AF588358

Multibody Dynamics

Computational Methods in Applied Sciences

Volume 28

Series Editor

E. Oñate
International Center for Numerical Methods in Engineering (CIMNE)
Technical University of Catalonia (UPC)
Edificio C-1, Campus Norte UPC
Gran Capitán, s/n
08034 Barcelona, Spain
onate@cimne.upc.edu
www.cimne.com

For further volumes:
www.springer.com/series/6899

Jean-Claude Samin • Paul Fisette
Editors

Multibody Dynamics

Computational Methods and Applications

Springer

Editors
Jean-Claude Samin
Institute for Mechanics, Materials and Civil Engineering
Université catholique de Louvain (UCL)
Louvain-la-Neuve, Belgium

Paul Fisette
Institute for Mechanics, Materials and Civil Engineering
Université catholique de Louvain (UCL)
Louvain-la-Neuve, Belgium

ISSN 1871-3033 Computational Methods in Applied Sciences
ISBN 978-94-007-5403-4 ISBN 978-94-007-5404-1 (eBook)
DOI 10.1007/978-94-007-5404-1
Springer Dordrecht Heidelberg New York London

Library of Congress Control Number: 2012950804

Printed on acid-free paper

Springer is part of Springer Science+Business Media (www.springer.com)

Preface

Multibody Dynamics is an exciting area of Computational Mechanics which merges and blends various disciplines in order to provide methods and tools for the virtual prototyping of complex mechanical systems. Multibody dynamics plays a central role these days in the modeling, analysis, simulation and optimization of mechanical and mechatronic systems in a variety of fields and for a wide range of scientific and industrial applications, some of which are illustrated below.

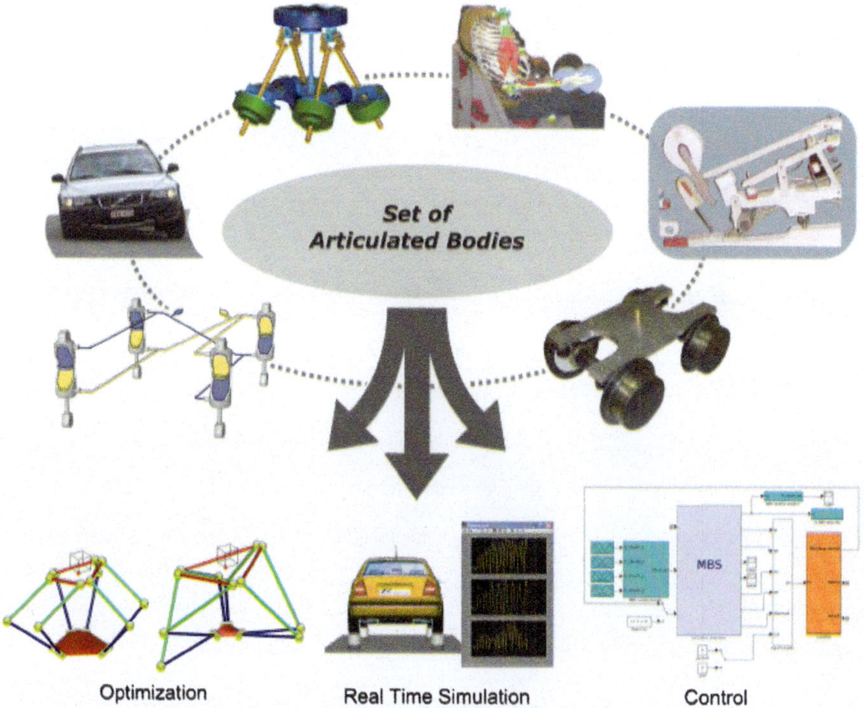

As new methods and procedures are being proposed at a fast pace in academia, research laboratories and industry, it is becoming important to provide researchers in multibody dynamics with appropriate venues for exchanging ideas and results. To answer these needs, the ECCOMAS Thematic Conference on Multibody Dynamics was initiated in Lisbon in 2003, and continued in Madrid in 2005, Milan in 2007 and Warsaw in 2009. Continuing this very successful series, the 2011 edition of the ECCOMAS Thematic Conference on Multibody Dynamics was held in Brussels, Belgium and organized by the Université catholique de Louvain, from 4th to 7th July 2011. More than 250 participants were attending the conference which provided a forum for fruitful discussion and technical exchanges.

This book contains the contributions of participants selected by the organizers that reflect the State-of-Art in the application of Multibody Dynamics to different areas of engineering. The chapters of this book are enlarged and revised versions of the communications, delivered at the conference, which were enhanced in terms of self-containment and tutorial quality by the authors. The result is a comprehensive text that constitutes a valuable reference for researchers and design engineers which helps to appraise the potential for the application of multibody dynamics methodologies to a wide range of areas of scientific and engineering relevance.

Louvain-la-Neuve, Belgium

Jean-Claude Samin
Paul Fisette

Contents

Speed Skating Modeling

A.L. Schwab, D.M. Fintelman, and O. den Braver

Abstract Advice about the optimal coordination pattern for an individual speed skater to reach their optimal performance, could well be addressed by simulation and optimization of a biomechanical model of speed skating. But before getting to this optimization approach one needs a model that matches observed behavior. In this chapter we present a simple 2-dimensional model of speed skating on the straights which mimics observed kinematic and force data. The primary features of the model are: the skater is modeled as three point masses, only motions in the horizontal plane are considered, air drag forces which are quadratic in the velocity and coulomb type ice friction forces at the skates are included, and idealized contact of the skate on the ice is modeled by a holonomic constraint in the vertical direction and a non-holonomic constraint in the lateral direction. Using the measured leg extension (relative motions of the skates with respect to the upper body) we are able to predict reasonable well the speed skater motions, even if we do not fit for that. The model seems to have the key terms for investigations of speed skating.

1 Introduction

The coordination pattern of speed skating appears to be completely different from all other types of human propulsion. In most patterns of human locomotion, humans generate forces by pushing against the environment in the opposite desired direction

A.L. Schwab (✉)
Laboratory for Engineering Mechanics, Delft University of Technology, Mekelweg 2, 2628 CD Delft, The Netherlands
e-mail: a.l.schwab@tudelft.nl

D.M. Fintelman
School of Sport and Exercise Sciences, University of Birmingham, Edgbaston, B15 2TT, Birmingham, UK
e-mail: dmf144@adf.bham.ac.uk

O. den Braver
BioMechanical Engineering, Delft University of Technology, Mekelweg 2, 2628 CD Delft, The Netherlands

J.-C. Samin, P. Fisette (eds.), *Multibody Dynamics*, Computational Methods in Applied Sciences 28,
DOI 10.1007/978-94-007-5404-1_1, © Springer Science+Business Media Dordrecht 2013

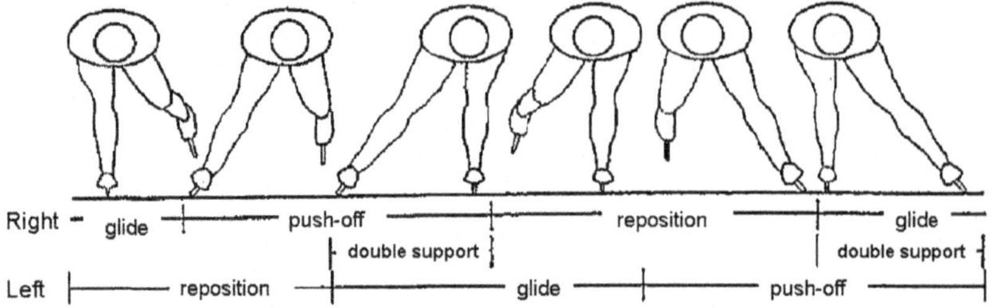

Fig. 1 Phases of a stroke: push-off phase, glide phase and reposition phase [1]

of motion. In speed skating humans generate forces by pushing in sideward direction. When we take a closer look at speed skating the straights we observe that a skating stroke can be divided in three phases: the glide, push-off and reposition phase, see Fig. 1. In the push-off phase the skate moves sidewards with respect to the center of mass (COM) of the body till near full leg extension. In the reposition phase the leg is retracted in the direction of the center of mass of the body. During the glide phase the body is supported over one leg that remains at nearly constant height (ankle to hip distance). Double support, where both skates are on the ice, only exists in the first part of the glide phase of one leg and in the second part of push-off phase of the other leg. This coordination pattern with sideward push-off results in a sinus-wave like trajectory of the upper body on the ice [4].

From these observations a number of questions arise. Of the many possible coordination patterns, that is position and orientation of the skates with respect to the upper body, why do skaters use this particular one? What is the optimal coordination pattern for an individual speed skater to reach their optimal performance? How do speed skaters create forward power on ice? Why are speed skaters steering back to their body at the end of the push-off? What is the effect of anthropometric differences on the coordination pattern of a speed skater (like the difference between a tall Dutch skater and a small Japanese skater)? All these questions are highly dependent on the coordination pattern of the speed skater and could well be addressed by simulation and optimization of a biomechanical model of speed skating. But before getting to this optimization approach one needs a model that reasonable matches observed behavior.

Currently, there exist three speed skating models [1, 6, 10]. The first models of speed skating were initiated by Gerrit Jan van Ingen Schenau [12] and further developed by researchers at the VU University Amsterdam [6]. By using power balances of the human and the environment useful information about the posture, athlete physiology and environmental parameters on the performance is obtained. Disadvantages of these models are that the validation is difficult and it is impossible to investigate differences in coordination pattern.

A more recent model was developed by Otten [10], in which forward and inverse dynamics are combined. The model is complex and includes up to 19 rigid bodies and 160 muscles. The model is able to simulate skating and can give insight in the forces/moments in the joints. Limitations of the model are that the kinematics in the model are manually tuned and that the model is not driven and validated with measurements of speed skaters. Unfortunately, no information about this model is available in the open literature, which makes it hard to review.

The most recent speed skater model is developed by Allinger and van den Bogert [1]. they developed a simple, one point mass, inverse dynamics model of a speed skater which is driven by individual strokes. The main limitations of the model are that the model is driven by a presumed leg function in time and that the model is not validated with force measurements. Furthermore, the effect of the assumptions on the model (e.g. constant height) are not investigated. On the other hand the model is possibly accurate and very useful for optimization the coordination pattern of speed skating.

Although three biomechanical models exist, none of these models is shown to accurately predict observed forces and motions. Which is partly due to the lack of experimental kinematic data and force data on stroke level.

In this chapter, we present a 2-dimensional inverse dynamics model on the straights which has minimal complexity. The model is based on three lumped masses and is validated with observed in-plane (horizontal) kinematics and forces at the skates. In the future, this model can be used to provide individual advice to elite speed skaters about their coordination pattern to reach their optimal performance.

2 Methods

We measured in time the 2-dimensional in-plane (horizontal) positions (x, y) of the two skates and the upper body, the normal forces and lateral forces at the two skates and lean angle of the skates. We developed a 2-dimensional inverse dynamic model of a skater. The model is driven by the measured leg extensions, that is relative motions of the skates with respect to the upper body and absolute orientation of the skates with respect to the ice. The upper body motions together with the forces exerted on the ice by the skates are calculated from the model.

A schematic of our 2-dimensional model is shown in Fig. 2. The model consists of three point masses: lumped masses at the body and the two skates. The total mass of the system is distributed over the three bodies by a constant mass distribution coefficient. The motions of the arms are neglected. We do not consider the vertical motion of the upper body, since experiments show that the upper body is at nearly constant height [3]. Air friction and ice friction are taken into account. Idealized contact of the skate on the ice is modeled by a holonomic constraint in the vertical direction and a non-holonomic constraint in the lateral direction.

Values for the mass distribution and air friction are found experimentally. The best agreement between the measurements and model can be achieved if we use accurate values for these parameters. Therefore we constructed an objective function J_{min} and minimized the error between the measurements and model. Details on the objective function can be found in Appendix 8.3.

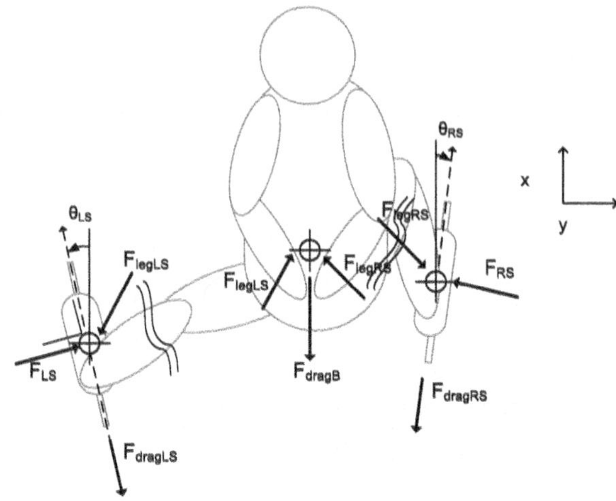

Fig. 2 Free body diagrams of the three point mass model (horizontal plane, top view). The masses are located at the COM of the body and at the COM of the skates. F_{ls} and F_{rs} are perpendicular with the skate blades, θ_{ls} and θ_{rs} are the steer angles of the skates with respect to the x-axis. The x- and y-axes are the inertial reference frame fixed to the ice rink

3 Model Analysis

In the model analysis for speed skating, three stages can be distinguished. First, the unconstrained equations of motion of the speed skater of a single stroke are derived. Secondly, the constraints are formulated and incorporated into the unconstrained equations of motion. Finally all equations are derived in terms of generalized coordinates and solved by numerical integration of these constrained equations of motion.

3.1 Equations of Motion

The equations of motion for each separate body (upper body, right skate and left skate) can be derived in x and y direction. Friction forces (air and ice friction) as well as the constraint forces are acting on the bodies. All constraints acting on the bodies will be explained in the next paragraph. The unconstrained equations of motions for all bodies are,

$$\begin{aligned} m\ddot{x}_i &= -F_{frictionX_i} + F_{constraintsX_i} \\ m\ddot{y}_i &= -F_{frictionY_i,} + F_{constraintsY_i} \end{aligned} \qquad i = B, LS, RS \tag{1}$$

where $F_{frictionX_i}$ is the component of the friction force in x direction and $F_{frictionY_i}$ the component of the friction force in y direction. $F_{constraintsX}$ are the constraint forces in x direction and $F_{constraintsY}$ the constraint forces in y direction.

3.2 Constraints

The first set of constraints are the leg extension constraints, they connect the skates to the upper body. The positions of the skates are prescribed by the position of the

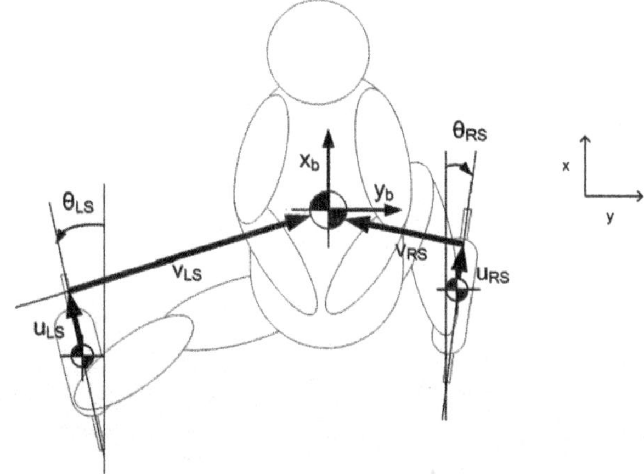

Fig. 3 Definition of generalized coordinates

upper body and the leg extension coordinates. The second set of constraints are at the skates. A holonomic constraint is applied in the vertical direction which establish that the skate is on the ice and a non-holonomic constraint in the lateral direction of the skate to express that there is no lateral slip of the skate on the ice.

3.3 Generalized Coordinates

The generalized coordinates of the skater model are chosen such that we can express the coordination of the motion of the skater in terms of the leg extensions and the skate orientations (steer angles). Therefore the configuration of the skater is expressed by the motion of the upper body and the leg extensions (relative motions of the skates with respect to the upper body, see Fig. 3) and can be described by the generalized coordinates,

$$\mathbf{q} = (xb, yb, u_{LS}, v_{LS}, \theta_{LS}, u_{RS}, v_{RS}, \theta_{RS})^T, \tag{2}$$

in which θ_{LS} and θ_{RS} are the steer angles of the skates with respect to the global x-axis. These steer angles, which are prescribed coordinates, are needed to apply the non-holonomic skate constraints. The equations of motion will be written in terms of the generalized coordinates. Detailed information on the transformation of the equations of motions in terms of the generalized coordinates can be found in Appendices 8.1, 8.7, and 8.8.

3.3.1 Leg Extension Constraints

The position of the right and left skate can be expressed as function of the generalized coordinates and will be incorporated into the equations of motion means by holonomic constraints. The left skate leg extension constraints are,

$$c_1 = x_{LS} - x_B + \cos(\theta_{LS})u_{LS} + \sin(\theta_{LS})v_{LS} = 0, \tag{3}$$

$$c_2 = y_{LS} - y_B - \sin(\theta_{LS})u_{LS} + \cos(\theta_{LS})v_{LS} = 0, \tag{4}$$

and the right skate leg extension constraints are,

$$c_3 = x_{RS} - x_B + \cos(\theta_{RS})u_{RS} + \sin(\theta_{RS})v_{RS} = 0, \tag{5}$$

$$c_4 = -y_{RS} + y_B - \sin(\theta_{RS})u_{RS} + \cos(\theta_{RS})v_{RS} = 0. \tag{6}$$

3.3.2 Skate Constraints

When the skate is on the ice we assume no lateral slip between the ice and skate, that is the lateral velocity of the skate is zero. This can be expressed by a non-holonomic constraint which are for the left and right skate respectively,

$$c_5 = -\sin(\theta_{LS})\dot{x}_{LS} - \cos(\theta_{LS})\dot{y}_{LS} = 0, \tag{7}$$

$$c_6 = -\sin(\theta_{RS})\dot{x}_{RS} + \cos(\theta_{RS})\dot{y}_{RS} = 0. \tag{8}$$

Since we do not consider vertical motions no constraints in the vertical direction are needed. Contact or no contact is described by on/off switching of the corresponding non-holonomic constraint.

3.4 Mass Distribution

The number of bodies in the model is based on an investigation of the shift in position of the center of mass on a complete anthropometric model of a speed skater during the gliding and the push-off phase of a stroke. A minimum of three bodies was shown to be necessary for describing the shift of the center of mass [8].

The total mass m of the skater is now distributed over the three point masses (body, left skate, right skate) by using a mass distribution coefficient α (Fig. 4). The distribution of the masses are given by $m_B = (1-\alpha)m$, $m_{LS} = (\alpha/2)m$, and $m_{RS} = (\alpha/2)m$.

3.5 Friction Forces

The total friction forces can be roughly divided in 80 % air friction and 20 % ice friction [5]. The ice friction in the model, following de Koning [7], is described by Coulomb's friction law,

$$F_{\text{ice}} = \mu F_N \tag{9}$$

where μ is the friction coefficient and F_N the normal force of the skate on the ice. Here we assume that the height of the skater is constant and that there is no double

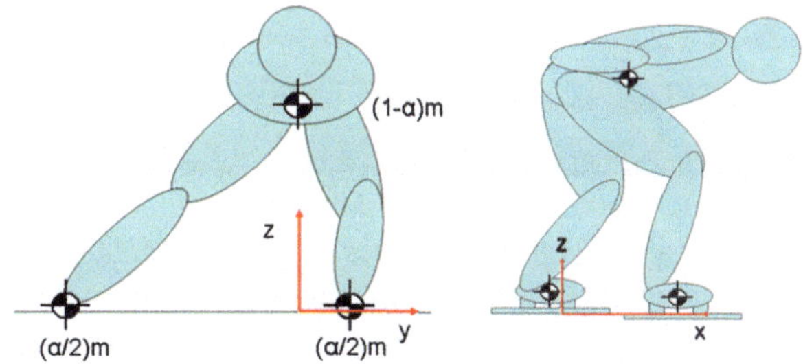

Fig. 4 Positions of the COM of the bodies during the push-off together with the mass distribution

stance phase. Therefore, the ice friction can be written as $F_{\text{ice}} = \mu m g$, in which m the total mass of the skater and g the earth gravity. The air friction can be described by,

$$F_{\text{air}} = \frac{1}{2}\rho C_d A v^2 = k_1 v^2 \tag{10}$$

where ρ represents the air density, C_d the drag coefficient, A the frontal projected area of the skater, and v the velocity of the air with respect to the skater. The air drag forces at each individual mass are calculated by multiplying the mass distribution coefficient of that mass by the total air drag. The drag coefficient k_1 can only be estimated experimentally. With an experimental method (see Appendix 8.5) both the drag coefficients μ and k_1 for every individual subject are estimated.

3.6 Model Summary

The equations of motion together with the constraint equations are completely defined by the state of skater. Combining the equations of motion for the individual masses (1) and including of the constraint forces and the constraints (3)–(6) on the acceleration level results in the constraint equations of motion for the system, $\mathbf{Au} = \mathbf{b}$, with

$$\mathbf{A} = \left[\begin{array}{cccccc|cccc} m(1-\alpha) & 0 & 0 & 0 & 0 & 0 & -1 & 0 & -1 & 0 \\ 0 & m(1-\alpha) & 0 & 0 & 0 & 0 & 0 & -1 & 0 & 1 \\ 0 & 0 & \frac{\alpha}{2}m & 0 & 0 & 0 & 1 & 0 & 0 & 0 \\ 0 & 0 & 0 & \frac{\alpha}{2}m & 0 & 0 & 0 & 1 & 0 & 0 \\ 0 & 0 & 0 & 0 & \frac{\alpha}{2}m & 0 & 0 & 0 & 1 & 0 \\ 0 & 0 & 0 & 0 & 0 & \frac{\alpha}{2}m & 0 & 0 & 0 & -1 \\ \hline -1 & 0 & 1 & 0 & 0 & 0 & 0 & 0 & 0 & 0 \\ 0 & -1 & 0 & 1 & 0 & 0 & 0 & 0 & 0 & 0 \\ -1 & 0 & 0 & 0 & 1 & 0 & 0 & 0 & 0 & 0 \\ 0 & 1 & 0 & 0 & 0 & -1 & 0 & 0 & 0 & 0 \end{array}\right] \tag{11}$$

$$\mathbf{u} = \begin{bmatrix} \ddot{x}_B & \ddot{y}_B & \ddot{x}_{LS} & \ddot{y}_{LS} & \ddot{x}_{RS} & \ddot{y}_{RS} & \lambda_1 & \lambda_2 & \lambda_3 & \lambda_4 \end{bmatrix}^T \tag{12}$$

$$\mathbf{b} = \begin{bmatrix} -F_{XfrictionB} & -F_{YfrictionB} & -F_{XfrictionLS} & -F_{YfrictionLS} & -F_{XfrictionRS} \\ -F_{YfrictionRS} & h_{c1} & h_{c2} & h_{c3} & h_{c4} \end{bmatrix}^T \tag{13}$$

where $h_{c1} \dots h_{c4}$ are the convective acceleration terms of the constraints (Appendix 8.8) and $\lambda_1 \dots \lambda_4$ are the constraint forces (Lagrange multipliers). Here λ_1 and λ_2 are the constraint forces in the left leg, and λ_3 and λ_4 the constraint forces in the right leg. The non-holonomic skate constraints are not yet included in this system, but will be in a later stage.

The model consists of 3 bodies with each 2 degrees of freedom, thus the unconstrained system has 6 degrees of freedom. However, there are 4 coordination constraints and 1 non-holonomic constraint of the skate on the ice (no double stance); therefore 1 degree of freedom remains. If there is a double stance phase then both skates are on the ice, the system is over-constrained and no degree freedom is left. Therefore for the model we will assume only single stance phases, and the model will alternatively switch between the right skate en left skate constraint. This assumption is validated by the experimental force data, where we see only a short period of double stance with load transfer.

We rewrite the equations of motion (11)–(13) (still without the non-holonomic skate constraints) in terms of the generalized coordinates (2), where the prescribed coordinates (leg extension coordinates $(u_{LS}, v_{LS}, \theta_{LS}, u_{RS}, v_{RS}, \theta_{RS})$) are pushed to the right-hand side (Appendix 8.1). Next, the constraint of the skate on the ice (left or right) is added to the equations. Finally the reduced constrained equations of motion are given by, for when the left skate is on the ice,

$$\begin{bmatrix} m & 0 & -s_{LS} \\ 0 & m & -c_{LS} \\ -s_{LS} & -c_{LS} & 0 \end{bmatrix} \begin{bmatrix} \ddot{x}_B \\ \ddot{y}_B \\ \lambda_5 \end{bmatrix} = \begin{bmatrix} \mathbf{T}_{,\mathbf{q}}^T(\mathbf{f} - \mathbf{M}\mathbf{h}) \\ h_{c5} \end{bmatrix}, \tag{14}$$

and for when the right skate is on the ice,

$$\begin{bmatrix} m & 0 & -s_{RS} \\ 0 & m & c_{RS} \\ -s_{RS} & c_{RS} & 0 \end{bmatrix} \begin{bmatrix} \ddot{x}_B \\ \ddot{y}_B \\ \lambda_6 \end{bmatrix} = \begin{bmatrix} \mathbf{T}_{,\mathbf{q}}^T(\mathbf{f} - \mathbf{M}\mathbf{h}) \\ h_{c6} \end{bmatrix}, \tag{15}$$

where λ_5 and λ_6 are the lateral constraint forces on the skate and h_{c5} and h_{c6} are the convective acceleration terms of the skate constraints, the latter are presented in Appendix 8.8. Clearly both systems have one degree of freedom left, one can think of it as being the forward motion.

3.7 Model Constants

Experimental data was obtained from four different riders. Listed in Table 1 are the values of the model parameters used in the simulations for these four riders. The total mass of the skater and gravity are a measured quantities. The other parameters are found by an optimization process as described in Appendix 8.3.

Table 1 Parameter values for the four riders

Variable	Description	Value
m	Mass skater	66, 80, 77, 84 [kg]
α	Mass distribution	0.604, 0.682, 0.607, 0.686 [–]
k_1	Drag coefficient	0.160, 0.153, 0.112, 0.299 [N/(m/s)^2]
g	Gravity	9.81 [m/s^2]

4 Model Analysis

4.1 Parametrization of the Coordination Body Functions

Input to the model are the measured motion coordinations, the leg extensions and the skate steer angles, and their velocities and accelerations. To determine these all measured positions have to be differentiated with respect to time. To get rid of model errors due to numerical differential and filtering errors (spikes), all positions are first parameterized by smooth functions. The required parametrization functions have to be twice differentiable. The combination of a linear and periodic functions satisfies this requirement. The used parametrization function is,

$$f = c_0 + c_1 t + \sum_{k=1}^{5} a_k \sin\left(2k\pi \frac{t}{T}\right) + b_k \cos\left(2k\pi \frac{t}{T}\right). \tag{16}$$

The fit is not accurate at the beginning and end of the stroke, which results in a mismatch of the initial conditions on the velocities and accelerations. Therefore the coordinates are fit at a somewhat longer time period and then cut off afterwards. We tried also other parametrization functions, like polynomial and cubic splines. The differentials of polynomial functions became unstable with increasing order, while piecewise cubic splines have no filtering which results in high frequent components in the positions. The measured positions of the body, left and right skate in x and y direction of a single stroke are parameterized according to (16) and by differentiating the equations of the fitted function the velocities and acceleration are calculated.

4.2 Integration of the Differential Equations

The differential algebraic equations (14), (15) describing the motion of the system cannot be solved analytically. Therefore, the equations will be numerically integrated, using the classic Runge-Kutta 4th order method (RK4). The stepsize h is taken constant during the whole simulation, and chosen identical to the sample time of the measurements $T_s = 1/100$ [sec]. After each numerical integration step the constraints are fulfilled by a projection method (Appendix 8.2).

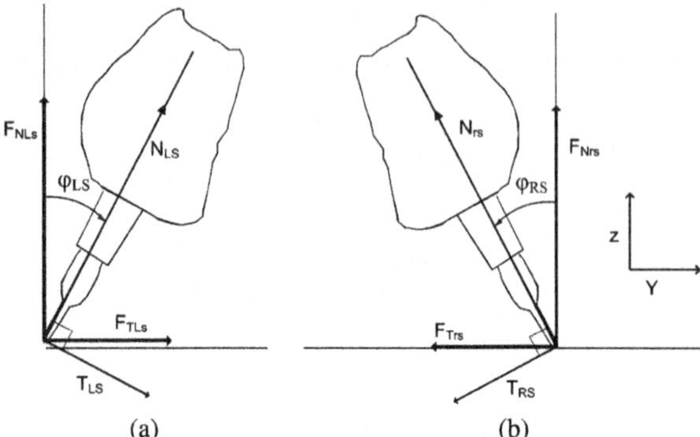

Fig. 5 Forces in local reference frame (N_{ls}, T_{ls}, N_{rs}, T_{rs}) and global reference frame (F_{Nls}, F_{Tls}, F_{Nrs}, F_{Trs}): (**a**) Left skate, (**b**) right skate

4.3 Data Collection

The data collection of the skater includes the 2-dimensional in-plane positions (x, y) of the two skates and the upper body, the normal and lateral forces at the two skates and lean angle of the skates. The global positions are measured by a radio frequency based so-called local position measurement system (LPM) from Inmotio.[1] This system is installed at the Thialf speed skate rink in Heerenveen, The Netherlands. The LPM system has been used for analysis of soccer matches, and can handle up to 22 active transponders at $1000/22$ Hz. The transponders are approximately placed at the positions of the point masses.

We have developed two instrumented clap skates to measure the normal and lateral forces (N_i, L_i) at the blades of the skates, see Fig. 5. To be able to compare these with the model output, which are the global lateral forces F_{Tls} and F_{Trs}, the lean angles of the skates, ϕ_i, has be measured too. These angels are measured using an inertial measurement unit from Xsens,[2] where only the lean angle is used.

For data acquisition a DAQ unit of National instruments[3] is used. All the force and orientation data is collected from the DAQ via a USB connection on a mini laptop which is carried by the skater in a backpack. The different measurement systems are synchronized by means of images from a high speed camera. See Appendix 8.4 for detailed description of the synchronization method.

Data sets of four trained speed skaters are used to validate the model. The data collection is performed with a standard measurement protocol which includes: skating two laps at an estimated 80 % of maximal performance level. The tests are repeated at least three times.

---

[1]http://www.inmotio.nl, Hettenheuvelweg 8, 1101 BN Amsterdam Zuidoost, The Netherlands.

[2]http://www.xsens.com.

[3]http://www.ni.com.

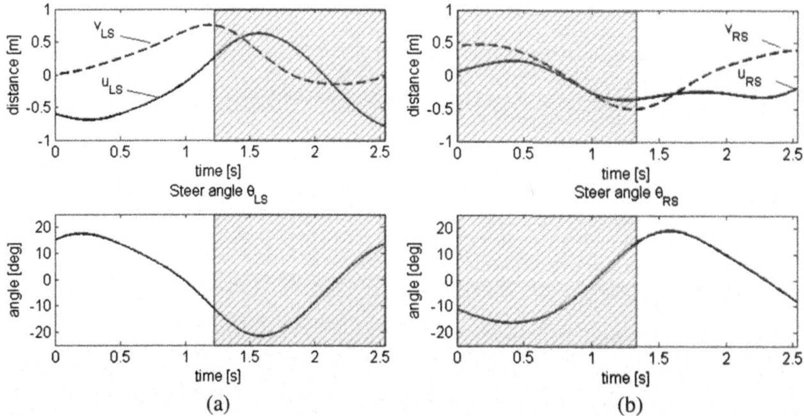

Fig. 6 Measured and parameterized leg extension coordinates u_i, v_i and θ_i as a function of time for a sequencing left and right stroke for rider 1 from Table 1. *Gray filled area* means that the skate is not active. (**a**) Left skate, (**b**) right skate

4.4 Fitting the Model to the Observed Data

The model is validated by showing how closely it can simulate the observed forces and motions. Quantification of the model errors are analyzed similar to that of McLean [9]. The measured data has different scales and units and therefore we constructed a measurement of error, $J_{\min}$, between the model and the measured data which includes the error of the upper body position, velocities and local normal forces (N_{ls} and N_{rs}). The measurement of error is dimensionless, reasonably scaled and independent of the number of time samples. See Appendix 8.3 for a detailed description of the measurement error function $J_{\min}$.

5 Results

Plots of the measured and simulated forces and motions (output of model) as a function of time for a sequencing left and right stroke are shown in Fig. 7 (the parameters are according to the first rider from Table 1). The corresponding measured and parameterized leg extensions (input of model) of the left and right leg are shown respectively in Fig. 6(a) and Fig. 6(b). At the beginning of the left stroke ($t = 0$) the skate is placed in front of the upper body, resulting in a negative u_{ls}. During the stroke the skate is moving sidewards and backwards, u_{ls} and v_{ls} increase. At the end of the stroke the skate is retracting to the upper body, u_{ls} and v_{ls} decrease. At the beginning of the right stroke ($t = 1.25$), the skate is again moving sidewards, v_{rs} increase. However the motion pattern of the u_{rs} is somewhat different in comparison with u_{ls}. The u_{rs} remains approximately constant during the stroke, which eventually will results in a different output motion of the upper body in y direction.

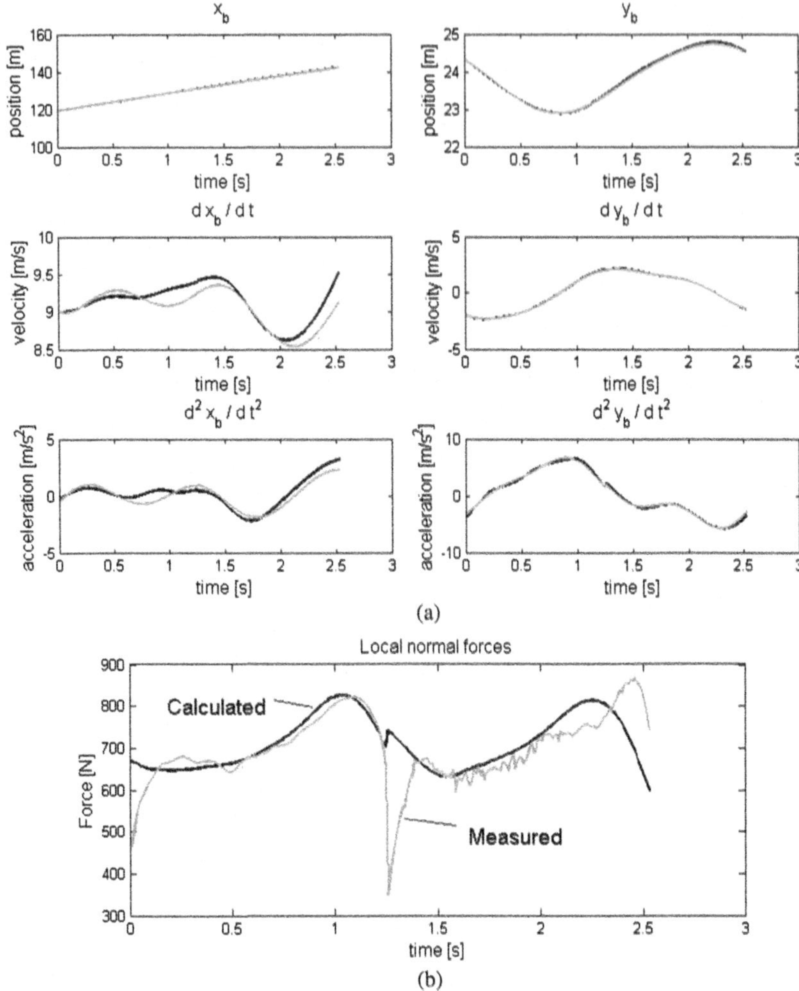

Fig. 7 Simulated (*black lines*) and measured (*gray lines*) upper body positions, velocities, accelerations and local normal forces on the skates (N_i), as a function of time for a sequencing left and right stroke, for rider 1 from Table 1 ($mg = 647$ N)

The skater has an average forward speed of ≈ 32 km/h. The upper body describes a sine-wave like trajectory with respect to the ice during speed skating the straights (Fig. 7(a), y_b), which has also been observed by de Boer [4]. The velocity pattern sidewards, $\dot{y}_b$, are alike for left and right stroke. However, the forward acceleration/deceleration pattern differ per stroke. This was observed for every rider.

The local normal forces N_{LS} and N_{RS} of the active skate are shown in Fig. 7(b), where the height of the body is assumed constant. At the large force drop in the measured force data a switch is made in the model from the left skate to the right

Table 2 Net error $J_{\min}$ per subject (average of all left straight strokes of all tests) divided by the number of optimization parameters

Skater	$J_{\min}$
1	0.0013
2	0.0015
3	0.0022
4	0.0013

skate. Note that the sum of the measured left and right force corresponds well to the calculated value. At the beginning of the stroke the normal force is rising above the body weight of the skater. Then a small force drop appears and at the end of stroke the normal forces rises again well above the body weight. The maximal normal force during push-off is approximately 150 % of the body weight.

Agreement exists between the measured and simulated positions and velocities. The largest error is in the force data, which mainly appears at the beginning and end of the stroke.

For all skaters the net error $J_{\min}$ (24) of all straight left strokes is calculated. This net error is divided by the number of optimization parameters being the upper body positions, upper body velocities and the local normal forces and presented in Table 2.

Averages of the magnitudes of the residuals are calculated similar to that of Cabrera [2] by $R_j = \sum_{i=1}^{N} |\tilde{y}_{ij} - y_{ij}|/N$. In which N the number of collected data points, y_i the measured value of the variable and $\tilde{y}_i$ the simulated value of the variable from the model. For all variables j the R_j is shown in Table 3. The residuals of the upper body are less than 0.10 m for the forward position, 0.031 m sidewards, 0.20 m/s in the forward velocity, 0.06 m/s sidewards, and 53 N for the local normal forces in the skate.

Table 3 Table of the residuals between measured and simulated values of the variables. Body position in x direction [m], body position in y direction [m], body velocity in x direction [m/s], body velocity in y direction [m/s], body acceleration in x direction [m/s^2], body acceleration in y direction [m/s^2], local normal forces [N]

Skater	R_{xb}	R_{yb}	$R_{\dot{x}b}$	$R_{\dot{y}b}$	$R_{\ddot{x}b}$	$R_{\ddot{y}b}$	R_{Nl}
1	0.0795	0.0165	0.1769	0.0464	0.5880	0.3836	22.01
2	0.0817	0.0245	0.1659	0.0491	0.5952	0.3379	34.30
3	0.1048	0.0314	0.2071	0.0626	1.0276	0.3244	53.91
4	0.0782	0.0186	0.1737	0.0401	0.8315	0.2380	26.45

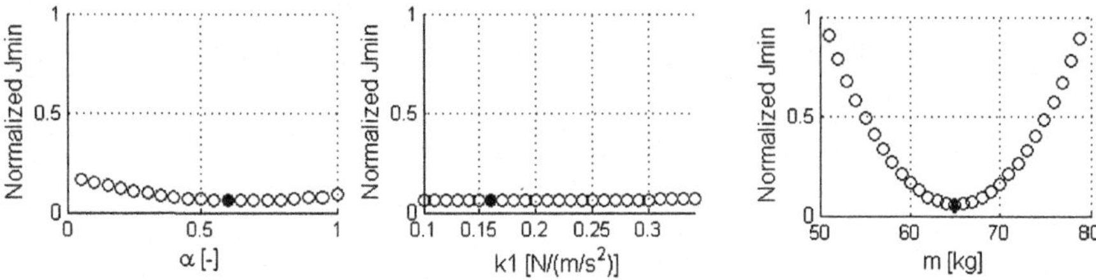

Fig. 8 Plots of $J_{\min}$ versus a single parameter value, mass distribution α, air friction coefficient k_1 and total mass of the skater m, as the parameter is varied about the nominal value for rider 1 from Table 1. The *filled circles* correspond to the value of $J_{\min}$ at the nominal parameter value

6 Discussion

6.1 Model Error

All position residuals are within the accuracy of the position measurement system (≈ 0.15 m). The accuracy of the LPM can be increased if two transponders, instead of one transponder are positioned at the skates and the upper body. The forward velocities $\dot{x}_B$ are less accurate than the sideward velocities $\dot{y}_B$, which is reasonable due to the fact that the forces are mainly in sideward direction instead of forward. Orientation errors have therefore more influence on the $\dot{x}_B$ than on the $\dot{y}_B$.

No total agreement exists between the measured forces and the forces calculated in the model, generally at the beginning and at the end of the stroke. There is no normal force drop in the calculated data which is a result of the simplification that there is no double stance phase, but the sum of the measured left and right force do correspond well with the calculated one. Conversion from global to local forces resulted in a force error, caused by the accuracy of the lean angle sensors. The accuracy of these sensors are < 2 deg root mean square, resulting in a local normal force error between $\approx 20/-20$ N. Besides conversion errors, crosstalk exists of ≈ 3 % of the lateral forces to the normal forces (max. $-9/9$ N). The maximal error due to inaccuracy of the measurement equipment is then approximately 29 N.

The net error $J_{\min}$ of all measurements are in the same magnitude, which shows that the model is valid for all subjects.

6.2 How Does the Fit Depend on Mechanical Constants

The sensitivity of the mechanical constants is obtained by minimizing the net error $J_{\min}$ (24). This net error is calculated by letting the upper body motions variable while fixing all other parameters to their optimal fit value, except for the wanted minimization parameter (mass distribution α, air friction coefficient k_1 or mass of the skater m). In Fig. 8 the normalized net error $J_{\min}$ are plotted as function of the minimization parameters. The minimal values in the figures correspond to the values of the parameters at the optimal fit.

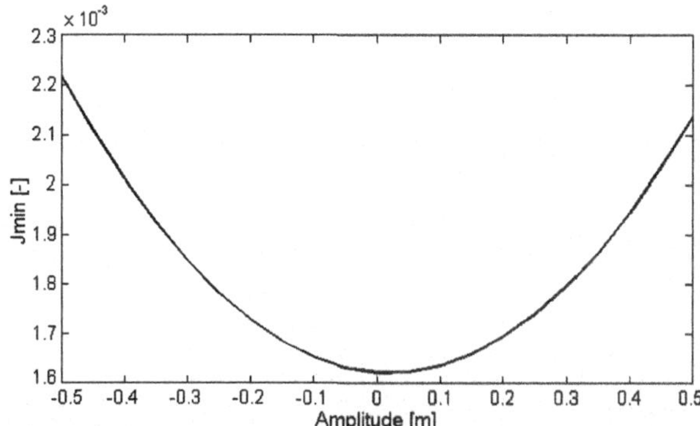

Fig. 9 Plot of error $J_{\min}$ versus the amplitude of the sine wave corrupting the velocity data of the upper body of the skater

The mass is the most sensitive mechanical parameter, however this parameter is measured accurately and therefore of no concern here. The value of the mass distribution α as well as the friction coefficient k_1 are more uncertain. The figure shows clearly that the fit depends little on these mechanical constant.

6.3 *Fitting False Data*

If the fits which are obtained are a result of good curve fitting, then it should be able to obtain good fits to false data. To test the model a pure sine function, $A\cos(2\pi t/T)$, with amplitude A, and stroke time T, is added to the measured velocity data of the upper body in either directions. In Fig. 9 the minimal error function versus the amplitude of the sinus wave is plotted. The total error between the model and the measured variables is minimal if the amplitude of the added function is zero. The model shows the best fit if there is not added corrupted data to the velocity data of the upper body. These results shows that the fits are not a result of good curve fitting, but rather the result of a good model.

6.4 *Kinematic Complexity*

The double stance phase was not included in the model. However, the sum of the measured left and right force during the short double stance phase do correspond well with the calculated forces (Fig. 7(b)), which demonstrates that there is little need for modeling this double stance phase.

Another major simplification of the model is that it was assumed that the center of mass remains at a constant height during skating, which was based on de Boer [3]. However, in accelerometer data of the upper body it was found that at the end of the stroke the upper body accelerates about 1.5 times gravity, which really influences the forces in the model. Therefore it seems beneficial to include the vertical motion of the body in the model.

7 Conclusions

We have constructed a simple 2-dimensional model of speed skating that does a reasonable job of imitating the forces and kinematics as observed in actual speed skating. The model reproduces these forces and motions reasonably well, even if we do not fit for that. The model is limited in accuracy due to the limited accuracy of the LPM position measurement system. Adding the (small) vertical motion of the upper body can increase the accuracy of the model.

The model seems promising for individual training advice. Coordination patterns of individual skaters can be optimized by using the model if psychological constraints of individual skater are added to the model. In Appendix 8.6 a detailed description of the needed constraints on the model is given. The model can also be used to give insight in the biomechanics of speed skating, like why speed skaters steer back to their body at the end of the stroke. Finally the effect of anthropometric differences between speed skaters can be determined.

Appendix

This appendix contains details on the modeling and the experimental validation and comments and remarks for future use of the model for optimization of speed skater performance.

8.1 Kinematic Transformation

This section describes the transformation of the equations of motion in terms of the generalized coordinates.

We start with the differential algebraic (constraint) equations of motion (DAEs), without the non-holonomic skate constraint, from (11), which can be written as,

$$\begin{bmatrix} \mathbf{M} & \mathbf{C}^T \\ \mathbf{C} & \emptyset \end{bmatrix} \begin{bmatrix} \ddot{\mathbf{x}} \\ \boldsymbol{\lambda} \end{bmatrix} = \begin{bmatrix} \mathbf{f} \\ \mathbf{h}_c \end{bmatrix}, \tag{17}$$

with the COM accelerations $\ddot{\mathbf{x}}$, the diagonal mass matrix $\mathbf{M}$, the applied forces $\mathbf{f}$ at the COM, the Jacobian $\mathbf{C} = \partial \mathbf{c}/\partial \mathbf{x}$ of the constraint equations $\mathbf{c}(\mathbf{x}) = 0$, the convective terms $\mathbf{h}_c = (\partial(\mathbf{C}\dot{\mathbf{x}})/\partial \mathbf{x})\dot{\mathbf{x}}$, and the Lagrange multipliers $\boldsymbol{\lambda}$, with respect to the constraints $\mathbf{c}$. The constrained equations of motion are,

$$\mathbf{M}\ddot{\mathbf{x}} = \mathbf{f} - \mathbf{C}^T \boldsymbol{\lambda}. \tag{18}$$

Next, we like to rewrite the equations in terms of the generalized coordinates $\mathbf{q}$. Therefore we introduce the coordinates of the COM $\mathbf{x}$ expressed in terms of the generalized coordinates $\mathbf{q}$,

$$\mathbf{x} = \mathbf{T}(\mathbf{q}). \tag{19}$$

Differentiate this twice with respect to time,

$$\dot{\mathbf{x}} = \mathbf{T}_{,\mathbf{q}}\dot{\mathbf{q}} \quad \text{and} \quad \ddot{\mathbf{x}} = \mathbf{T}_{,\mathbf{q}}\ddot{\mathbf{q}} + \mathbf{h}. \tag{20}$$

The subscript comma followed by one or more variables denotes the partial derivatives with respect to these variables, and with the convective terms $\mathbf{h} = (\mathbf{T}_{,\mathbf{q}}\dot{\mathbf{q}})_{,\mathbf{q}}\dot{\mathbf{q}}$. Substitution of these accelerations in (18) and pre-multiplying with the transposed Jacobian $\mathbf{T}_{,\mathbf{q}}^T$ gives,

$$\mathbf{T}_{,\mathbf{q}}^T\mathbf{M}(\mathbf{T}_{,\mathbf{q}}\ddot{\mathbf{q}} + \mathbf{h}) = \mathbf{T}_{,\mathbf{q}}^T(\mathbf{f} - \mathbf{C}^T\boldsymbol{\lambda}). \tag{21}$$

Since the generalized coordinates fulfill the constraints, $\mathbf{T}_{,\mathbf{q}}^T\mathbf{C}^T$ is identical to zero, that is the constraint forces $\boldsymbol{\lambda}$ drop out of the equations. The result is the equations of motion expressed in terms of the generalized coordinates $\mathbf{q}$,

$$\mathbf{T}_{,\mathbf{q}}\mathbf{M}\mathbf{T}_{,\mathbf{q}}^T\ddot{\mathbf{q}} = \mathbf{T}_{,\mathbf{q}}^T(\mathbf{f} - \mathbf{M}\mathbf{h}). \tag{22}$$

Finally the skate constraint can be added to these equations of motion, which results in the constraint equations of motion (14) and (15).

8.2 State Projection

After numerical integration of the equations of motion for one time increment, the state variables in general do not fulfill the constraints. This can be solved by formulating a minimization problem such that the distance from the predicted solution $\tilde{\mathbf{q}}_{n+1}$ to the solution which is on the constraint surface $\mathbf{q}_{n+1}$ is minimal: $\|\tilde{\mathbf{q}}_{n+1} - \mathbf{q}_{n+1}\|_2 = \min_{\mathbf{q}_{n+1}}$ and where all $\mathbf{q}_{n+1}$ have to fulfill the constraints $\mathbf{c}(\mathbf{q}_{n+1}) = 0$. This non-linear constraint least-square problem is in general solved with a Gauss-Newton method after every numerical integration step. However, here we have to deal with non-holonomic constraints only, which are linear in the speeds $\mathbf{C}(\mathbf{q}_{n+1})\dot{\mathbf{q}}_{n+1} = 0$. The optimization problem then reduces to a linear constraint least-square problem which can be solved in one step.

8.3 Objective Function $J_{\min}$

The best agreement between simulation and measurements can be achieved if we use accurate values for the air friction coefficient and the mass distribution. This is solved by minimizing the error between the model and the measurements. The objective function is defined by equation:

$$E_j = \frac{1}{N}\sum_{i=1}^{N}(\tilde{y}_i - y_i)^2 \tag{23}$$

where N the number of collected data points, y_i the measured value of the variable and $\tilde{y}_i$ the simulated value of the variable from the model. This is a constrained multivariable minimization problem: $\min_x f(x)$ with the constraint: $\text{lb} \le \text{x} \le \text{ub}$ in which x are the air friction coefficient k_1 and mass distribution constant α. The upper and lower limit of α are defined as 0 and 1 while the limits of k_1 are defined as 0.1 and 0.3. With the optimization function *fmincon* of Matlab the optimal combination of α and k_1 are found. The optimization function uses an interior point algorithm and starts at the initial guess of the minimum x_0. For each measured variable the optimal mechanical parameters can be fit.

Besides calculating the optimal values by minimizing one variable, the net error is calculated including the error of the upper body position, velocities and local normal forces (N_i). The net error is calculated with:

$$J_{\min} = \frac{\sum_{j=1}^{M} w_j \left(\frac{1}{N} \sum_{i=1}^{N} \frac{(\tilde{y}_{ij} - y_{ij})^2}{\bar{y}_j^2}\right)}{\sum_{j=1}^{M} w_j \bar{y}_j^2} \tag{24}$$

in which $\tilde{y}_{ij}$ is the simulated value of a variable, y_{ij} the measured value of a variable, w_j is the weighting factor of a variable and $\bar{y}_j$ is the characteristic value of the variable. The peak to peak values of the x and y upper body positions, average value of the body velocity in forward direction, peak to peak value of the velocity in sideward direction and the local measured normal peak force are used as the characteristic values of the parameters. Equal weights are used ($w_j = 1$) for all j in the error function.

8.4 Synchronization Method

The LPM position measurement system and the DAQ data acquisition unit are synchronized with video frames from a high speed (300 Hz) digital photo camera.[4]

8.4.1 Synchronization LPM and Video

The LPM is synchronized by using an extra static transponder, which was placed in line with the start line on the ice. During the synchronization test the line is filmed with the high speed camera (300 Hz). The moment of crossing the start line can be found in the LPM data and video.

8.4.2 DAQ and Video

In order to synchronize the DAQ and video frames a reset button with LED, which lights up when pressed, is used. The reset button is connected with the DAQ. At the start and the end of the measurement the subject has to push the reset button in view of the high speed camera, such that the video frames at which the LED reset button when pushed lights up can be easily determined.

[4]Casio Exilim EX-F1.

8.5 Friction Estimation

In order to estimate the friction coefficients, the subjects got the following instruction: after skating two laps immediately stop skating and glide along the line of the lanes in the same skating posture as you were skating before for 100 m.

In order to determine the coefficients of friction from the glide exercise, it is assumed that the friction coefficients are constant during the glide.

From this estimation the conclusion is drawn that the speed should decreases linearly. A first order polynomial is fitted through the velocity profile of the LPM data of the COM of the skater during gliding.

$$\dot{y}(t) = -at + b. \tag{25}$$

The gradient a of the line is the decelerations of the skater during gliding. The total friction force F_{friction} is the total mass m times the deceleration a of the skater during gliding; $F_{\text{friction}} = ma$. The air friction coefficient is assumed to behave like,

$$F_{\text{air}} = k_1 v^2 = \beta F_{\text{friction}}, \tag{26}$$

with the velocity v of the center of mass of the skater, k_1 the air drag factor and β the friction distribution factor which is assumed to be 0.8. The ice friction is assumed to behave like Coulomb friction as in

$$F_{\text{ice}} = \mu m g = (1 - \beta) F_{\text{friction}}, \tag{27}$$

where μ is the Coulomb friction coefficient between the ice and the skate.

8.6 Optimization

For future application of the model where one wants to find coordination patterns which result in optimal performance, additional constraints are required. The model has to be constrained by the physiology of the skater during optimization of his coordination pattern. The physiology constraints of a skater are given by: the leg length, average power and maximal power of the speed skater.

8.6.1 Maximal Power Constraint

The maximal power during a single stroke must not exceed the maximal possible power from a leg extension motion of a skater. The maximal power constraint value could be based on either literature or experimentally determined. First, the maximal power can be determined from the push-off force and velocity of leg extension in the horizontal plane, since no work is done in the vertical plane. The maximal possible power of single leg extension as a function of the leg extension velocity can be based on force-velocity data extracted from leg press results of Vandervoort et al. [11]. The

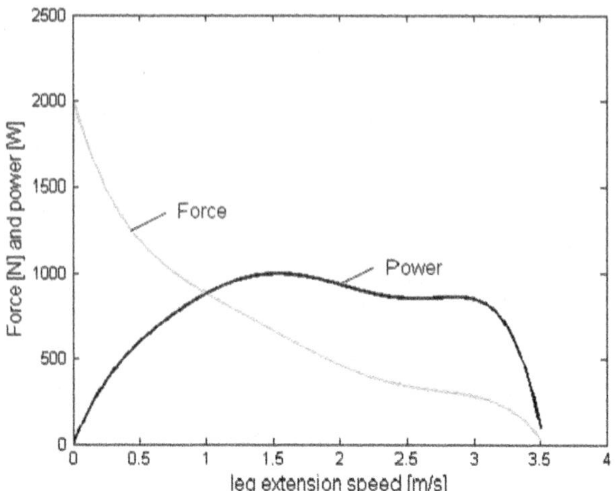

Fig. 10 Force velocity relation [1]

power is estimated by multiplying the force with the extension velocity of the leg [1] (Fig. 10).

Secondly, the maximal power of a single leg extension could be experimentally determined per subject. In power models the maximal and average power of an athlete is measured with an ergometer test [6]. The measured ergometer power is then multiplied by an experimentally determined constant to determine the power during skating. However, the uncertainty on the constant is large so there is room for a better method to determine subject individual skating power.

8.6.2 Average Power Constraint

The average power of a stroke must not exceed the available aerobic power of a leg extension motion. The average power of a stroke can be calculated by:

$$P_{avg} = \frac{1}{t_{stroke}} \int_0^{t_{stroke}} P\, dt \tag{28}$$

in which t_{stroke} is the stroke time and P is the available aerobic power for skating. For this equation it is assumed that the skater is running at steady state speed, which results in zero anaerobic power.

The average power exerted during skating can be either measured with oxygen measurements during the speed skating measurements or calculated by Eq. (28).

8.6.3 Leg Power Calculation

For optimization the power exerted by the skater of stroke has to fulfill the constraints. The power of a stroke can be determined from the push-off force of the

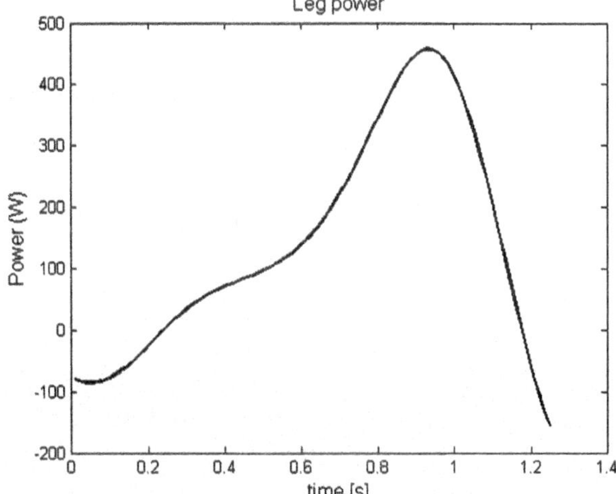

Fig. 11 Leg power in horizontal plane

skate on the ice and the leg extension in the horizontal plane, since no work is done in the vertical plane (Eq. (29)).

$$P_{leg} = F_{Tls} \dot{v}_{LS}. \tag{29}$$

An example of the leg power during a stroke can be seen in Fig. 11. At the beginning of the stroke the leg power becomes negative, which is caused by the negative direction of the lean angle. During the stroke the leg power increases to approximately 500 W. At the end of the stroke the extension speed decreases and the leg power becomes smaller.

8.7 Generalized Coordinates

The positions of the bodies (B, LS, RS) written in the generalized coordinates:

$$\begin{bmatrix} x_B \\ y_B \\ x_{LS} \\ y_{LS} \\ x_{RS} \\ y_{RS} \end{bmatrix} = \begin{bmatrix} x_B \\ y_B \\ x_B - \cos(\theta_{LS})u_{LS} - \sin(\theta_{LS})v_{LS} \\ y_B + \sin(\theta_{LS})u_{LS} - \cos(\theta_{LS})v_{LS} \\ x_B - \cos(\theta_{RS})u_{RS} - \sin(\theta_{RS})v_{RS} \\ y_B - \sin(\theta_{RS})u_{RS} + \cos(\theta_{RS})v_{RS} \end{bmatrix} \tag{30}$$

in which θ_i are the steer angles. These planar angular rotations can be calculated, since the velocity data of the skate in plane (x, y) are obtained. The skate steer angles are calculated by:

$$\begin{bmatrix} \theta_{LS} \\ \theta_{RS} \end{bmatrix} = \begin{bmatrix} -\tan^{-1}\left(\frac{\dot{y}_{LS}}{\dot{x}_{LS}}\right) \\ \tan^{-1}\left(\frac{\dot{y}_{RS}}{\dot{x}_{RS}}\right) \end{bmatrix}. \tag{31}$$

8.8 Convective Acceleration Terms

The convective acceleration terms of the leg extension constraints are:

$$\begin{aligned} h_{c1} = {} & \ddot{\theta}_{LS}\big(v_{LS}\cos(\theta_{LS}) - u_{LS}\sin(\theta_{LS})\big) + \ddot{u}_{LS}\cos(\theta_{LS}) + \ddot{v}_{LS}\sin(\theta_{LS}) \\ & - \dot{\theta}_{LS}\big(\dot{\theta}_{LS}\big(u_{LS}\cos(\theta_{LS}) + v_{LS}\sin(\theta_{LS})\big) - \dot{v}_{LS}\cos(\theta_{LS}) + \dot{u}_{LS}\sin(\theta_{LS})\big) \\ & + \dot{\theta}_{LS}\dot{v}_{LS}\cos(\theta_{LS}) - \dot{\theta}_{LS}\dot{u}_{LS}\sin(\theta_{LS}) \end{aligned} \tag{32}$$

$$\begin{aligned} h_{c2} = {} & \ddot{v}_{LS}\cos(\theta_{LS}) - \ddot{\theta}_{LS}\big(u_{LS}\cos(\theta_{LS}) + v_{LS}\sin(\theta_{LS})\big) - \ddot{u}_{LS}\sin(\theta_{LS}) \\ & - \dot{\theta}_{LS}\big(\dot{\theta}_{LS}\big(v_{LS}\cos(\theta_{LS}) - u_{LS}\sin(\theta_{LS})\big) + \dot{u}_{LS}\cos(\theta_{LS}) + \dot{v}_{LS}\sin(\theta_{LS})\big) \\ & - \dot{\theta}_{LS}\dot{u}_{LS}\cos(\theta_{LS}) - \dot{\theta}_{LS}\dot{v}_{LS}\sin(\theta_{LS}) \end{aligned} \tag{33}$$

$$\begin{aligned} h_{c3} = {} & \ddot{u}_{RS}\cos(\theta_{RS}) - \ddot{\theta}_{RS}\big(-v_{RS}\cos(\theta_{RS}) + u_{RS}\sin(\theta_{RS})\big) + \ddot{v}_{RS}\sin(\theta_{RS}) \\ & - \dot{\theta}_{RS}\big(\dot{\theta}_{RS}\big(u_{RS}\cos(\theta_{RS}) + v_{RS}\sin(\theta_{RS})\big) - \dot{v}_{RS}\cos(\theta_{RS}) + \dot{u}_{RS}\sin(\theta_{RS})\big) \\ & - \dot{\theta}_{RS}\dot{-}v_{RS}\cos(\theta_{RS}) - \dot{\theta}_{RS}\dot{u}_{RS}\sin(\theta_{RS}) \end{aligned} \tag{34}$$

$$\begin{aligned} h_{c4} = {} & \dot{\theta}_{RS}\big(\dot{\theta}_{RS}\big(-v_{RS}\cos(\theta_{RS}) + u_{RS}\sin(\theta_{RS})\big) - \dot{u}_{RS}\cos(\theta_{RS}) + \dot{v}_{RS}\sin(\theta_{RS})\big) \\ & + \ddot{v}_{RS}\cos(\theta_{RS}) - \ddot{u}_{RS}\sin(\theta_{RS}) - \ddot{\theta}_{RS}\big(u_{RS}\cos(\theta_{RS}) + v_{RS}\sin(\theta_{RS})\big) \\ & - \dot{\theta}_{RS}\dot{u}_{RS}\cos(\theta_{RS}) - \dot{\theta}_{RS}\dot{v}_{RS}\sin(\theta_{RS}). \end{aligned} \tag{35}$$

The convective acceleration terms of the skate constraints are:

$$\begin{aligned} h_{c5} &= \dot{\theta}_{LS}\dot{u}_{LS} - \ddot{v}_{LS} + \ddot{\theta}_{LS}u_{LS} + \dot{\theta}_{LS}\dot{x}_B\cos(\theta_{LS}) - \dot{\theta}_{LS}\dot{y}_B\sin(\theta_{LS}) \\ h_{c6} &= \dot{\theta}_{RS}\dot{u}_{RS} - \ddot{v}_{RS} + \ddot{\theta}_{RS}u_{RS} + \dot{\theta}_{RS}\dot{x}_B\cos(\theta_{RS}) + \dot{\theta}_{RS}\dot{y}_B\sin(\theta_{RS}). \end{aligned} \tag{36}$$

References

1. Allinger, T.L., van den Bogert, A.J.: Skating technique for the straights, based on the optimization of a simulation model. Med. Sci. Sports Exerc. **29**(2), 279–286 (1997)
2. Cabrera, D., Ruina, A., Kleshnev, V.: A simple 1+ dimensional model of rowing mimics observed forces and motions. Hum. Mov. Sci. **25**, 192–220 (2006)
3. de Boer, R.W., Nislen, K.L.: The gliding and push-off technique of male and female Olympic speed skaters. Int. J. Sport Biomech. **5**, 119–134 (1989)
4. de Boer, R.W., Schermerhorn, P., Gademan, J., de Groot, G., van Ingen Schenau, G.J.: Characteristic stroke mechanics of elite and trained male speed skaters. Int. J. Sport Biomech. **2**, 175–185 (1986)
5. de Koning, J.J.: Biomechanical aspects of speed skating. PhD thesis, Free University of Amsterdam (1991). ISBN 90-9003956-2
6. de Koning, J.J., van Ingen Schenau, G.J.: Performance-determining factors in speed skating. In: Zatsiorsky, V. (ed.) Biomechanics in Sport: Performance Enhancement and Injury Prevention. Olympic Encyclopedia of Sports Medicine, vol. IX (2000). ISBN 978-0-632-05392-6
7. de Koning, J.J., de Groot, G., van Ingen Schenau, G.J.: Ice friction during speed skating. J. Biomech. **25**(6), 565–571 (1992)

8. Fintelman, D.M.: Literature study: biomechanical models for speed skating. MSc report, BioMechanical Engineering, Delft University of Technology (2010)
9. McLean, S.G., Su, A., van den Bogert, A.J.: Development and validation of a 3-D model to predict knee joint loading during dynamic movement. J. Biomech. Eng. **125**, 864–874 (2003)
10. Otten, E.: Inverse and forward dynamics: models of multi-body systems. Philos. Trans. R. Soc. Lond. **358**, 1493–1500 (2003)
11. Vandervoort, A.A., Sale, D.G., Moroz, J.: Comparison of motor unit activation during unilateral and bilateral leg extension. J. Sport Sci. **13**, 153–170 (1995)
12. van Ingen Schenau, G.J.: The influence of air friction in speed skating. J. Biomech. **15**(6), 449–458 (1982)

Contact Modelling in Multibody Systems by Means of a Boundary Element Co-simulation and a DIRICHLET-to-NEUMANN Algorithm

János Zierath and Christoph Woernle

Abstract The present contribution introduces the modelling of elastic contacts by coupled multibody an boundary element systems. Compared to contacts modelled by impact laws, physically more accurate results can be obtained. Due to the use of boundary element systems, the contact stresses are obtained within the contact calculation.

A new three-dimensional contact element for boundary element systems is developed. The mortar element uses the mixed formulation of boundary element formulations. The algorithm for the iteration of contact states is based on an DIRICHLET-to-NEUMANN algorithm. Herein, both contacting bodies are calculated serially. In the first calculation step one of the contacting bodies represents a rigid obstacle for the other elastic one. The resulting reaction forces on the elastic body are partially transferred on the other one, which is for the second calculation step no longer rigid. As a result, the obstacle is deformed and the next iteration starts. The algorithm converges if the numerical equilibrium in the contact interface is reached.

1 Introduction

Simulation of elastic multibody systems based on a finite element analysis has become usual in commercial multibody programs, such as MSC.Adams or SIMPACK. To model contacts between elastic bodies, the classical impact laws used in rigid-body dynamics, such as NEWTON's kinematical approach or POISSON's impact law [29] based on integrated forces, are not suitable. Instead contact models derived from the elastic body description have to be applied. Typically elastic body models are formulated by means of a modal approach like the Craig-Bampton method [12].

J. Zierath (✉)
W2E Wind to Energy GmbH, Strandstrasse 96, Rostock, Germany
e-mail: jzierath@wind-to-energy.de

C. Woernle
Chair of Technical Dynamics, University of Rostock, Justus-von-Liebig-Weg 6, Rostock, Germany
e-mail: woernle@uni-rostock.de

J.-C. Samin, P. Fisette (eds.), *Multibody Dynamics*,
Computational Methods in Applied Sciences 28,
DOI 10.1007/978-94-007-5404-1_2,

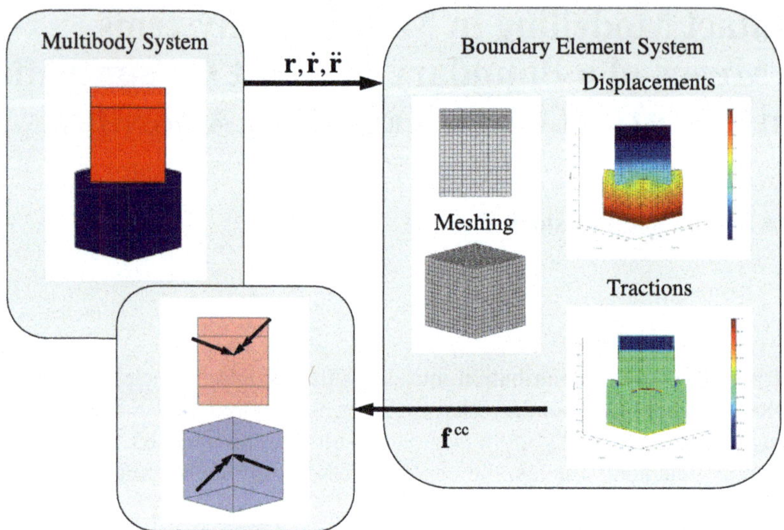

Fig. 1 Coupling of multibody system and boundary element models

Modal functions describe the elastic deformations by a superposition of global mode shapes with static mode shapes or frequency response modes. However, the global mode shapes are not able to describe the elastic deformations of the bodies in the contact area. Consequently, the number of modes used for the contacting elastic bodies has to be increased. Another method to describe elastic contacts is the coupling between a finite element (FE) model and a multibody system (MBS), see [15]. As a FE body has typically a higher degree of freedom (DOF) than a complete MBS, the coupling of FE models and MBS results in a large computational amount.

The aim of this chapter is to present a method to couple multibody systems with boundary element systems for contact simulation as shown in Fig. 1. This approach is seen as an alternative to model contact by coupled multibody/finite element simulations as presented by [2]. In this chapter three-dimensional problems are considered only. Due to the fact that three-dimensional contact algorithms for boundary element systems are sparely described in literature, a new contact element based on mortar methods has been developed, see [37].

This contribution is organised as follows. In Sect. 2 the fundamentals of multibody systems are introduced. The description restricts on differential algebraic equations. Based on the boundary integral formulation for elastostatics introduced in Sect. 3, the DIRICHLET-to-NEUMANN algorithm for contact calculation in boundary element systems is described in Sect. 4. This algorithm is based on mortar methods which are presented in Sect. 5. In Sect. 6 the numerical implementation of the mortar methods is described. The contact stresses are obtained from the contact calculation in boundary element systems. The calculation of the resulting contact force from the stresses is described in Sect. 7. As an example the dynamic simulation of two impacting spheres is presented in Sect. 8.

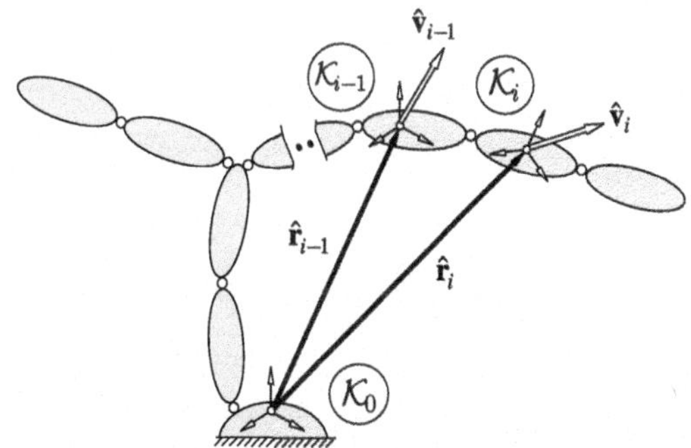

Fig. 2 Kinematics of rigid multibody systems

2 Fundamentals of Multibody Systems

In this section the theoretical background of multibody systems is shortly described. Due to the fact that the contact acts as a force element inside the multibody environment an introduction to constrained systems of rigid bodies is given only. The equations of motion are formulated as differential-algebraic equations (DAE). This formulation is typically used in commercial multibody simulation packages. A detailed description of multibody system dynamics is given in [32] and [35].

The position of the ith body of a constrained multibody system is described by a spatial vector $\hat{\mathbf{r}}_i$ consisting of the position vector $\mathbf{r}_i$ from an inertial system $\mathcal{K}_0$ to a body-fixed coordinate system $\mathcal{K}_i$ and $m_i \geq 3$ coordinates γ_i describing the spatial orientation of $\mathcal{K}_i$ relative to $\mathcal{K}_0$ (Fig. 2),

$$\hat{\mathbf{r}}_i = \begin{bmatrix} \mathbf{r}_i \\ \gamma_i \end{bmatrix}. \tag{1}$$

Examples for the definition of γ_i are EULER angles or EULER parameters (unit quaternions). Accordingly, the absolute velocity of the ith body is described by the velocity $\mathbf{v}_i$ and the angular velocity ω_i of $\mathcal{K}_i$ put together in the six-dimensional spatial vector

$$\hat{\mathbf{v}}_i = \begin{bmatrix} \mathbf{v}_i \\ \omega_i \end{bmatrix}. \tag{2}$$

To describe the position and velocity of the overall system with N bodies, the vectors

$$\hat{\mathbf{r}} = \begin{bmatrix} \hat{\mathbf{r}}_1 \\ \vdots \\ \hat{\mathbf{r}}_N \end{bmatrix}, \qquad \hat{\mathbf{v}} = \begin{bmatrix} \hat{\mathbf{v}}_1 \\ \vdots \\ \hat{\mathbf{v}}_N \end{bmatrix} \tag{3}$$

are introduced. The relation between the time derivatives of $\hat{\mathbf{r}}_i$ and $\hat{\mathbf{v}}_i$ is given by kinematic differential equations of the form

$$\begin{aligned} &\dot{\hat{\mathbf{r}}}_i = \widehat{\mathbf{H}}_i(\hat{\mathbf{r}}_i)\hat{\mathbf{v}}_i, \quad i = 1, \ldots, N \\ &\text{or} \\ &\dot{\hat{\mathbf{r}}} = \widehat{\mathbf{H}}(\hat{\mathbf{r}})\hat{\mathbf{v}}, \quad \widehat{\mathbf{H}} = \mathrm{diag}(\widehat{\mathbf{H}}_1 \ldots \widehat{\mathbf{H}}_N). \end{aligned} \tag{4}$$

The constraints between the rigid bodies at the position, velocity, and acceleration levels are formulated in the implicit form,

$$\mathbf{g}(\hat{\mathbf{r}}, t) = \mathbf{0}, \tag{5}$$

$$\dot{\mathbf{g}} \equiv \mathbf{G}(\hat{\mathbf{r}}, t)\hat{\mathbf{v}} + \bar{\mathbf{g}}(\hat{\mathbf{r}}, t) = \mathbf{0}, \tag{6}$$

$$\ddot{\mathbf{g}} \equiv \mathbf{G}(\hat{\mathbf{r}}, t)\dot{\hat{\mathbf{v}}} + \bar{\bar{\mathbf{g}}}(\hat{\mathbf{r}}, \hat{\mathbf{v}}, t) = \mathbf{0}. \tag{7}$$

The kinetic differential equations are derived from the principles of linear momentum and angular momentum. Using spatial force vectors $\hat{\mathbf{f}}_i$ containing a pair of a force and a torque the kinetic equations of the ith body are

$$\widehat{\mathbf{M}}_i \dot{\hat{\mathbf{v}}}_i = \hat{\mathbf{f}}_i^{\mathrm{c}} + \hat{\mathbf{f}}_i^{\mathrm{a}} + \hat{\mathbf{f}}_i^{\mathrm{r}} + \hat{\mathbf{f}}_i^{\mathrm{cc}}, \quad i = 1, \ldots, N, \tag{8}$$

with the symmetric, positive definite $(6, 6)$ mass matrix $\widehat{\mathbf{M}}_i$, gyroscopic and CORIOLIS forces $\hat{\mathbf{f}}_i^{\mathrm{c}}$, applied forces $\hat{\mathbf{f}}_i^{\mathrm{a}}$, and reaction forces (constraint forces) $\hat{\mathbf{f}}_i^{\mathrm{r}}$. Additionally, the vector $\hat{\mathbf{f}}_i^{\mathrm{cc}}$ is introduced to represent the contact forces. Inside the multibody environment, the contact force is represented by an applied force. The overall kinetic equations can be written as

$$\widehat{\mathbf{M}}\dot{\hat{\mathbf{v}}} = \hat{\mathbf{f}}^{\mathrm{c}} + \hat{\mathbf{f}}^{\mathrm{a}} + \hat{\mathbf{f}}^{\mathrm{r}} + \hat{\mathbf{f}}^{\mathrm{cc}}$$
$$\text{with} \quad \widehat{\mathbf{M}} = \mathrm{diag}(\widehat{\mathbf{M}}_1 \ldots \widehat{\mathbf{M}}_N), \quad \hat{\mathbf{f}}^{\mathrm{c/a/r/cc}} = \begin{bmatrix} \hat{\mathbf{f}}_1^{\mathrm{c/a/r/cc}} \\ \vdots \\ \hat{\mathbf{f}}_N^{\mathrm{c/a/r/cc}} \end{bmatrix}. \tag{9}$$

The reaction forces $\hat{\mathbf{f}}_i^{\mathrm{r}}$ have components in the constrained spatial directions only, given by the row vectors of the constraint matrix $\mathbf{G}$ from (6). Accordingly, they can be expressed by means of the explicit reaction force equations

$$\hat{\mathbf{f}}^{\mathrm{r}} = \mathbf{G}^{\mathrm{T}}\boldsymbol{\lambda} \tag{10}$$

with reaction force coordinates (LAGRANGIAN multipliers) $\boldsymbol{\lambda}$. If the positions $\hat{\mathbf{r}}$ and the velocities $\hat{\mathbf{v}}$, which have to be consistent with the constraints (5) and their first-order time derivatives (6), are given, (9) with (10) and (7) together represent a set of linear equations to determine uniquely the accelerations $\dot{\hat{\mathbf{v}}}$ and the reaction force coordinates $\boldsymbol{\lambda}$,

$$\begin{bmatrix} \widehat{\mathbf{M}} & \mathbf{G}^{\mathrm{T}} \\ \mathbf{G} & \mathbf{0} \end{bmatrix} \cdot \begin{bmatrix} \dot{\hat{\mathbf{v}}} \\ -\boldsymbol{\lambda} \end{bmatrix} = \begin{bmatrix} \hat{\mathbf{f}}^{\mathrm{c}} + \hat{\mathbf{f}}^{\mathrm{a}} + \hat{\mathbf{f}}^{\mathrm{cc}} \\ -\bar{\bar{\mathbf{g}}} \end{bmatrix}. \tag{11}$$

Numerical integration of the velocity $\hat{\mathbf{v}}$ obtained from the kinematic differential equation (4) and of the acceleration $\dot{\hat{\mathbf{v}}}$ obtained from (11) yields the motion of the system, described by the position $\hat{\mathbf{r}}(t)$ and the velocity $\hat{\mathbf{v}}(t)$. The reaction force coordinates $\boldsymbol{\lambda}$ obtained from (11) yield the reaction forces $\hat{\mathbf{f}}^{\mathrm{r}}$ by means of the explicit reaction force equations (10). During numerical integration, the constraints at the position and velocity levels may be violated necessitating a constraint stabilisation [4].

3 Boundary Integral Formulation in Elastostatics

The boundary integral formulation starts from the partial differential equation representing the quasistatic equilibrium inside the volume V,

$$\sigma_{ij,j} + b_i = 0 \quad \text{in } V \quad \text{for } i = x, y, z. \tag{12}$$

Herein the variable σ_{ij} represents the stress tensor inside the volume V and b_i the body forces due to the acceleration of the body. In addition to the equilibrium inside the volume V, the equilibrium conditions on the surface S,

$$t_i = \sigma_{ij} n_j \quad \text{on } S, \tag{13}$$

have to be fulfilled. Herein t_i is the stress vector, also called traction vector, representing the equilibrium on the surface S with the outward normal vector n_j.

The surface is divided into regions with different types of boundary conditions. There are regions on the surface S where the DIRICHLET boundary conditions

$$u_i = \bar{u}_i \quad \text{on } S_u \tag{14}$$

have to be fulfilled, and regions with NEUMANN boundary conditions,

$$t_i = \bar{t}_i \quad \text{on } S_t. \tag{15}$$

Due to the fact that each point on the surface S has three degrees of freedom, each direction $i = x, y, z$ has to be treated separately. Thus, the requirements $S_i = S_{ui} \cup S_{ti}$ and $S_{ui} \cap S_{ti} = 0$ have to be fulfilled for each coordinate direction i.

By using the weighted residual formulation

$$\int_V (\sigma_{ij,j} + b_i)\, w_i \,\mathrm{d}\tilde{V} + \int_{S_u} (u_i - \bar{u}_i)\tilde{w}_i \,\mathrm{d}\tilde{S} + \int_{S_t} (t_i - \bar{t}_i)\tilde{w}_i \,\mathrm{d}\tilde{S} = 0 \tag{16}$$

and introducing the fundamental solutions u^*_{ij} and t^*_{ij} as weighting functions, the SOMIGLIANA identity

$$u_i(\mathbf{x}) = \int_S u^*_{ij}(\mathbf{x},\mathbf{y}) t_j(\mathbf{y}) \,\mathrm{d}\tilde{S} - \int_S t^*_{ij}(\mathbf{x},\mathbf{y}) u_j(\mathbf{y}) \,\mathrm{d}\tilde{S} + \int_V u^*_{ij}(\mathbf{x},\mathbf{y}) b_j(\mathbf{y}) \,\mathrm{d}\tilde{V} \tag{17}$$

is obtained which is valid for an arbitrary load point $\mathbf{x}$ inside the volume V. The fundamental solutions for elastostatics have been developed by Lord KELVIN, see [33]. They represent the response of the system at any arbitrary field point $\mathbf{y}$ due to a unit load at a chosen load point $\mathbf{x}$. In this chapter, three-dimensional boundary element formulations are treated only. Because each point has three degrees of freedom, the coordinate direction has also to be considered. A volume integral with the specific body forces b_j is given on the left hand side of (17).

According to [1], the displacement fundamental solution, also called DIRICHLET fundamental solution, at a field point $\mathbf{y}$ in direction j due to a unit load at a load point $\mathbf{x}$ in direction i is given by

$$u^*_{ij}(\mathbf{x},\mathbf{y}) = \frac{1+\nu}{8\pi E(1-\nu) r}\left[(3-4\nu)\delta_{ij} + r_{,i}\, r_{,j}\right]. \tag{18}$$

Herein, the variable E represents YOUNG's modulus, ν POISSON's ratio, and δ_{ij} the KRONECKER delta. The distance r between a load point $\mathbf{x}$ and a field point $\mathbf{y}$ is given by the EUCLIDIAN norm

$$r = \|\mathbf{y} - \mathbf{x}\|. \tag{19}$$

The partial derivatives of the EUCLIDIAN distance r with respect to the coordinate direction i can easily be obtained by

$$r_{,i} = \frac{y_i - x_i}{r}. \tag{20}$$

The stress fundamental solution, also known as NEUMANN fundamental solution, at a field point $\mathbf{y}$ in direction j due to a unit load at a load point $\mathbf{x}$ in direction i is given by

$$t^*_{ij}(\mathbf{x}, \mathbf{y}) = -\frac{1}{8\pi(1-\nu)r^2}\left[\frac{\partial r}{\partial n}\left[(1-2\nu)\delta_{ij} + 3r_{,i}\, r_{,j}\right] - (1-2\nu)(n_i r_{,i} - n_j r_{,j})\right], \tag{21}$$

where n_i represents the ith component of the outward normal vector $\mathbf{n}$. Herein, the derivative in normal direction is calculated by

$$\frac{\partial r}{\partial n} = r_{,i}\, n_i. \tag{22}$$

According to the third summand in (17), a remaining volume integral with the specific body forces b_j has to be taken into account. There are different procedures to calculate body forces. A method is the discretisation of the volume by using cells, see [1]. For a boundary element formulation this method is less appropriate because an additional volume mesh has to be introduced. A proposal for the transformation of the body forces from the volume V on the surface S was made in [13]. In the present chapter, quasistatic body forces are taken into account only. By this, elastic vibrations of the body are neglected.

Starting from SOMIGLIANA's identity (17) for a load point $\mathbf{x}$ on the surface S, the integral equation under consideration of the boundary factor c_{ij} is given by

$$c_{ij} u_j(\mathbf{x}) = \int_S u^*_{ij}(\mathbf{x}, \mathbf{y}) t_j(\mathbf{y})\, \mathrm{d}\tilde{S} - \int_S t^*_{ij}(\mathbf{x}, \mathbf{y}) u_j(\mathbf{y})\, \mathrm{d}\tilde{S} + B_i(\mathbf{x}). \tag{23}$$

The boundary factor c_{ij} is a correction factor to consider load points on the surface S of the volume V. Due to the introduction of surface elements, the surface integrals in (23) are evaluated element-wise (summation over all elements e),

$$c_{ij} u_j(\mathbf{x}) = \sum_e \int_{S_e} u^*_{ij}(\mathbf{x}, \mathbf{y}) t_j(\mathbf{y})\, \mathrm{d}\tilde{S} - \sum_e \int_{S_e} t^*_{ij}(\mathbf{x}, \mathbf{y}) u_j(\mathbf{y})\, \mathrm{d}\tilde{S} + \sum_e B_{ie}(\mathbf{x}). \tag{24}$$

The continuous functions $u_j(\mathbf{y})$ for the displacements and $t_j(\mathbf{y})$ for the tractions can be replaced by discrete nodal values which leads to

$$u_j(\mathbf{y}) = N_k\big(\mathbf{y}(\xi, \eta)\big) u_{kj} \quad \text{and} \quad t_j(\mathbf{y}) = N_k\big(\mathbf{y}(\xi, \eta)\big) t_{kj}. \tag{25}$$

Herein, ξ and η are the local element coordinates. The index k represents the kth node of element e and j represents one of the coordinate directions x, y, or z. Caused by the fact that the values u_{kj} and t_{kj} can belong to more than one element, it is necessary to replace the summation over all elements e in (24) by an assembly,

$$\begin{aligned} c_{ij} N_k\big(\mathbf{x}(\xi,\eta)\big) u_{kj} = & \bigcup_e \int_{S_e} u^*_{ij}\big(\mathbf{x},\mathbf{y}(\xi,\eta)\big) N_k\big(\mathbf{y}(\xi,\eta)\big)\, \mathrm{d}\tilde{S}\, t_{kj} \\ & - \bigcup_e \int_{S_e} t^*_{ij}\big(\mathbf{x},\mathbf{y}(\xi,\eta)\big) N_k\big(\mathbf{y}(\xi,\eta)\big)\, \mathrm{d}\tilde{S}\, u_{kj} + \sum_e B_{ie}(\mathbf{x}). \end{aligned} \tag{26}$$

Within this chapter, collocation methods are applied to obtain a discretised system of equations from (26). Assuming that the body is discretised by n nodes, the load point $\mathbf{x}$ is set to the location of each node $\mathbf{x}_l$ with $l = 1,\ldots,n$. That means, instead of the arbitrary load point $\mathbf{x}$ in (26), the nodal coordinates $\mathbf{x}_l$ of the discretised geometry are used as collocation points leading to

$$\begin{aligned} c_{ij} u_{1j} = & \bigcup_e \int_{S_e} u^*_{ij}\big(\mathbf{x}_1,\mathbf{y}(\xi,\eta)\big) N_k\big(\mathbf{y}(\xi,\eta)\big)\, \mathrm{d}\tilde{S}\, t_{kj} \\ & - \bigcup_e \int_{S_e} t^*_{ij}\big(\mathbf{x}_1,\mathbf{y}(\xi,\eta)\big) N_k\big(\mathbf{y}(\xi,\eta)\big)\, \mathrm{d}\tilde{S}\, u_{kj} + \sum_e B_{ie}(\mathbf{x}_1), \\ & \vdots \\ c_{ij} u_{lj} = & \bigcup_e \int_{S_e} u^*_{ij}\big(\mathbf{x}_l,\mathbf{y}(\xi,\eta)\big) N_k\big(\mathbf{y}(\xi,\eta)\big)\, \mathrm{d}\tilde{S}\, t_{kj} \\ & - \bigcup_e \int_{S_e} t^*_{ij}\big(\mathbf{x}_l,\mathbf{y}(\xi,\eta)\big) N_k\big(\mathbf{y}(\xi,\eta)\big)\, \mathrm{d}\tilde{S}\, u_{kj} + \sum_e B_{ie}(\mathbf{x}_l), \\ & \vdots \\ c_{ij} u_{nj} = & \bigcup_e \int_{S_e} u^*_{ij}\big(\mathbf{x}_n,\mathbf{y}(\xi,\eta)\big) N_k\big(\mathbf{y}(\xi,\eta)\big)\, \mathrm{d}\tilde{S}\, t_{kj} \\ & - \bigcup_e \int_{S_e} t^*_{ij}\big(\mathbf{x}_n,\mathbf{y}(\xi,\eta)\big) N_k\big(\mathbf{y}(\xi,\eta)\big)\, \mathrm{d}\tilde{S}\, u_{kj} + \sum_e B_{ie}(\mathbf{x}_n). \end{aligned} \tag{27}$$

The integrals from (27) can be evaluated with respect to the element surface S_e. By assembling (27) over all elements e, the system of equations

$$(\mathbf{C} + \mathbf{H}^*)\mathbf{u} = \mathbf{G}\mathbf{t} + \mathbf{B} \tag{28}$$

is obtained. Herein, the matrix $\mathbf{C}$ contains the boundary factors c_{ij}. According to (27), this matrix is block diagonal. Due to the use of the fundamental solutions u^*_{ij} and t^*_{ij}, the matrices $\mathbf{H}^*$ and $\mathbf{G}$ are fully populated, unsymmetrical and not necessarily positive definite. The vector $\mathbf{B}$ contains the projected body forces. The system of equations (28) can be written as

$$\mathbf{H}\mathbf{u} = \mathbf{G}\mathbf{t} + \mathbf{B} \tag{29}$$

which is the form known from literature [1, 5, 17].

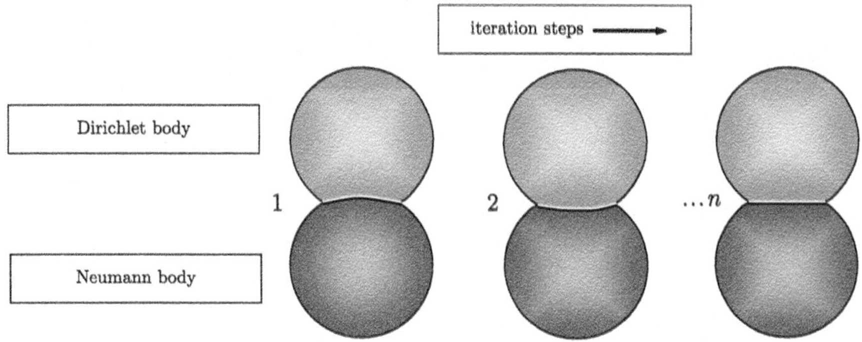

Fig. 3 Principle of DIRICHLET-to-NEUMANN algorithm

4 Contact Calculation by a DIRICHLET-to-NEUMANN Algorithm

This contribution presents an extension to the LAGRANGIAN multiplier approach introduced in [42]. The algorithm is introduced to reduce the amount of random access memory and to decrease the calculation time for large systems. The algorithm presented here is based on [25] and [38]. The principle of this DIRICHLET-to-NEUMANN algorithm which is called a nonlinear GAUSS-SEIDEL block in [24] is shown in Fig. 3.

The two contacting bodies are divided into a NEUMANN body and a DIRICHLET body. In this chapter the mortar and non-mortar are equivalent to the NEUMANN and DIRICHLET body, respectively. Within each iteration, a linear NEUMANN problem and a nonlinear DIRICHLET problem is solved. Starting from (29), the system of equations for the NEUMANN body can be expressed by

$$\mathbf{H}^{\mathrm{N}} \Delta \mathbf{u}^{\mathrm{N}(j)} = \mathbf{G}^{\mathrm{N}} \Delta \mathbf{t}^{\mathrm{N}(j)} + \mathbf{B}^{\mathrm{N}(j)}, \tag{30}$$

where the index j represents the iteration step within the DIRICHLET-to-NEUMANN algorithm and the upper right index N denotes the NEUMANN body. The system of equations can be partitioned into degrees of freedom that are in contact and such that are not,

$$\begin{bmatrix} \mathbf{H}^{\mathrm{N}}_{\mathrm{nc}} & \mathbf{H}^{\mathrm{N}}_{\mathrm{cc}} \end{bmatrix} \begin{bmatrix} \Delta \mathbf{u}^{\mathrm{N}(j)}_{\mathrm{nc}} \\ \Delta \mathbf{u}^{\mathrm{N}(j)}_{\mathrm{cc}} \end{bmatrix} = \begin{bmatrix} \mathbf{G}^{\mathrm{N}}_{\mathrm{nc}} & \mathbf{G}^{\mathrm{N}}_{\mathrm{cc}} \end{bmatrix} \begin{bmatrix} \Delta \mathbf{t}^{\mathrm{N}(j)}_{\mathrm{nc}} \\ \Delta \mathbf{t}^{\mathrm{N}(j)}_{\mathrm{cc}} \end{bmatrix} + \mathbf{B}^{\mathrm{N}(j)}. \tag{31}$$

Herein, the degrees of freedom which are not in contact are denoted by the lower right indices nc and those which are in contact by cc, respectively. The body forces $\mathbf{B}^{\mathrm{N}(j)}$ are typically applied within the first iteration, that means $\mathbf{B}^{\mathrm{N}(j)} \stackrel{!}{=} \mathbf{0}$ for $j > 1$. As indicated by the name of the algorithm, the tractions caused by the contact are applied to the NEUMANN body, where for the first iteration step, thus $j = 1$, the contact tractions $\Delta \mathbf{t}^{\mathrm{N}(1)}_{\mathrm{cc}}$ are assumed to be zero. Considering also the other boundary conditions in the non-contact area and sorting the system of equations (31) with respect to known and unknown values leads to

$$\begin{bmatrix} -\mathbf{G}^{\mathrm{N}}_{\mathrm{ncD}} & \mathbf{H}^{\mathrm{N}}_{\mathrm{ncN}} & \mathbf{H}^{\mathrm{N}}_{\mathrm{cc}} \end{bmatrix} \begin{bmatrix} \Delta\mathbf{t}^{\mathrm{N}(j)}_{\mathrm{ncD}} \\ \Delta\mathbf{u}^{\mathrm{N}(j)}_{\mathrm{ncN}} \\ \Delta\mathbf{u}^{\mathrm{N}(j)}_{\mathrm{cc}} \end{bmatrix} = \begin{bmatrix} -\mathbf{H}^{\mathrm{N}}_{\mathrm{ncD}} & \mathbf{G}^{\mathrm{N}}_{\mathrm{ncN}} & \mathbf{G}^{\mathrm{N}}_{\mathrm{cc}} \end{bmatrix} \begin{bmatrix} \Delta\mathbf{u}^{\mathrm{N}(j)}_{\mathrm{ncD}} \\ \Delta\mathbf{t}^{\mathrm{N}(j)}_{\mathrm{ncN}} \\ \Delta\mathbf{t}^{\mathrm{N}(j)}_{\mathrm{cc}} \end{bmatrix} + \mathbf{B}^{\mathrm{N}(j)}. \tag{32}$$

Equation (32) represents a fully populated quadrangular system of equation of the form $\mathbf{A}^{\mathrm{N}}\mathbf{x}^{\mathrm{N}} = \mathbf{b}^{\mathrm{N}}$ which can be solved by GAUSS or *LU* decomposition. The deformed mesh of the NEUMANN body is stored after each iteration and acts as rigid obstacle for the DIRICHLET body. In order to obtain the overall deformations and tractions the incremental deformations $\Delta\mathbf{u}^{\mathrm{N}(j)}$ and tractions $\Delta\mathbf{t}^{\mathrm{N}(j)}$ are summarised over all iterations j. According to [25] a numerical damping parameter δ_{D} with $0 < \delta_{\mathrm{D}} \leq 1$ is introduced, resulting in

$$\mathbf{u}^{\mathrm{N}(j)} = \mathbf{u}^{\mathrm{N}(j-1)} + \delta_{\mathrm{D}}\Delta\mathbf{u}^{\mathrm{N}(j)} \quad \text{and} \quad \mathbf{t}^{\mathrm{N}(j)} = \mathbf{t}^{\mathrm{N}(j-1)} + \delta_{\mathrm{D}}\Delta\mathbf{t}^{\mathrm{N}(j)}. \tag{33}$$

The actual nodal coordinates of the NEUMANN body are obtained by

$$\mathbf{x}^{\mathrm{N}(j)} = \mathbf{x}^{\mathrm{N}(j-1)} + \delta_{\mathrm{D}}\Delta\mathbf{u}^{\mathrm{N}(j)}. \tag{34}$$

The second step within each iteration is the solution of a nonlinear DIRICHLET problem. The formulation starts from the system of equations

$$\mathbf{H}^{\mathrm{D}}\mathbf{u}^{\mathrm{D}(j)} = \mathbf{G}^{\mathrm{D}}\mathbf{t}^{\mathrm{D}(j)} + \mathbf{B}^{\mathrm{D}}, \tag{35}$$

where the upper right index D denotes the DIRICHLET body and j the number of iteration. Comparing (30) and (35) differences can be determined between the systems of equations of the NEUMANN and the DIRICHLET body. The calculation of the DIRICHLET body always starts from the reference configuration $\mathbf{x}^{\mathrm{D}(1)}$ within each iteration step j, which means in general the undeformed configuration. Hence, no incremental update has to be done for the displacements $\mathbf{u}^{\mathrm{D}(j)}$ and $\mathbf{t}^{\mathrm{D}(j)}$. The body forces $\mathbf{B}^{\mathrm{D}}$ have to be applied within each iteration step or by superposition once in a preprocessing step.

The system of equations (35) can be partitioned and sorted leading to

$$\begin{bmatrix} -\mathbf{G}^{\mathrm{D}}_{\mathrm{ncD}} & \mathbf{H}^{\mathrm{D}}_{\mathrm{ncN}} & \mathbf{H}^{\mathrm{D}}_{\mathrm{cc}} & -\mathbf{G}^{\mathrm{D}}_{\mathrm{cc}} \end{bmatrix} \begin{bmatrix} \mathbf{t}^{\mathrm{D}(j)}_{\mathrm{ncD}} \\ \mathbf{u}^{\mathrm{D}(j)}_{\mathrm{ncN}} \\ \mathbf{u}^{\mathrm{D}(j)}_{\mathrm{cc}} \\ \mathbf{t}^{\mathrm{D}(j)}_{\mathrm{cc}} \end{bmatrix} = \begin{bmatrix} -\mathbf{H}^{\mathrm{D}}_{\mathrm{ncD}} & \mathbf{G}^{\mathrm{D}}_{\mathrm{ncN}} \end{bmatrix} \begin{bmatrix} \mathbf{u}^{\mathrm{D}(j)}_{\mathrm{ncD}} \\ \mathbf{t}^{\mathrm{D}(j)}_{\mathrm{ncN}} \end{bmatrix} + \mathbf{B}^{\mathrm{D}}, \tag{36}$$

which represents a system of linear equations of the form $\mathbf{A}^{\mathrm{D}}\mathbf{x}^{\mathrm{D}(j)} = \mathbf{b}^{\mathrm{D}(j)}$. Due to the fact that within the potential contact area the displacements $\mathbf{u}^{\mathrm{D}(j)}_{\mathrm{cc}}$ and the tractions $\mathbf{t}^{\mathrm{D}(j)}_{\mathrm{cc}}$ are unknown, the system of equations (36) is rectangular, that means the number of equations is lower than the number of unknown nodal values. This results in an infinite number of possible solutions.

To solve (36), a quadratic optimisation problem with equality and inequality constraints is formulated,

$$\frac{1}{2}\mathbf{x}^{\mathrm{D}(j)\mathrm{T}}\mathbf{Q}^{\mathrm{D}(j)}\mathbf{x}^{\mathrm{D}(j)} + \mathbf{x}^{\mathrm{D}(j)\mathrm{T}}\mathbf{p}^{\mathrm{D}(j)} = \min_{\mathbf{x}^{\mathrm{D}(j)}}$$
$$\text{subject to} \quad \mathbf{A}^{\mathrm{D}}\mathbf{x}^{\mathrm{D}(j)} = \mathbf{b}^{\mathrm{D}(j)} \quad \text{and} \quad \mathbf{N}^{\mathrm{D}(j)}\mathbf{x}^{\mathrm{D}(j)} + \mathbf{d}_0^{\mathrm{D}(j)} \geq \mathbf{0}. \tag{37}$$

Herein $\mathbf{Q}^{\mathrm{D}(j)}$ represents the objective function matrix containing the weighting factors of the nodal values $\mathbf{x}^{\mathrm{D}(j)}$, and $\mathbf{p}^{\mathrm{D}(j)}$ contains the coefficients of the linear part of the quadratic objective function. Typically, $\mathbf{Q}^{\mathrm{D}(j)}$ is a unit matrix, and the elements of $\mathbf{p}^{\mathrm{D}(j)}$ are zero. If the objective function matrix $\mathbf{Q}^{\mathrm{D}(j)}$ is equal to the identity matrix, the result of such an optimisation is identical to the MOORE-PENROSE inverse of the left-hand side matrix $\mathbf{A}^{\mathrm{D}}$ given in (36). As equality condition the system of equations (36) is used. The inequality equations $\mathbf{N}^{\mathrm{D}(j)}\mathbf{x}^{\mathrm{D}(j)} + \mathbf{d}_0^{\mathrm{D}(j)} \geq \mathbf{0}$ represent the one-sided constraints which are built up with respect to the deformed surface of the NEUMANN body representing the mortar side of the contacting bodies. The matrix $\mathbf{N}^{\mathrm{D}(j)}$ containing the normal vectors and the vector $\mathbf{d}_0^{\mathrm{D}(j)}$ containing the initial gaps have to be built up within each iteration step j.

The theoretical background and computational implementation of the quadratic programming is not explained here. A description and implementation on that topic is given by [26].

If the optimisation is successful, the reaction forces in form of the tractions $\mathbf{t}_{\mathrm{cc}}^{\mathrm{D}(j)}$ are obtained. The sum of the tractions inside the contact area has to vanish,

$$\mathbf{t}_{\mathrm{cc}}^{\mathrm{D}(j)} + \mathbf{t}_{\mathrm{cc}}^{\mathrm{N}(j)} = \mathbf{0}. \tag{38}$$

Because of the possibility of nonconforming meshes of the contacting bodies, dynamic constraints based on a mortar method not explained so far are introduced. This leads to a reformulation of (38),

$$\mathbf{M}_t^{\mathrm{D}(j)}\mathbf{t}_{\mathrm{cc}}^{\mathrm{D}(j)} + \mathbf{M}_t^{\mathrm{N}(j)}\mathbf{t}_{\mathrm{cc}}^{\mathrm{N}(j)} = \mathbf{0}, \quad \mathbf{0} \in \mathbb{R}^{\mathrm{N}}, \tag{39}$$

where the mortar matrices $\mathbf{M}_t^{\mathrm{D}(j)}$ and $\mathbf{M}_t^{\mathrm{N}(j)}$ have to be built up within each iteration based on the actual geometry of the NEUMANN body as it will be shown in Sect. 5. Due to the fact that mortar matrices are formed with respect to the degrees of freedom in contact on the NEUMANN body representing the mortar side, the matrix $\mathbf{M}_t^{\mathrm{N}(j)}$ is quadratic and has a full rank. Inversion of the matrix $\mathbf{M}_t^{\mathrm{N}(j)}$ in (39) leads to

$$\mathbf{t}_{\mathrm{cc}}^{\mathrm{N}(j)} = -\mathbf{M}_t^{\mathrm{N}(j)-1}\mathbf{M}_t^{\mathrm{D}(j)}\mathbf{t}_{\mathrm{cc}}^{\mathrm{D}(j)}, \tag{40}$$

which represents a mapping of the tractions in the contact area from the DIRICHLET body to the NEUMANN body. Because the system of equations (30) is given in an incremental form, the incremental tractions $\Delta\mathbf{t}_{\mathrm{cc}}^{\mathrm{N}(j)}$ are calculated by

$$\Delta\mathbf{t}_{\mathrm{cc}}^{\mathrm{N}(j)} = \delta_{\mathrm{N}}\left(\mathbf{t}_{\mathrm{cc}}^{\mathrm{N}(j)} - \mathbf{t}_{\mathrm{cc}}^{\mathrm{N}(j-1)}\right), \tag{41}$$

where δ_{N} is an additional numerical damping parameter with $0 < \delta_{\mathrm{N}} \leq 1$. For further information on building up mortar matrices see Sect. 5. A summary of the global solution algorithm is given in Algorithm 1.

input : A geometric set comprising *elements* and *nodes*,
System matrices $\mathbf{H}^{\mathrm{D}}, \mathbf{H}^{\mathrm{N}}, \mathbf{G}^{\mathrm{D}}, \mathbf{G}^{\mathrm{N}}$,
Boundary conditions $\Delta\mathbf{u}_{\mathrm{ncD}}^{\mathrm{N}(j)}, \Delta\mathbf{t}_{\mathrm{ncN}}^{\mathrm{N}(j)}, \mathbf{u}_{\mathrm{ncD}}^{\mathrm{D}(j)}, \mathbf{t}_{\mathrm{ncN}}^{\mathrm{D}(j)}$

output: Displacements $(\mathbf{u}^{\mathrm{D}}, \mathbf{u}^{\mathrm{N}})$ and tractions $(\mathbf{t}^{\mathrm{D}}, \mathbf{t}^{\mathrm{N}})$ caused by contact

for $j = 1, 2, \ldots,$ *until convergence* **do**
  do collision detection
    → List of element pairs which are potentially in contact
  Build-up of mortar matrices for tractions according to Sect. 5
    → $\mathbf{M}_t^{\mathrm{D}(j)}, \mathbf{M}_t^{\mathrm{N}(j)}$
  Transfer tractions: $\mathbf{t}_{\mathrm{cc}}^{\mathrm{N}(j)} = -\mathbf{M}_t^{\mathrm{N}(j)-1}\mathbf{M}_t^{\mathrm{D}(j)}\mathbf{t}_{\mathrm{cc}}^{D(j-1)}$, (40)
  Remark: $\mathbf{t}_{\mathrm{cc}}^{D(j-1)} = \mathbf{0}$ for $j = 1$
  Solve linear NEUMANN problem: $\mathbf{H}^{\mathrm{N}}\Delta\mathbf{u}^{\mathrm{N}(j)} = \mathbf{G}^{\mathrm{N}}\Delta\mathbf{t}^{\mathrm{N}(j)} + \mathbf{B}^{\mathrm{N}(j)}$, (30)
  Update geometry of NEUMANN body: $\mathbf{x}^{\mathrm{N}(j)} = \mathbf{x}^{\mathrm{N}(j-1)} + \delta_{\mathrm{D}}\Delta\mathbf{u}^{\mathrm{N}(j)}$, (34)
  Project nodes from DIRICHLET body on actual geometry of the
    NEUMANN body
    → normal vector matrix $\mathbf{N}^{\mathrm{D}(j)}$, initial gap vector $\mathbf{d}^{\mathrm{D}(j)}$
  Solve nonlinear DIRICHLET problem:
    $\frac{1}{2}\mathbf{x}^{\mathrm{D}(j)\mathrm{T}}\mathbf{Q}^{\mathrm{D}(j)}\mathbf{x}^{\mathrm{D}(j)} + \mathbf{x}^{\mathrm{D}(j)\mathrm{T}}\mathbf{p}^{\mathrm{D}(j)} \rightarrow \mathrm{MIN}$
    subject to: $\mathbf{A}^{\mathrm{D}}\mathbf{x}^{\mathrm{D}(j)} = \mathbf{b}^{\mathrm{D}(j)}$ and $\mathbf{N}^{\mathrm{D}(j)}\mathbf{x}^{\mathrm{D}(j)} + \mathbf{d}^{\mathrm{D}(j)} \geq \mathbf{0}$, (37)
    → contact tractions $\mathbf{t}_{\mathrm{cc}}^{\mathrm{D}(j)}$
  Check for convergence: $\|\mathbf{t}_{\mathrm{cc}}^{\mathrm{N}(j)} - \mathbf{t}_{\mathrm{cc}}^{\mathrm{N}(j-1)}\| \leq TOL \rightarrow$ *exit loop*
end

Algorithm 1: DIRICHLET-to-NEUMANN algorithm

Note that if the damping parameters δ_{D} in (36) and δ_{N} in (41) are equal to 1, no convergence will be achieved. In [25] it is suggested to choose 0.7 for δ_{D} and δ_{N}. A value of 1.0 for the parameter δ_{D} and a value of 0.5 for the parameter δ_{N} for the algorithm described above is recommended. Due to the fact that the surface tractions are calculated, fast convergence is achieved if the half of the tractions are transferred from the DIRICHLET body to the NEUMANN body.

5 Build-up of Mortar Matrices

In this section the theoretical background of the mortar methods for contact and the numerical implementation to create the mortar matrices $\mathbf{M}_t$ and $\mathbf{M}_u$ from (39) are described. The aim of the mortar methods is to find a mapping matrix between the degrees of freedom of the contacting bodies. The basic concept presented here was published by [30] for finite elements. It is here adapted for boundary elements. Similar to Fig. 3, the contacting bodies are divided into a mortar and non-mortar body, denoted by the indices m and nm, respectively, that are equivalent to the NEUMANN body and the DIRICHLET body, respectively.

As mentioned before, mortar contact belongs to the group of contacts which is based on LAGRANGIAN multipliers. The unilateral constraint for contact formulation can be defined using the KUHN-TUCKER-KARUSH conditions,

$$\begin{array}{llll} \text{1-dimensional} & d \geq 0 & \lambda \leq 0 & \lambda d = 0, \\ \text{2- or 3-dimensional} & \mathbf{d} \geq \mathbf{0} & \boldsymbol{\lambda} \leq \mathbf{0} & \boldsymbol{\lambda}^{\mathrm{T}}\mathbf{d} = 0. \end{array} \tag{42}$$

The first line of (42) presents the constraints for one-dimensional problems. As the gap cannot be negative, $d \geq 0$, and only pressure forces can be transferred, $\lambda \leq 0$, the product of the gap d and the LAGRANGIAN multiplier λ always vanishes.

For two- and three-dimensional problems typically a distinction has to be made between normal and tangential contact. For normal contact, the same constraints as for one-dimensional contact can be used where vectors of the gap $\mathbf{d}$ and of the LAGRANGIAN multiplier $\boldsymbol{\lambda}$ have to be projected in normal direction. Stiction and sliding may occur in tangential direction. In case of sliding, the tangential forces act as applied forces in opposite directions of the relative motion so that force laws such as COULOMB's law can be applied. In case of stiction, no motion in tangential direction occurs. In that case, the tangential contact can also be treated as a constraint and the resulting tangential tractions have to be treated as reaction stresses. Hence, the normal and tangential contact can be summarised according to the second line of (42).

The starting point for building up three-dimensional mortar matrices is the weak form of the KUHN-TUCKER-KARUSH condition,

$$\Pi_{\mathrm{c}}^{\mathrm{LM}} = \int_{S_{\mathrm{cc}}} \boldsymbol{\lambda}^{\mathrm{T}}\mathbf{d}\,\mathrm{d}S. \tag{43}$$

Due to the weak form, the constraints are fulfilled in an integral meaning over the potential contact area S_{cc} only. The variation of the weak form from (43) leads to

$$\delta\Pi_{\mathrm{c}}^{\mathrm{LM}} \equiv C_{\mathrm{c}}^{\mathrm{LM}} = \int_{S_{\mathrm{cc}}} \delta\boldsymbol{\lambda}^{\mathrm{T}}\mathbf{d}\,\mathrm{d}S + \int_{S_{\mathrm{cc}}} \boldsymbol{\lambda}^{\mathrm{T}}\delta\mathbf{d}\,\mathrm{d}S = 0. \tag{44}$$

The first integral in (44) represents the fulfilment of the gap function $\mathbf{d}$ choosing an arbitrary LAGRANGIAN multiplier vector $\boldsymbol{\lambda}$. The second integral represents the fulfilment of the reaction forces for an arbitrary chosen gap function $\mathbf{d}$. In case of contact the gap between the two contacting bodies is closed, and the gap function $\mathbf{d}$ can be replaced by the displacements of the mortar body ${}^{m}\mathbf{u}^{m}$ and of the non-mortar body ${}^{m}\mathbf{u}^{nm}$, and the initial gap ${}^{m}\mathbf{d}_{0}^{m}$,

$$C_{\mathrm{c}}^{\mathrm{LM}} = \int_{S_{\mathrm{cc}}} \delta\boldsymbol{\lambda}^{\mathrm{T}}({}^{m}\mathbf{u}^{nm} - {}^{m}\mathbf{u}^{m} - {}^{m}\mathbf{d}_{0}^{m})\,\mathrm{d}S + \int_{S_{\mathrm{cc}}} ({}^{m}\mathbf{t}^{nm} + {}^{m}\mathbf{t}^{m})\delta\mathbf{d}^{\mathrm{T}}\,\mathrm{d}S = 0. \tag{45}$$

The upper left index m denotes that the displacement vectors are expressed with respect to the body-fixed mortar reference frame $\mathcal{K}_{m}$. Additionally, the LAGRANGIAN multiplier $\boldsymbol{\lambda}$ in the second integral can be replaced by the sum of the tractions of the mortar body ${}^{m}\mathbf{t}^{m}$ and of the non-mortar body ${}^{m}\mathbf{t}^{nm}$. Hence, the equilibrium in the contact area is fulfilled by an integral meaning only.

The LAGRANGIAN multipliers $\boldsymbol{\lambda}$ and their variations $\delta\boldsymbol{\lambda}$ can be discretised in the same way as it was done for displacements and tractions. This leads to

$$\boldsymbol{\lambda}(\xi,\eta)=\sum_i M_i^{\lambda}(\xi,\eta)\boldsymbol{\lambda}_i, \qquad \delta\boldsymbol{\lambda}(\xi,\eta)=\sum_i M_i^{\lambda}(\xi,\eta)\delta\boldsymbol{\lambda}_i, \tag{46}$$

where $M_i^{\lambda}(\xi,\eta)$ represents the general shape functions over one element e and $\boldsymbol{\lambda}_i$ and $\delta\boldsymbol{\lambda}_i$ the corresponding nodal values. The same procedure can be applied to the gap function $\mathbf{d}$ and the corresponding variation $\delta\mathbf{d}$,

$$\mathbf{d}(\xi,\eta)=\sum_i M_i^{d}(\xi,\eta)\mathbf{d}_i, \qquad \delta\mathbf{d}(\xi,\eta)=\sum_i M_i^{d}(\xi,\eta)\delta\mathbf{d}_i. \tag{47}$$

Inserting (46) and (47) into (45) leads to the constraint equations for one pair of contact elements,

$$\begin{aligned}
C_{\mathrm{ec}}^{\mathrm{LM}}=\sum_i \delta\boldsymbol{\lambda}_i^{\mathrm{T}}\int_{S_{ecc}}\Bigg(&\sum_j M_i^{\lambda}(\xi_{nm},\eta_{nm})N_j(\xi_{nm},\eta_{nm})\,{}^{m}\mathbf{u}_j^{nm}\\
&-\sum_j M_i^{\lambda}(\xi_m,\eta_m)N_j(\xi_m,\eta_m)\,{}^{m}\mathbf{u}_j^{m}\\
&-\sum_j M_i^{\lambda}(\xi_m,\eta_m)N_j(\xi_m,\eta_m)\,{}^{m}\mathbf{d}_{0j}^{m}\Bigg)\,\mathrm{d}S\\
+\sum_i \delta\mathbf{d}_i^{\mathrm{T}}\int_{S_{ecc}}\Bigg(&\sum_j M_i^{d}(\xi_{nm},\eta_{nm})N_j(\xi_{nm},\eta_{nm})\,{}^{m}\mathbf{t}_j^{nm}\\
&+\sum_j M_i^{d}(\xi_m,\eta_m)N_j(\xi_m,\eta_m)\,{}^{m}\mathbf{t}_j^{m}\Bigg)\,\mathrm{d}S=0.
\end{aligned} \tag{48}$$

The nodal values ${}^{m}\mathbf{u}_j^{nm}$ and ${}^{m}\mathbf{t}_j^{nm}$ for the degrees of freedom on the non-mortar side are given with respect to the body-fixed reference frame $\mathcal{K}_m$, see (48). Typically the system matrices $\mathbf{H}^{nm}$ and $\mathbf{G}^{nm}$ are calculated with respect to the body-fixed reference frame $\mathcal{K}_{nm}$. The corresponding displacements and tractions are also given with respect to $\mathcal{K}_{nm}$. An additional transformation for the current orientation between $\mathcal{K}_m$ and $\mathcal{K}_{nm}$ is introduced so that

$${}^{m}\mathbf{u}_i^{nm}={}^{m\,nm}\mathbf{T}\,{}^{nm}\mathbf{u}_i^{nm} \quad\text{and}\quad {}^{m}\mathbf{t}_i^{nm}={}^{m\,nm}\mathbf{T}\,{}^{nm}\mathbf{t}_i^{nm}, \tag{49}$$

where ${}^{m\,nm}\mathbf{T}$ is the transformation matrix between body-fixed reference frames on mortar body m and the non-mortar body nm. If (48) vanishes, the constraints for displacements and tractions are fulfilled. Considering (49) and using the fact that

(48) vanishes for arbitrary variations of the LAGRANGIAN multipliers $\delta\boldsymbol{\lambda}$ and gap functions $\delta\mathbf{d}$, the contact constraints for one pair of contact elements become

$$\begin{aligned}
&\int_{S_{ecc}} \Big(\sum_j M_i^{\lambda}(\xi_{nm}, \eta_{nm}) N_j(\xi_{nm}, \eta_{nm})^{m\,nm}\mathbf{T}^{nm}\mathbf{u}_j^{nm} \\
&\quad - \sum_j M_i^{\lambda}(\xi_m, \eta_m) N_j(\xi_m, \eta_m)^{m}\mathbf{u}_j^{m} \\
&\quad - \sum_j M_i^{\lambda}(\xi_m, \eta_m) N_j(\xi_m, \eta_m)^{m}\mathbf{d}_{0j}^{m} \Big)\, \mathrm{d}S = \mathbf{0}, \\
&\int_{S_{ecc}} \Big(\sum_j M_i^{d}(\xi_{nm}, \eta_{nm}) N_j(\xi_{nm}, \eta_{nm})^{m\,nm}\mathbf{T}^{nm}\mathbf{t}_j^{nm} \\
&\quad + \sum_j M_i^{d}(\xi_m, \eta_m) N_j(\xi_m, \eta_m)^{m}\mathbf{t}_j^{m} \Big) \mathrm{d}S = \mathbf{0}.
\end{aligned} \tag{50}$$

For a pair of bilinear elements, the first equation of (50) results for the displacements on the non-mortar side to the matrix

$$\mathbf{M}_u^{nm\,e} = \begin{bmatrix}
\int_{S_{ecc}} M_1^{\lambda} N_1\, \mathrm{d}S^{m\,nm}\mathbf{T} & \int_{S_{ecc}} M_1^{\lambda} N_2\, \mathrm{d}S^{m\,nm}\mathbf{T} & \int_{S_{ecc}} M_1^{\lambda} N_3\, \mathrm{d}S^{m\,nm}\mathbf{T} & \int_{S_{ecc}} M_1^{\lambda} N_4\, \mathrm{d}S^{m\,nm}\mathbf{T} \\
\int_{S_{ecc}} M_2^{\lambda} N_1\, \mathrm{d}S^{m\,nm}\mathbf{T} & \int_{S_{ecc}} M_2^{\lambda} N_2\, \mathrm{d}S^{m\,nm}\mathbf{T} & \int_{S_{ecc}} M_2^{\lambda} N_3\, \mathrm{d}S^{m\,nm}\mathbf{T} & \int_{S_{ecc}} M_2^{\lambda} N_4\, \mathrm{d}S^{m\,nm}\mathbf{T} \\
\int_{S_{ecc}} M_3^{\lambda} N_1\, \mathrm{d}S^{m\,nm}\mathbf{T} & \int_{S_{ecc}} M_3^{\lambda} N_2\, \mathrm{d}S^{m\,nm}\mathbf{T} & \int_{S_{ecc}} M_3^{\lambda} N_3\, \mathrm{d}S^{m\,nm}\mathbf{T} & \int_{S_{ecc}} M_3^{\lambda} N_4\, \mathrm{d}S^{m\,nm}\mathbf{T} \\
\int_{S_{ecc}} M_4^{\lambda} N_1\, \mathrm{d}S^{m\,nm}\mathbf{T} & \int_{S_{ecc}} M_4^{\lambda} N_2\, \mathrm{d}S^{m\,nm}\mathbf{T} & \int_{S_{ecc}} M_4^{\lambda} N_3\, \mathrm{d}S^{m\,nm}\mathbf{T} & \int_{S_{ecc}} M_4^{\lambda} N_4\, \mathrm{d}S^{m\,nm}\mathbf{T}
\end{bmatrix} \tag{51}$$

and for the displacements on the mortar side to the matrix

$$\mathbf{M}_u^{m\,e} = \begin{bmatrix}
\int_{S_{ecc}} M_1^{\lambda} N_1\, \mathrm{d}S\mathbf{E} & \int_{S_{ecc}} M_1^{\lambda} N_2\, \mathrm{d}S\mathbf{E} & \int_{S_{ecc}} M_1^{\lambda} N_3\, \mathrm{d}S\mathbf{E} & \int_{S_{ecc}} M_1^{\lambda} N_4\, \mathrm{d}S\mathbf{E} \\
\int_{S_{ecc}} M_2^{\lambda} N_1\, \mathrm{d}S\mathbf{E} & \int_{S_{ecc}} M_2^{\lambda} N_2\, \mathrm{d}S\mathbf{E} & \int_{S_{ecc}} M_2^{\lambda} N_3\, \mathrm{d}S\mathbf{E} & \int_{S_{ecc}} M_2^{\lambda} N_4\, \mathrm{d}S\mathbf{E} \\
\int_{S_{ecc}} M_3^{\lambda} N_1\, \mathrm{d}S\mathbf{E} & \int_{S_{ecc}} M_3^{\lambda} N_2\, \mathrm{d}S\mathbf{E} & \int_{S_{ecc}} M_3^{\lambda} N_3\, \mathrm{d}S\mathbf{E} & \int_{S_{ecc}} M_3^{\lambda} N_4\, \mathrm{d}S\mathbf{E} \\
\int_{S_{ecc}} M_4^{\lambda} N_1\, \mathrm{d}S\mathbf{E} & \int_{S_{ecc}} M_4^{\lambda} N_2\, \mathrm{d}S\mathbf{E} & \int_{S_{ecc}} M_4^{\lambda} N_3\, \mathrm{d}S\mathbf{E} & \int_{S_{ecc}} M_4^{\lambda} N_4\, \mathrm{d}S\mathbf{E}
\end{bmatrix} \tag{52}$$

with the 3×3 identity matrix $\mathbf{E}$. The compatibility matrices for the tractions $\mathbf{M}_t^{m\,e}$ and $\mathbf{M}_t^{nm\,e}$ are created with the same procedure. Assembling the matrices of all element pairs which are possibly in contact leads to

$$\begin{aligned}
\mathbf{M}_u^{m} &= \bigcup_e \mathbf{M}_u^{m\,e}, & \mathbf{M}_u^{nm} &= \bigcup_e \mathbf{M}_u^{nm\,e}, \\
\mathbf{M}_t^{m} &= \bigcup_e \mathbf{M}_t^{m\,e}, & \mathbf{M}_t^{nm} &= \bigcup_e \mathbf{M}_t^{nm\,e}.
\end{aligned} \tag{53}$$

Concluding some remarks on the shape functions $M_i(\xi, \eta)$ used for the interpolation of the gap $\mathbf{d}$ and the LAGRANGIAN multipliers $\boldsymbol{\lambda}$ are given. According to [18, 37, 39], the shape functions M_i^{λ} for the interpolation of the LAGRANGIAN multipliers $\boldsymbol{\lambda}$ in (46) and M_i^{d} for the interpolation of gap $\mathbf{d}$ according to (47) have to fulfil the BABUŠKA-BREZZI or *inf-sup* condition. A detailed description on that topic is provided by [9]. This condition ensures a unique solution and the maximum rank of the mortar matrices $\mathbf{M}_u^{nm}$ and $\mathbf{M}_t^{m}$. To realise this condition, the function

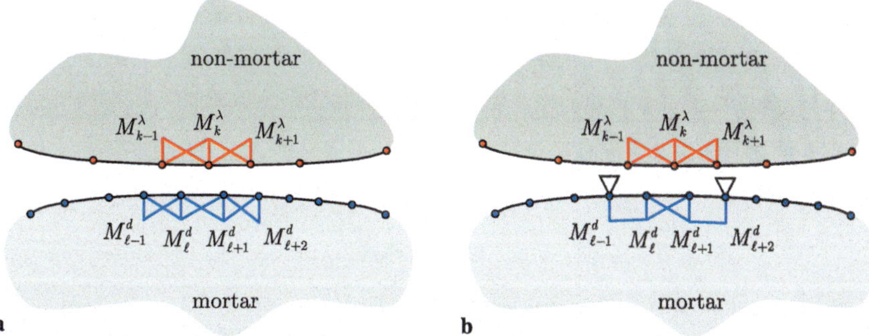

Fig. 4 Approximation of the LAGRANGIAN multiplier $\boldsymbol{\lambda}$ and the gap function **d** within the contact area. **a** Without boundary conditions. **b** With DIRICHLET boundary conditions

space of the shape functions M_i^{λ} and M_i^d has to be sufficiently rich. Many research work has been done to obtain an optimal set of shape functions used for interpolation of LAGRANGIAN multipliers for domain decomposition within finite element calculations, see [36]. A dual LAGRANGIAN multiplier formulation was proposed by [36] which leads to diagonal mortar matrices. In the present work the original approach as described by [6] is implemented. The same interpolation functions are used for the approximation of the LAGRANGIAN multiplier $\boldsymbol{\lambda}$ and the gap function **d** within the contact area as used for the interpolation of the displacements **u** and the tractions **t**. The use of that shape functions typically ensures the fulfilment of the BABUŠKA-BREZZI conditions. For two-dimensional contact problems, the interpolation functions are schematically shown in Fig. 4.

For the interpolation of the LAGRANGIAN multipliers $\boldsymbol{\lambda}$ the mesh on the non-mortar side is used, where for the interpolation of the gap function **d** the mesh on the mortar side is used, as presented by M_i^{λ} and M_i^d in Fig. 4. Due to that formulation, a construction of an intermediate mesh as described by [34] is not necessary. Nevertheless to obtain mortar matrices with a maximum rank, the shape functions have to fulfil the BABUŠKA-BREZZI condition, which leads according to [34] to the simple requirement

$$\min(N^m, N^{nm}) \leq N^{\lambda}, N^d \leq \max(N^m, N^{nm}), \tag{54}$$

where N^m and N^{nm} represent the number of contact nodes of the mortar and non-mortar body, respectively. The corresponding numbers of nodes for the interpolation of the LAGRANGIAN multiplier and the gap function **d** are given by N^{λ} and N^d. Due to the facts that the LAGRANGIAN multipliers are approximated using the potential contacting elements on the non-mortar side and the gap function is approximated using contacting elements on the mortar side, the requirement (54) is always fulfilled if no boundary conditions are applied at the boundary of the contact area. If DIRICHLET boundary conditions at the boundary of the contact area are applied, the shape functions M_i^{λ} and M_i^d at the boundary have to be modified, see Fig. 4b. According to [18, 34], the shape functions are kept constant to ensure the maximum rank of $\mathbf{M}_u^{nm}$ and $\mathbf{M}_t^m$.

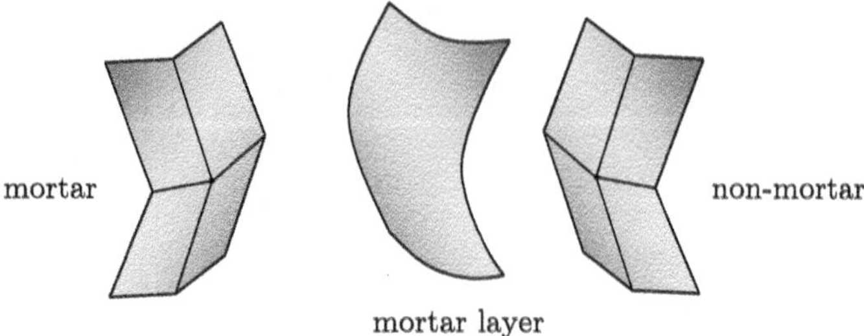

Fig. 5 Introduction of a mortar layer

6 Numerical Implementation of Mortar Matrices

First, a mortar layer is introduced between the mortar and non-mortar side, see Fig. 5.

The mortar layer can either be located as an intermediate surface between the mortar and the non-mortar surface as presented by [27, 28, 31], or one of the contacting surfaces can be chosen as shown in [25, 36]. According to [30], the numerical integration scheme is shown in Fig. 6 and in Algorithm 2.

First, the element facets are projected onto the common mortar layer, see Fig. 6a. Therefore, only the corner nodes are taken into account. For bilinear elements, this consideration is sufficient because two-dimensional bilinear elements consist of four nodes only, but in case of biquadratic elements where the edges could be curved errors may occur. For simplification and an easier handle of the projected elements, this error is neglected. This approach is valid if the differences between the geometry of the biquadratic element and the corresponding 4-node element facet are not too large.

In a second step, the common area due to the overlap of the projected elements is determined, see Fig. 6b. Therefore, a modified version of the SUTHERLAND-HODGMAN polygon clipping algorithm is used as presented in [16]. The algorithm described in [16] is valid for an axis parallel clipping only. The algorithm modifications presented here are valid for clipping with arbitrary 4-point polygons as presented in Fig. 7a.

The principle of the algorithm is to clip the parts *outside* of one polygon with respect to another one. This is done by dividing one of the elements into four clipping edges represented by the four projected edges of the element facet. According to Fig. 7a, the element 1234 is chosen to be the clipping element. The element $1'2'3'4'$ is clipped with respect to the element 1234. Therefore, a flow direction according to the arrow in Fig. 7a is chosen. In case of the chosen direction the area of the element $1'2'3'4'$ located on the right side of the clipping edges is cut off, because they are *outside* the element 1234. To create new corner points of the overlap polygon, the edges of the element $1'2'3'4'$ are cut off with respect to the clipping edges.

A calculation whether one point is lying *inside* or *outside* with respect to the clipping edge is necessary. According to Fig. 7b, this is done by a perpendicular projection of the point with respect to the clipping edge. The distance vector $\mathbf{r}_{12}$

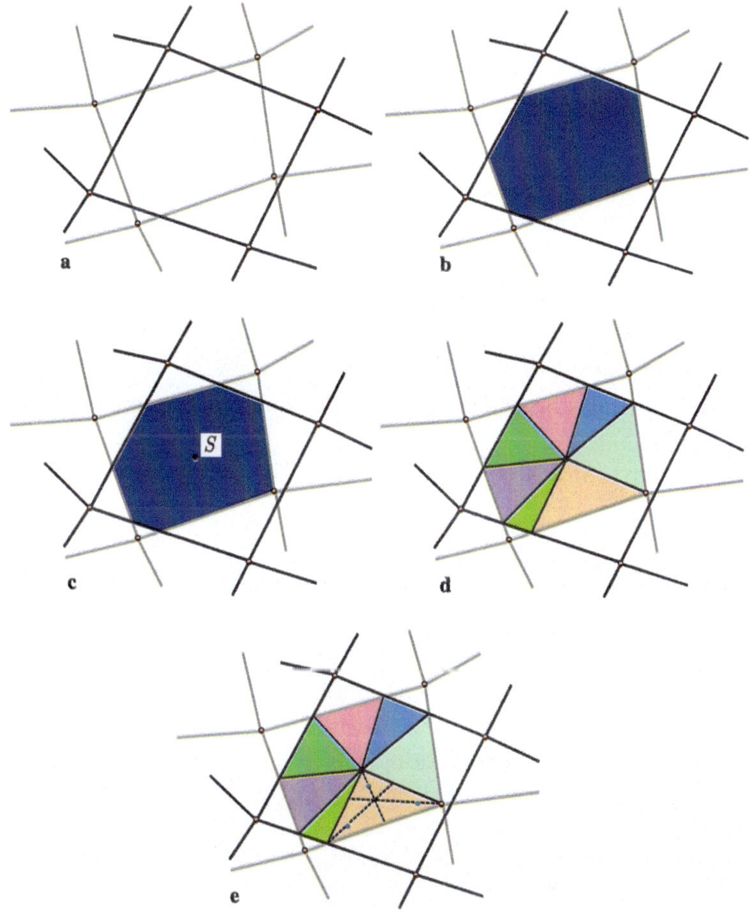

Fig. 6 Integration scheme for 3D mortar elements. **a** Projections of the element facets onto the mortar layer. **b** Common area of the projected facets. **c** Centre of area S. **d** Division of the polygon into triangles. **e** Locating GAUSS-RADAU integration points on the triangles

of the two points of the clipping edge and the corresponding normal vector $\mathbf{n}_{12}$ are given by

$$\mathbf{r}_{12} = \begin{bmatrix} x_2 - x_1 \\ y_2 - y_1 \end{bmatrix} \quad \text{and} \quad \mathbf{n}_{12} = \begin{bmatrix} y_2 - y_1 \\ x_1 - x_2 \end{bmatrix}. \tag{55}$$

The resulting clipping test is given by

$$\mathbf{n}_{12}^{\mathrm{T}} \begin{bmatrix} x_P - x_1 \\ y_P - y_1 \end{bmatrix} \begin{cases} > 0 \rightarrow \textit{outside} \\ \leq 0 \rightarrow \textit{inside}. \end{cases} \tag{56}$$

The other parts of the algorithm remain unchanged, compare to [16]. Thereafter the centre of the area S of the resulting polygon, see Fig. 6c, is calculated by the

input : List of element pairs $Elp(i)$, which is possibly in contact
output: Mortar matrices $\mathbf{M}_u^{nm}$, $\mathbf{M}_u^{m}$, $\mathbf{M}_t^{nm}$, $\mathbf{M}_t^{m}$

forall the *element pairs* $Elp(i)$ **do**

1. projection of the element facets on a common mortar layer, see Fig. 6a
2. use clipping algorithm to form polygon of the overlap of the projected element facets, see Fig. 6b
3. locate geometric centre of the polygon, see Fig. 6c
4. divide polygon into n_t triangles, see Fig. 6d

forall the n_t *triangles* **do**

- locate GAUSS-RADAU integration points on the triangle, see Fig. 6e
- project GAUSS integration points on the mortar and non-mortar element to obtain ξ_m, η_m and ξ_{nm}, η_{nm}
- compute integrals over triangles, see (57)
- sum integral contributions, (58), to obtain $\mathbf{M}_u^{nm\,e}$, $\mathbf{M}_u^{m\,e}$, $\mathbf{M}_t^{nm\,e}$, $\mathbf{M}_t^{m\,e}$ according to (51)

end

end

assemble the element matrices according to (53)

Algorithm 2: Build-up of mortar matrices

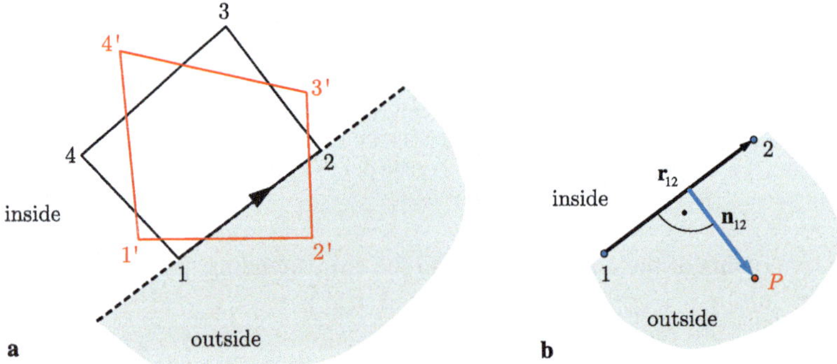

Fig. 7 Scheme of clipping algorithm. **a** Clipping with 4-point polygons. **b** Projection of an edge point P onto the normal of the clipping edge

arithmetic middle of the corner coordinates of the polygon. This results in a division of the polygon into triangles according to Fig. 6d. In the next step the GAUSS-RADAU integration points can be determined for each of the triangles according to

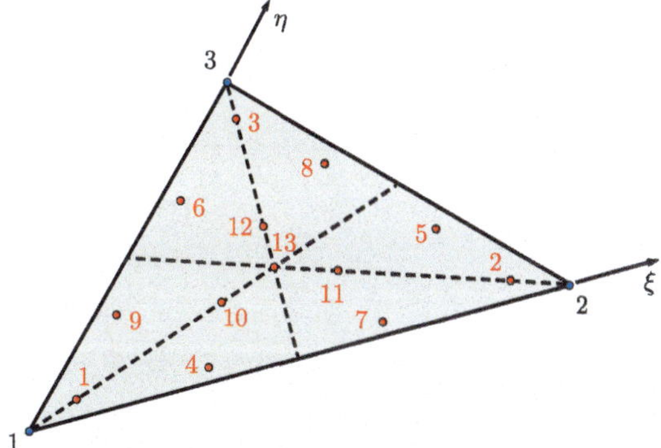

Fig. 8 GAUSS-RADAU integration points for triangles

[11, 14]. In [30] a thirteen point GAUSS-RADAU rule is recommended, see Fig. 8, to overcome the problem that warped meshes can not be integrated exactly.

In contrast to typical numerical integration, not the local but the absolute coordinates are necessary. These coordinates are projected back on the mortar and the non-mortar elements, respectively. As a result, the local coordinates ξ_m, η_m and ξ_{nm}, η_{nm} will be obtained. The integrals over one triangle from (51) can be expressed by

$$\int_{S_{et}} M_i(\xi_k, \eta_k) N_j(\xi_k, \eta_k)\,\mathrm{d}S = \sum_{l=1}^{n_g} M_i(\xi_{kl}, \eta_{kl}) N_j(\xi_{kl}, \eta_{kl}) W_l, \tag{57}$$

where S_{et} is the surface of a triangle t of the element e while k represents the projection on the mortar m and non-mortar side nm, respectively. The variable n_g represents the number of integration points and W_l the corresponding weighting factors. The integral over the element surface S_{ecc} can be evaluated by

$$\int_{S_{ecc}} M_i(\xi_k, \eta_k) N_j(\xi_k, \eta_k)\,\mathrm{d}S = \sum_{t=1}^{n_t} \int_{S_{et}} M_i(\xi_k, \eta_k) N_j(\xi_k, \eta_k)\,\mathrm{d}S, \tag{58}$$

which represents the sum over all triangles. Finally, the integrals can be assembled according to (53).

7 From Contact Tractions to Applied Forces

If the contact calculation inside the boundary element environment has converged, the tractions and displacements within the contact area are given. However, within the multibody environment only forces and torques can be treated. Therefore, the contact tractions $\mathbf{t}_j$ have to be integrated with respect to the contact area to form the resulting wrench, consisting of the force $\mathbf{f}_T^{cc}$ and torque $\mathbf{f}_R^{cc}$, see Fig. 9.

According to Fig. 9, the contact tractions have to be integrated for each contacting body with respect to the body-fixed reference frame $\mathcal{K}_i$. The upper right index i denotes the mortar body m and non-mortar body nm, respectively. The body-fixed

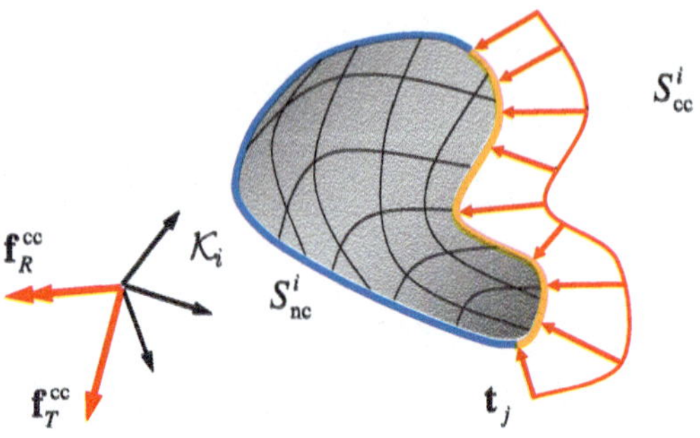

Fig. 9 From contact tractions to resulting wrench

reference frame $\mathcal{K}_i$ within the boundary element system is equal to the force application point within the multibody system. The resulting contact force $\mathbf{f}_T^{cc}$ can be calculated by

$$\mathbf{f}_T^{cc} = \sum_e^{n_c} \left[\sum_k \int_{S_e} N_k \mathbf{t}_k \, d\tilde{S} \right], \tag{59}$$

where $\mathbf{t}_i$ represents the surface traction vector of the ith node, N_i the corresponding shape functions, and S_e the element surface. The integral in (59) has to be evaluated over all n_c contacting elements. The resulting contact torque $\mathbf{f}_R^{cc}$ built up by the tractions $\mathbf{t}_i$ can be calculated by

$$\mathbf{f}_R^{cc} = \sum_e^{n_c} \left[\sum_k \sum_l \int_{S_e} \tilde{x}_{kl} N_l \mathbf{t}_l \, d\tilde{S} \right], \tag{60}$$

where $\tilde{x}_{kl}$ are the components of the skew symmetric matrix representing the location of a point on the surface S_e with respect to the body-fixed reference frame $\mathcal{K}_i$. The integrals can be evaluated using standard GAUSS integration. The overall contact wrench $\hat{\mathbf{f}}^{cc}$ from (11) is obtained by assembling the calculated contact force $\mathbf{f}_T^{cc}$ and torque $\mathbf{f}_R^{cc}$.

8 Dynamic Simulation of Two Contacting Spheres

To test the functionality of the algorithm, a simple multibody model is considered. A schematic representation of the dynamic model is given in Fig. 10a.

The one-dimensional model consists of two identical spheres, where the lower sphere is fixed at its lower side. The upper sphere is suspended at its upper side by a spring-damper element in vertical direction. The properties of the spring-damper element are defined by the stiffness coefficient k_{spring} and the damping coefficient b_{damper}. Because of the gravity g_{grav}, the upper sphere moves downwards until the initial gap d_0 is closed. Then elastic contact between both spheres based on the elastic behaviour of both curves occurs. The multibody model consists of two rigid spheres, see Fig. 10b. The multibody data of the spheres such as mass are calculated by the density ρ and the diameter d_{sph}. The calculation of the contact forces

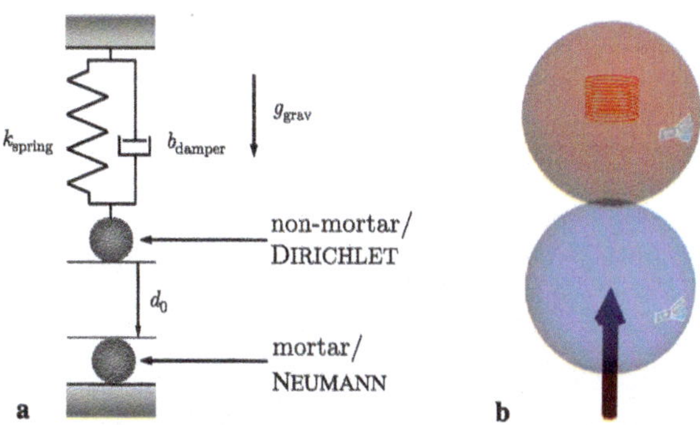

Fig. 10 Model for dynamic simulation of two spheres. **a** Schematic representation. **b** Multibody model

Table 1 Properties of the dynamic model

Property	Value
YOUNG's modulus E	210,000 MPa
POISSON's ratio ν	0.3
density ρ	$7{,}850.0\ \frac{\text{kg}}{\text{m}^3}$
diameter d_{sph}	1 m
initial gap d_0	0.2 m
gravity g_{grav}	$9.81\ \frac{\text{m}}{\text{s}^2}$
stiffness coefficient k_{spring}	$40{,}000\ \frac{\text{N}}{\text{m}}$
damping coefficient b_{damper}	$110\ \frac{\text{N s}}{\text{m}}$

depends mainly on the motion of the reference frames of the spheres, which are located at the centres of the spheres. The data of the model are summarised in Table 1.

For comparison, a reference model with existing force elements of the multibody program SIMPACK™ was created. Here, the force element containing the developed BEM contact model is replaced by a HERTZIAN pressure element. Here, the bodies remain undeformed and body forces are also not considered in the reference model. The transition between no contact and contact is realised by an event function. The event function depends on the relative position and on the geometry of the contacting bodies. If the gap between the spheres is closed, the integrator is stopped and restarted with new initial conditions. The *Sodarst2* integrator from the multibody program SIMPACK™ is chosen. This integrator is based on an implicit formulation with automatic step size calculation.

The co-simulation is done by the use of the explicit EULER integrator. The constant step size is equal to 0.0002 s. This step size is chosen to overcome the impact problem of both spheres. The simulation time is 1 s. In contrast to the reference model, the body forces are considered. The contact gap of both simulations is shown in Fig. 11.

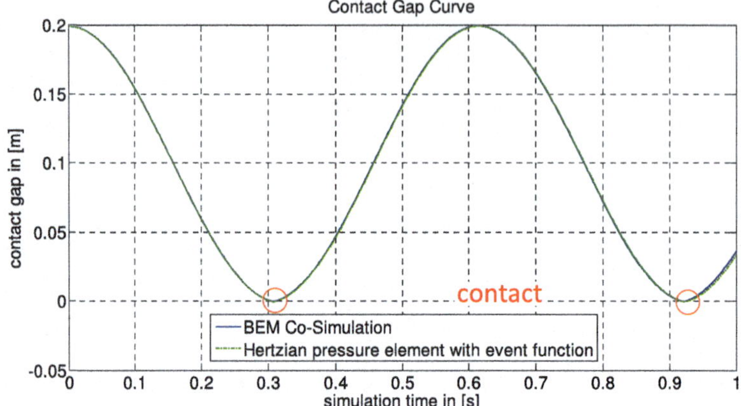

Fig. 11 Contact gap vs. time for BEM co-simulation and HERTZIAN pressure element

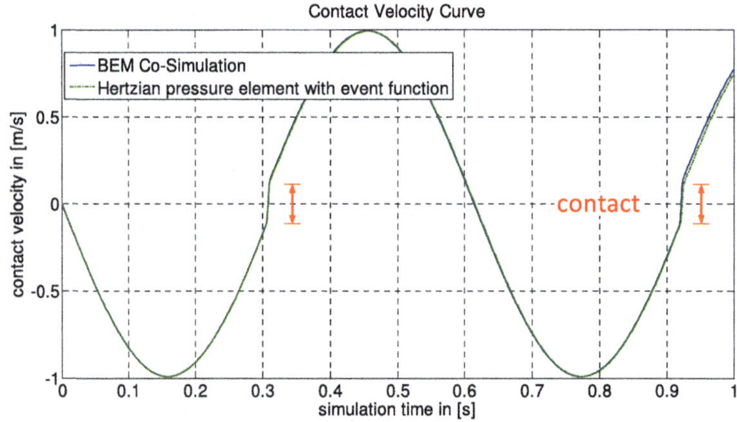

Fig. 12 Contact velocity vs. time for BEM co-simulation and HERTZIAN pressure element

The two curves agree well. Only small differences can be denoted at the end of the simulation time. The corresponding relative contact velocity of the upper contact point with respect to the lower contact point is shown in Fig. 12.

Herein also small differences between the two curves can be noticed at the end of the simulation time. The contact forces of both simulations are shown in Fig. 13.

The peaks of the two contact force curves differ from each other. The first peak of the HERTZIAN reference model reaches a value of −244.942 kN. The corresponding contact force value of the BEM co-simulation is obtained with −276.003 kN, which means a difference of −12.68 % compared to the reference model. Higher pressure force are to be expected due to the consideration of body forces. Additional differences occur due to the integrators chosen. These differences can be seen at the second peak of both curves. The contact force values are obtained as −223.215 kN

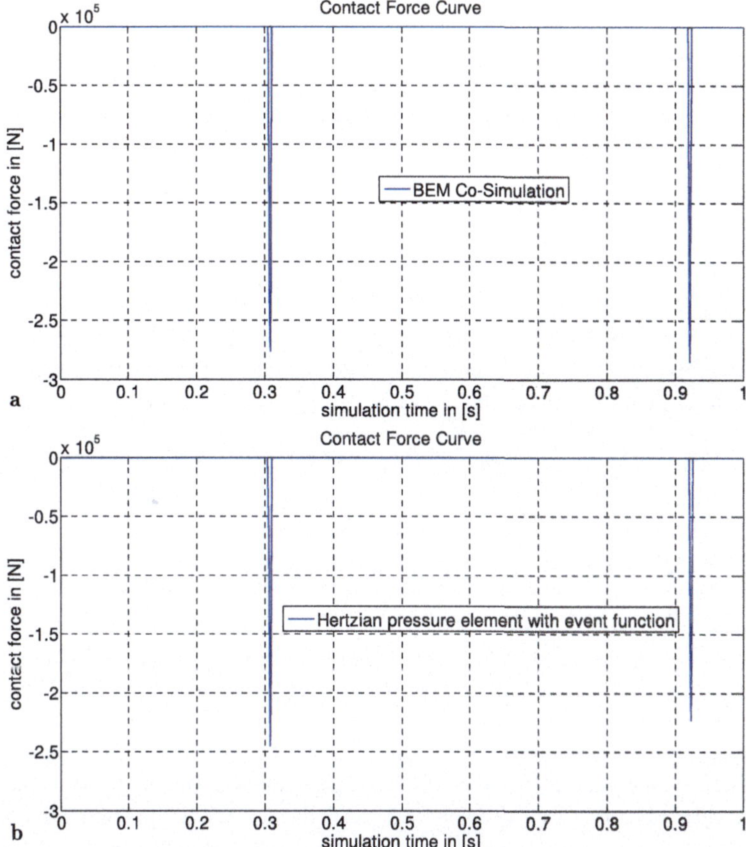

Fig. 13 Contact force vs. time. **a** BEM co-simulation. **b** HERTZIAN pressure element

and −285.308 kN for the reference model and the BEM co-simulation, respectively. The second peak of the reference model is lower than the first one because of the modelled damping d_{damper}. Additional damping occurs due to the implicit integrator. In contrast the second peak of the co-simulation is higher. This results from the explicit EULER integrator which leads to an excitation of the numerical solution.

9 Conclusions and Outlook

The present chapter introduces the modelling of elastic contacts by coupled multibody and boundary element systems. Compared to contacts modelled by impact laws, physically more accurate results can be obtained. Due to the use of boundary element systems, the contact stresses are obtained within the contact calculation.

The contact formulation is based on mortar methods, which enables the contact calculation of non-conforming meshes. A new three-dimensional contact element for boundary element systems is developed. The mortar element uses the mixed formulation of boundary element formulations. Constraints in a weak form are defined for the displacements in the contact interface. This principle is also known from mortar formulations of finite elements. Additionally, a weak equilibrium is introduced for the tractions in the contact interface. The algorithm for the iteration of contact states is based on a DIRICHLET-to-NEUMANN algorithm. Herein, both contacting bodies are calculated serially. In the first calculation step, one of the contacting bodies represents a rigid obstacle for the other elastic one. The resulting reaction forces on the elastic body are partially transferred on the other one, which is for the second calculation step no longer rigid. As a result the obstacle is deformed and the next iteration starts. The algorithm converges if the numerical equilibrium in the contact interface is reached.

The incorporation of the multibody and boundary element program is realised by interprocess communication. Further details are described in [41]. The multibody program stops during the calculation of the contact forces by the boundary element program. Unix domain sockets are used for communication of both programs. This coupling scheme is applied because both programs works under the same operating system on the same computer. The applied communication protocol can easily be extended to network sockets. In contrast to Unix domain sockets, network sockets allow both programs working on different computers. This property is very important for the developed BEM co-simulation because the boundary element program needs more computational resources than the multibody program. The position data provided by the multibody program are used by the boundary element program to calculate the displacements and tractions on the contacting bodies. The resulting contact tractions are summarised to a contact wrench by numerical integration.

Altogether the MBS-BEM co-simulation is an appropriate way for contact calculation in multibody systems. The main advantage is that contact stresses are obtained within the dynamic calculation. These data can be used for strength, fatigue and durability analyses. Despite to the fact that the calculation of complex models needs a large calculation time the coupled simulation of multibody and boundary element systems offers various applications.

As an application for the BEM contact algorithm, the femoral-patellar joint with sliding contact between a human patella and a femur bone is under investigation. First tests show a good convergence of the BEM contact algorithm, see Fig. 14.

A future task should be the speed-up of the co-simulation. The main advantage of the developed BEM contact algorithm is the possibility to integrate fast boundary element methods, because the two contacting bodies are treated separately. Fast boundary methods are developed to overcome large calculation times caused by fully populated system matrices. The panel clustering method described in [19] approximates the matrix-vector multiplication so that instead of a system matrix $\mathbf{A}$ of the size $n \times n$ only two vectors of the size n are calculated. Thus, the memory requirements are strongly reduced. The build-up of a binary tree or

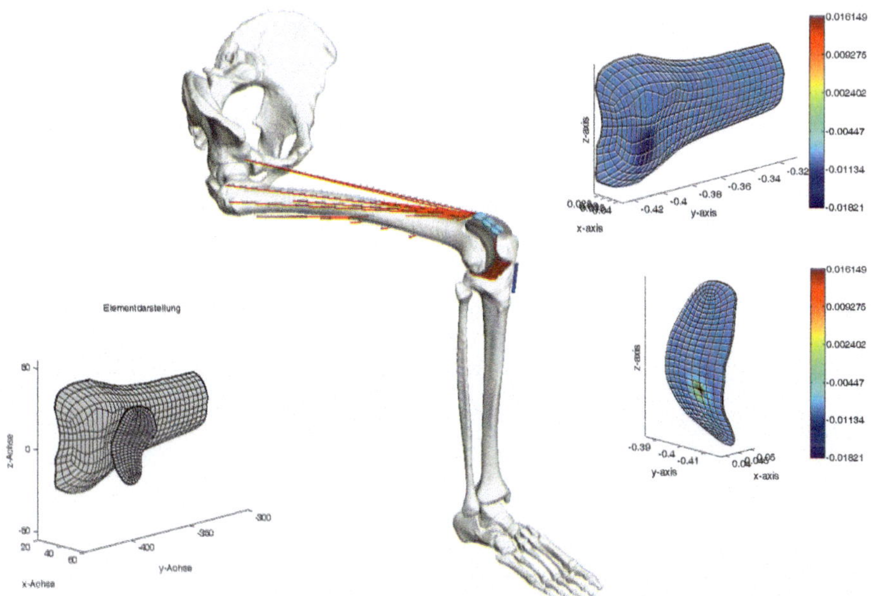

Fig. 14 Simulation of a patellar joint (in cooperation with Department of Orthopaedics, University Medicine, Rostock)

octree for the boundary elements is necessary, so that the collision detection algorithms described in [40, 41] can be used. The panel clustering method is extended to three-dimensional elastostatics in [20, 21] where the integral free term c_{ij} from (23) is assumed to be equal to $\frac{1}{2}\delta_{ij}$. From theoretical background this is valid for flat surfaces only, which is not typical for physical shapes in mechanical engineering. Therefore, these algorithms have to be checked carefully. Panel clustering uses hierarchical matrices which are explained in detail by [7]. First implementations of the panel clustering method for temperature distribution problems is shown in [22].

In the present contribution, quasistatic boundary element formulations are considered only. To take local vibrations and wave propagation of the elastic bodies into account dynamic co-simulation is recommended. A possible implementation is the dual reciprocity method described in [8]. The result of that method are dynamic system matrices, whereby also a system matrix in analogy to the mass matrix of finite element systems is obtained. The coupling of multibody and boundary element systems based on dual reciprocity methods to model flexible bodies is described by [3].

The integration of dynamic boundary element formulations leads to better results for elastic impact problems because the energy loss due to wave propagation is taken into account. The system matrices for that formulation are fully populated and not necessarily positive definite so that complex eigenfrequencies are obtained. To overcome these problem internal nodes are included to represent the mass distribution.

In [23] it is shown that lower eigenfrequencies become real if a sufficient number of internal nodes is used. For dynamic co-simulation a separate integrator has to be implemented within the boundary element formulation. In addition other coupling schemes have to be implemented because the two simulations have to be synchronised. Special coupling schemes for co-simulation of multibody and finite element systems are described in [10] which are also applicable for the co-simulation of multibody and boundary element systems.

Acknowledgements Thanks to Prof. Klaus Schittkowski from the University of Bayreuth for providing the source code of quadratic programming.

References

1. Aliabadi, F.: The Boundary Element Method: Applications in Solids and Structures. Wiley, New York (2002)
2. Ambrósio, J., Rauter, F., Pereira, M., Pombo, J.: A flexible multibody pantograph model for the analysis of the catenary-pantograph contact problem. In: Fracek, J., Wojtyra, M. (eds.) ECCOMAS Thematic Conference: Multibody Dynamics, Warsaw (2009)
3. Amirouche, F.: Fundamentals of Multibody Dynamics. Theory and Applications. Birkhäuser, Boston (2005)
4. Baumgarte, J.: Stabilization of constraints and integrals of motion in dynamical systems. Comput. Methods Appl. Mech. Eng. **1**, 1–16 (1972)
5. Beer, G., Smith, I., Duenser, C.: The Boundary Element Method with Programming: For Engineers and Scientists. Springer, Berlin (2008)
6. Bernardi, C., Maday, Y., Patera, A.: A new conforming approach to domain decomposition: the mortar element. In: Brezis, H., Lions, J.L. (eds.) Nonlinear Partial Differential Equations and Their Applications, pp. 13–51. Wiley, New York (1992)
7. Börm, S., Grasedyck, L., Hackbusch, W.: Hierarchical matrices. Lecture note No. 21, Max-Planck-Institut für Mathematik in den Naturwissenschaften, Leipzig (2003)
8. Brebbia, C.A., Wrobel, L.C., Partridge, P.W.: Dual Reciprocity Boundary Element Method. Springer, Berlin (1991)
9. Brezzi, F., Fortin, M.: Mixed and Hybrid Finite Element Methods. Springer, Berlin (1991)
10. Busch, M., Schweizer, B.: Numerical stability and accuracy of different co-simulation techniques: analytical investigations based on a 2-DOF test model. In: Mikkola, A. (ed.) Proceedings of the 1st Joint International Conference on Multibody System Dynamics—IMSD, Lappeenranta, pp. 71–83 (2010)
11. Cowper, G.R.: Gaussian quadrature formulas for triangles. Int. J. Numer. Methods Eng. **7**, 405–408 (1973)
12. Craig, R.R., Bampton, M.C.C.: Coupling of substructures for dynamic analyses. J. Aircr. **6**(7), 1313–1319 (1968)
13. Danson, D.J.: A boundary element formulation of problems in linear isotropic elasticity with body forces. In: 3rd International Conference on Boundary Element Methods, pp. 105–122 (1981)
14. Dunavant, D.A.: High degree efficient symmetrical Gaussian quadrature rules for the triangle. Int. J. Numer. Methods Eng. **21**, 1129–1148 (1985)
15. Eberhard, P.: Kontaktuntersuchungen durch hybride Mehrkörpersystem/Finite Elemente Simulationen. Shaker, Aachen (2000)
16. Foley, J.D., Van Dam, A., Feiner, S.K., Hughes, J.F., Phillips, R.L.: Introduction to Computer Graphics. Addison-Wesley Longman, Amsterdam (1993)

17. Gaul, L., Fiedler, C.: Methode der Randelemente in Statik und Dynamik. Vieweg, Braunschweig (1998)
18. Gaul, L., Fischer, M.: Large-scale simulations of acoustic-structure interaction using the fast multipole BEM. Z. Angew. Math. Mech. **86**, 4–17 (2006)
19. Hackbusch, W., Nowak, Z.P.: On the fast matrix multiplication in the boundary element method by panel clustering. Numer. Math. **54**, 463–491 (1989)
20. Hayami, K., Sauter, S.: Application of the panel clustering method to the three-dimensional elastostatic problem. In: Boundary Elements XIX, pp. 625–635. Computational Mechanics Publications, Southampton (1997)
21. Hayami, K., Sauter, S.: Panel clustering for 3-D elastostatics using spherical harmonics. In: Boundary Elements XX, pp. 289–299. Computational Mechanics Publications, Southampton (1998)
22. Koch, A.: Implementierung eines direkten Randelementverfahrens und einer Panel-Clustering-Methode zur Berechnung von Temperaturproblemen. Master thesis, Universität Rostock (2011)
23. Kögl, M.: A boundary element method for dynamic analysis of anisotropic elastic, piezoelectric, and thermoelastic solids. Ph.D. thesis, Institut A für Mechanik (Bericht 6/2000), Universität Stuttgart (2000)
24. Krause, R.H.: Monotone multigrid methods for Signorini's problem with friction. Ph.D. thesis, FU Berlin (2001)
25. Krause, R.H., Wohlmuth, B.I.: A Dirichlet-Neumann type algorithm for contact problems with friction. Comput. Vis. Sci. **5**, 139–148 (2002)
26. Lehmann, T., Schittkowski, K., Spickenreuther, T.: MIQL: a Fortran code for convex mixed-integer quadratic programming—user's guide. Tech. Rep. 1, Department of Computer Science, University of Bayreuth (2009)
27. McDevitt, T.W., Laursen, T.A.: A mortar-finite element formulation for frictional contact problems. Int. J. Numer. Methods Eng. **48**, 1525–1547 (2000)
28. Park, K., Felippa, C., Rebel, G.: A simple algorithm for localized construction of non-matching structural interfaces. Int. J. Numer. Methods Eng. **53**, 2117–2142 (2002)
29. Pfeiffer, F., Glocker, C.: Multibody Dynamics with Unilateral Contacts. Wiley, New York (1996)
30. Puso, M.A.: A 3D mortar method for solid mechanics. Int. J. Numer. Methods Eng. **59**, 315–336 (2004)
31. Rebel, G., Park, K., Felippa, C.: A contact formulation based on localized Lagrange multipliers: formulation and application to two-dimensional problems. Int. J. Numer. Methods Eng. **54**, 263–298 (2002)
32. Schiehlen, W., Eberhard, P.: Technische Dynamik. Teubner, Leipzig (2004)
33. Thomson, W.: Note on the integration of the equations of equilibrium of an elastic solid. Cambridge and Dublin Mathematical Journal **3**, 87–89 (1848)
34. Vodička, R., Mantič, V., Paris, F.: SGBEM with Lagrange multipliers applied to elastic domain decomposition problems with curved interfaces using non-matching meshes. Int. J. Numer. Methods Eng. **83**, 91–128 (2010)
35. Woernle, C.: Mehrkörpersysteme: Eine Einführung in die Kinematik und Dynamik von Systemen starrer Körper. Springer, Berlin (2011)
36. Wohlmuth, B.I.: A mortar finite element method using dual spaces for the Lagrange multiplier. SIAM J. Numer. Anal. **38**, 989–1012 (2000)
37. Wohlmuth, B.I.: Discretization Methods and Iterative Solvers Based on Domain Decomposition. Springer, Berlin (2001)
38. Wohlmuth, B.I., Krause, R.H.: Monotone methods on non-matching grids for nonlinear contact problems. SIAM J. Sci. Comput. **25**, 324–347 (2003)
39. Wriggers, P.: Computational Contact Mechanics. Springer, Berlin (2006)
40. Zachmann, G.: Virtual reality in assembly simulation—collision detection, simulation algorithms, and interaction techniques. Ph.D. thesis, Technische Universität Darmstadt (2000)

41. Zierath, J.: Modelling of Three-Dimensional Contacts by Coupled Multibody- and Boundary-Element-Systems. Verlag Dr. Hut, München (2011)
42. Zierath, J., Woernle, C.: Elastic contact analyses with hybrid boundary element/multibody systems. In: Mikkola, A. (ed.) Proceedings of the 1st Joint International Conference on Multibody System Dynamics, Lappeenranta, pp. 327–336 (2010)

Trajectory Control of Serial and Parallel Flexible Manipulators Using Model Inversion

Robert Seifried, Markus Burkhardt, and Alexander Held

Abstract Traditional manipulator designs are based on maximized stiffness to suppress undesired elastic vibrations. This results in high accuracy in end-effector trajectory tracking, while it usually includes a drastic mass increase, a poor weight-to-payload ratio and high energy consumption. In contrast, modern light weight designs result in low energy consumption and allow often high working speeds. However, due to the light weight design the bodies have a significant flexibility which yields undesired vibrations. Therefore, in the control design these flexibilities must be taken into account. In this chapter feedforward control designs based on inverse models are presented and applied to serial and parallel flexible manipulators. Thereby, for a given system output the inverse model provides the control input for exact reproduction of the desired output trajectory and the trajectories of the generalized coordinates.

1 Introduction

In order to achieve in modern machines low energy consumption and allowing high working speeds light weight designs are increasingly often used. However, due to the light weight design the bodies of the manipulators have a significant flexibility which yields undesired vibrations. Therefore, the manipulators must be modeled as flexible multibody system and in the control design these flexibilities must be taken into account. Flexible manipulators are typical examples of underactuated multibody systems, since they generally possess less control inputs than degrees of freedom. In order to obtain a good performance in end-effector trajectory tracking

R. Seifried (✉) · M. Burkhardt · A. Held
Institute of Engineering and Computational Mechanics, University of Stuttgart, Pfaffenwaldring 9, 70569 Stuttgart, Germany
e-mail: robert.seifried@itm.uni-stuttgart.de

M. Burkhardt
e-mail: markus.burkhardt@itm.uni-stuttgart.de

A. Held
e-mail: alexander.held@itm.uni-stuttgart.de

J.-C. Samin, P. Fisette (eds.), *Multibody Dynamics*,
Computational Methods in Applied Sciences 28,
DOI 10.1007/978-94-007-5404-1_3, © Springer Science+Business Media Dordrecht 2013

an accurate and efficient feedforward control is necessary. This is then supplemented by additional feedback control to account for small disturbances and uncertainties. In this chapter the feedforward control design based on inverse models is presented and applied to serial and parallel flexible manipulators. Thereby for a given system output the inverse model provides the control input for exact reproduction of the desired output trajectory. In addition the inverse model provides the trajectories for all generalized coordinates, which can be used in additional feedback control.

In this chapter, firstly, an exact inverse model using concepts from differential geometric control theory [8, 12] is presented and applied to serial and parallel flexible manipulators. The starting point is the explicit symbolic transformation of the equations of motion into the nonlinear input-output normal-form. From this the inverse model is derived, consisting of a chain of differentiators, the driven internal dynamics and an algebraic part. The stability properties of the internal dynamics determine the complexity of the feedforward control design. If the internal dynamics are stable they can be solved by forward time integration. Otherwise, bounded solutions for the internal dynamics must be found by the solution of a two-sided boundary value problem [4, 17]. In order to avoid this, optimization based output relocation is proposed to obtain a system with stable internal dynamics, while keeping the tracking error of the flexible manipulator small.

In addition an alternative approach for feedforward control based on servo-constraints is presented. This yields at first a set of differential-algebraic equations. By using numerical projection into the unconstrained subspace the description of the internal dynamics is obtained, while its differentiation index is reduced. Using the methods available for the first approach, the internal dynamics are then solved in a similar way. This feedforward control concept is applied to a parallel flexible manipulator, where the loop closing constraints and servo-constraints are treated concurrently.

2 Feedforward Control Design by Symbolic Coordinate Transformation

For modeling flexible light weight manipulators the method of flexible multibody systems is often most suitable to represent large nonlinear working motions coupled with elastic vibrations. Since for manipulators the elastic deformations are comparatively small, the floating frame of reference approach can be used [16]. The symbolic multibody system research software Neweul-M^2 [9] is used to derive the equations of motion in minimal coordinates based on the Newton-Euler equations and D'Alembert's principle. The availability of the symbolic equations of motion is very convenient for nonlinear controller design.

With the vector of generalized coordinates $\boldsymbol{q} \in \mathbb{R}^f$ the equations of motion in minimal coordinates are obtained as

$$\boldsymbol{M}(\boldsymbol{q})\ddot{\boldsymbol{q}} + \boldsymbol{k}(\boldsymbol{q},\dot{\boldsymbol{q}}) = \boldsymbol{g}(\boldsymbol{q},\dot{\boldsymbol{q}}) + \boldsymbol{B}(\boldsymbol{q})\boldsymbol{u}. \tag{1}$$

Thereby $\boldsymbol{M}$ is the mass matrix, $\boldsymbol{k}$ the vector of generalized Coriolis-, centrifugal- and gyroscopic-forces and $\boldsymbol{g}$ the vector of applied forces and inner forces due to the body elasticity. The input matrix $\boldsymbol{B}$ distributes the control inputs $\boldsymbol{u} \in \mathbb{R}^m$ onto the directions of the f generalized coordinates. The vector of generalized coordinates consists of the rigid coordinates $\boldsymbol{q}_r \in \mathbb{R}^{f_r}$ representing the rigid body motion and the elastic coordinates $\boldsymbol{q}_e \in \mathbb{R}^{f_e}$. Then, the equations of motion can be partitioned into

$$\begin{bmatrix} \boldsymbol{M}_{rr}(\boldsymbol{q}) & \boldsymbol{M}_{re}(\boldsymbol{q}) \\ \boldsymbol{M}_{re}^T(\boldsymbol{q}) & \boldsymbol{M}_{ee}(\boldsymbol{q}) \end{bmatrix} \begin{bmatrix} \ddot{\boldsymbol{q}}_r \\ \ddot{\boldsymbol{q}}_e \end{bmatrix} + \begin{bmatrix} \boldsymbol{k}_r(\boldsymbol{q}, \dot{\boldsymbol{q}}) \\ \boldsymbol{k}_e(\boldsymbol{q}, \dot{\boldsymbol{q}}) \end{bmatrix} = \begin{bmatrix} \boldsymbol{g}_r(\boldsymbol{q}, \dot{\boldsymbol{q}}) \\ \boldsymbol{g}_e(\boldsymbol{q}, \dot{\boldsymbol{q}}) \end{bmatrix} + \begin{bmatrix} \boldsymbol{B}_r \\ \boldsymbol{B}_e \end{bmatrix} \boldsymbol{u}. \tag{2}$$

For serial flexible manipulators the actuation occurs only at the joints of the systems. Then, using a tangent frame for the elastic bodies, the control inputs $\boldsymbol{u}$ do not directly effect the elastic coordinates and it is $\boldsymbol{B}_e = \boldsymbol{0}$. In addition, using relative coordinates the inputs act directly on the rigid coordinates and it is $\boldsymbol{B}_r = \boldsymbol{I}$. Since serial manipulators are considered first, this special choice is used in the remainder of this section.

Flexible multibody systems are typical underactuated multibody systems, since they have less control inputs $\boldsymbol{u} \in \mathbb{R}^m$ than generalized coordinates $\boldsymbol{q} \in \mathbb{R}^f$ with $m < f$. The control goal is tracking of a system output $\boldsymbol{y} \in \mathbb{R}^m$, e.g. the end-effector point. In this section an exact inverse model for feedforward control design is derived using concepts from differential geometric control theory [8, 12]. Thereby the starting point is the transformation of the flexible multibody system into the nonlinear input-output normal-form. Then, from this the inverse model is derived.

2.1 Coordinate Transformation into Input-Output Normal-Form

For control design it is often helpful to transform the nonlinear system into the so-called nonlinear input-output normal-form by a diffeomorphic coordinate transformation $\boldsymbol{z} = \boldsymbol{\Phi}(\boldsymbol{x})$. Thereby $\boldsymbol{x} = [\boldsymbol{q}^T, \dot{\boldsymbol{q}}^T]^T$ are the original coordinates and $\boldsymbol{z}$ are the new coordinates of the input-output normal-form, which are derived partially from the system output. The first two derivatives of the system output are

$$\dot{\boldsymbol{y}} = \boldsymbol{H}(\boldsymbol{q})\dot{\boldsymbol{q}}, \qquad \ddot{\boldsymbol{y}} = \boldsymbol{H}(\boldsymbol{q})\ddot{\boldsymbol{q}} + \boldsymbol{h}''(\boldsymbol{q}, \dot{\boldsymbol{q}}), \tag{3}$$

where $\boldsymbol{H}$ is the Jacobian matrix of the system output and $\boldsymbol{h}'' = \dot{\boldsymbol{H}}\dot{\boldsymbol{q}}$ is the local acceleration. In (3) the second derivative of the generalized coordinates $\ddot{\boldsymbol{q}}$ can be replaced by the equations of motion (1), yielding

$$\begin{aligned} \ddot{\boldsymbol{y}} &= \boldsymbol{H}\ddot{\boldsymbol{q}} + \boldsymbol{h}'' = \boldsymbol{H}\boldsymbol{M}^{-1}[\boldsymbol{g} - \boldsymbol{k} + \boldsymbol{B}\boldsymbol{u}] + \boldsymbol{h}'' \\ &= \boldsymbol{H}\boldsymbol{M}^{-1}\boldsymbol{B}\boldsymbol{u} + \boldsymbol{H}\boldsymbol{M}^{-1}[\boldsymbol{g} - \boldsymbol{k}] + \boldsymbol{h}''. \end{aligned} \tag{4}$$

If the matrix $\boldsymbol{H}\boldsymbol{M}^{-1}\boldsymbol{B}$ is nonsingular (4) can be solved for the control inputs $\boldsymbol{u}$. In this case the matrix $\boldsymbol{H}\boldsymbol{M}^{-1}\boldsymbol{B}$ is called decoupling matrix and the system is said to have vector relative degree $\boldsymbol{r} = \{r_1, \ldots, r_m\} = \{2, \ldots, 2\}$. Following [8] the relative degree is defined as the minimal number of derivatives of each system output $h_i(\boldsymbol{q})$,

$i = 1, \ldots, m$ until the inputs $\boldsymbol{u}$ can be computed. Then, no further derivatives are necessary and the first part of the coordinate transformation is found, which is typical for many flexible manipulators. Then, the nonlinear coordinate transformation is given by

$$z = \boldsymbol{\Phi}(\boldsymbol{x}) = \boldsymbol{\Phi}(\boldsymbol{q}, \dot{\boldsymbol{q}}) = \begin{bmatrix} z_1 \\ z_2 \\ z_3 \end{bmatrix} \quad \text{with} \quad \begin{aligned} z_1 &= \boldsymbol{y} = \boldsymbol{h}(\boldsymbol{q}) \in \mathbb{R}^m, \\ z_2 &= \dot{\boldsymbol{y}} = \boldsymbol{H}(\boldsymbol{q})\dot{\boldsymbol{q}} \in \mathbb{R}^m, \\ z_3 &= \boldsymbol{\Phi}_3(\boldsymbol{q}, \dot{\boldsymbol{q}}) \in \mathbb{R}^{2(f-m)}. \end{aligned} \tag{5}$$

Thereby the coordinates z_3 are determined such that (5) forms at least a local diffeomorphic coordinate transformation, which requires that the Jacobian matrix $\boldsymbol{J} = \partial\boldsymbol{\Phi}(\boldsymbol{x})/\partial\boldsymbol{x}$ is nonsingular.

Applying the coordinate transformation (5) to the equations of motion (1) yields the nonlinear input-output normal-form. However, the complete symbolic transformation is often quite difficult. In the following it is shown, that the input-output normal-form can be established efficiently using a linearly combined system output and the partitioned equations of motion (2). For flexible manipulators a linearly combined output $\boldsymbol{y} = \boldsymbol{q}_r + \boldsymbol{\Gamma}\boldsymbol{q}_e$ of rigid coordinates $\boldsymbol{q}_r$ and elastic coordinates $\boldsymbol{q}_e$ is often a suitable choice [11, 15]. With such a system output the end-effector position of serial flexible manipulators may be approximated such that $\boldsymbol{r}^{ef}(\boldsymbol{q}_r, \boldsymbol{q}_e) \approx \boldsymbol{r}(\boldsymbol{y})$. The determination of the weighting matrix $\boldsymbol{\Gamma}$ is discussed in Sect. 3. For the special case of $\boldsymbol{\Gamma} = \boldsymbol{0}$ the output reduces to $\boldsymbol{y} = \boldsymbol{q}_r$, which is the so-called collocated output. Thus, for flexible manipulators with linearly combined output a suitable coordinate transformation is given by

$$z_1 = \boldsymbol{y} = \boldsymbol{q}_r + \boldsymbol{\Gamma}\boldsymbol{q}_e, \qquad z_2 = \dot{\boldsymbol{y}} = \dot{\boldsymbol{q}}_r + \boldsymbol{\Gamma}\dot{\boldsymbol{q}}_e, \qquad z_3 = \left[\boldsymbol{q}_e^T, \dot{\boldsymbol{q}}_e^T\right]^T. \tag{6}$$

In order to derive the input-output normal-form the rigid coordinates $\boldsymbol{q}_r$ are expressed in terms of the output $\boldsymbol{y}$ and the elastic coordinates $\boldsymbol{q}_e$. Then, after symbolic manipulations the nonlinear input-output normal-form is obtained

$$\widetilde{\boldsymbol{M}}\ddot{\boldsymbol{y}} = \widetilde{\boldsymbol{g}} - \widetilde{\boldsymbol{k}} + \boldsymbol{u} \tag{7a}$$

$$\left(\boldsymbol{M}_{ee} - \boldsymbol{M}_{re}^T\boldsymbol{\Gamma}\right)\ddot{\boldsymbol{q}}_e = \boldsymbol{g}_e - \boldsymbol{k}_e - \boldsymbol{M}_{re}^T\widetilde{\boldsymbol{M}}^{-1}\left(\widetilde{\boldsymbol{g}} - \widetilde{\boldsymbol{k}} + \boldsymbol{u}\right), \tag{7b}$$

with the terms summarized according to the convention $\widetilde{\boldsymbol{M}} = \boldsymbol{M}_{rr} - (\boldsymbol{M}_{re} - \boldsymbol{M}_{rr}\boldsymbol{\Gamma})(\boldsymbol{M}_{ee} - \boldsymbol{M}_{re}^T\boldsymbol{\Gamma})^{-1}\boldsymbol{M}_{re}^T$ and $\widetilde{\boldsymbol{g}} = \boldsymbol{g}_r - (\boldsymbol{M}_{re} - \boldsymbol{M}_{rr}\boldsymbol{\Gamma})(\boldsymbol{M}_{ee} - \boldsymbol{M}_{re}^T\boldsymbol{\Gamma})^{-1}\boldsymbol{g}_e$ and $\widetilde{\boldsymbol{k}} = \boldsymbol{k}_r - (\boldsymbol{M}_{re} - \boldsymbol{M}_{rr}\boldsymbol{\Gamma})(\boldsymbol{M}_{ee} - \boldsymbol{M}_{re}^T\boldsymbol{\Gamma})^{-1}\boldsymbol{k}_e$. For details on the performed symbolic computations see e.g. [15]. Equation (7a) has dimension m and describes the relationship between the inputs $\boldsymbol{u}$ and outputs $\boldsymbol{y}$. Equation (7b) has in this case dimension $f - m$ and is called internal dynamics. Its behavior is crucial for control design and has to be analyzed carefully. For this task the concept of zero-dynamics is often very useful. The zero-dynamics of a nonlinear system are the internal dynamics of the system under the constraint that the output is kept exactly constant, e.g. $\boldsymbol{y} = \boldsymbol{0}$, $\forall t$. A nonlinear system is called asymptotically minimum phase if the equilibrium point of the zero-dynamics is asymptotically stable. Otherwise the system is called non-minimum phase [8, 12].

2.2 Inverse Model

The inverse model follows directly from the input-output normal-form (7a)–(7b). The inverse model provides the inputs $\boldsymbol{u}_d$ which are required for exact reproduction of a desired system output trajectory $\boldsymbol{y} = \boldsymbol{y}_d$. In addition the corresponding trajectories of all generalized coordinates $\boldsymbol{q}_d$ are obtained. The inputs $\boldsymbol{u}_d$ follow from Eq. (7a) as

$$\boldsymbol{u}_d = \widetilde{\boldsymbol{M}}(\boldsymbol{y}_d, \boldsymbol{q}_e)\ddot{\boldsymbol{y}}_d - \widetilde{\boldsymbol{g}}(\boldsymbol{y}_d, \boldsymbol{q}_e, \dot{\boldsymbol{y}}_d, \dot{\boldsymbol{q}}_e) + \widetilde{\boldsymbol{k}}(\boldsymbol{y}_d, \boldsymbol{q}_e, \dot{\boldsymbol{y}}_d, \dot{\boldsymbol{q}}_e). \tag{8}$$

The computation of the inputs $\boldsymbol{u}_d$ depends on the desired outputs $\boldsymbol{y}_d, \dot{\boldsymbol{y}}_d$ and the elastic states $\boldsymbol{q}_e, \dot{\boldsymbol{q}}_e$. The latter are the solution of the internal dynamics given by Eq. (7b) which is driven by $\boldsymbol{y}_d, \dot{\boldsymbol{y}}_d$ and $\boldsymbol{u}_d$. Replacing $\boldsymbol{u}_d$ in the internal dynamics (7b) by Eq. (8) yields for the values of the elastic states $\boldsymbol{q}_e, \dot{\boldsymbol{q}}_e$ the differential equation

$$\begin{aligned}&\big[\boldsymbol{M}_{ee}(\boldsymbol{y}_d, \boldsymbol{q}_e) - \boldsymbol{M}_{re}^T(\boldsymbol{y}_d, \boldsymbol{q}_e)\boldsymbol{\Gamma}\big]\ddot{\boldsymbol{q}}_e \\ &\quad = \boldsymbol{g}_e(\boldsymbol{y}_d, \boldsymbol{q}_e, \dot{\boldsymbol{y}}_d, \dot{\boldsymbol{q}}_e) - \boldsymbol{k}_e(\boldsymbol{y}_d, \boldsymbol{q}_e, \dot{\boldsymbol{y}}_d, \dot{\boldsymbol{q}}_e) - \boldsymbol{M}_{re}^T(\boldsymbol{y}_d, \boldsymbol{q}_e)\ddot{\boldsymbol{y}}_d.\end{aligned} \tag{9}$$

Several methods for model inversion exist which differ in the solution of the internal dynamics (9):

Classical Inversion In classical inversion [6] the $\boldsymbol{q}_e, \dot{\boldsymbol{q}}_e$ variables can be found through forward integration of the internal dynamics (9) from the starting time point t_0 to the final time point t_f, using the initial values $\boldsymbol{q}_e(t_0) = \boldsymbol{q}_{e0}, \dot{\boldsymbol{q}}_e(t_0) = \dot{\boldsymbol{q}}_{e0}$. However, depending on the stability of the internal dynamics forward integration might yield unbounded $\boldsymbol{q}_e, \dot{\boldsymbol{q}}_e$ values and thus unbounded inputs $\boldsymbol{u}_d$, which cannot be used as feedforward control. Therefore, this approach can only be used for feedforward control design if the internal dynamics (9) remain bounded, which implies that the system is minimum phase. An example is a flexible manipulator with collocated output $\boldsymbol{y} = \boldsymbol{q}_r$, i.e. $\boldsymbol{\Gamma} = \boldsymbol{0}$.

Stable Inversion Flexible manipulators with the end-effector point as system output turn out to be often non-minimum phase and classical inversion cannot be used. However, using stable inversion [4] the inversion problem can be solved, such that the trajectories $\boldsymbol{q}_e, \dot{\boldsymbol{q}}_e$ of the internal dynamics (9) and the control inputs $\boldsymbol{u}_d$ remain bounded. However, the solution might be non-causal. The solution of the stable inversion is formulated as a two-sided boundary value problem, where the boundary conditions are described by the unstable and stable eigenspaces $\boldsymbol{E}_0^u$, $\boldsymbol{E}_f^s$ at the corresponding equilibrium points of the internal dynamics. These are local approximations of the unstable manifold $\boldsymbol{W}_0^u$ and stable manifold $\boldsymbol{W}_f^s$ at the starting and ending equilibrium point, respectively, see [12]. This yields for the internal dynamics bounded trajectories $\boldsymbol{q}_e, \dot{\boldsymbol{q}}_e$ which start at time t_0 on the unstable manifold $\boldsymbol{W}_0^u$ and reach the stable manifold $\boldsymbol{W}_f^s$ at time t_f. Thus the initial conditions $\boldsymbol{q}_{u0}$, $\dot{\boldsymbol{q}}_{u0}$ at time t_0 cannot exactly be pre-designated. A pre-actuation phase $[t_{pr}, t_0]$ is

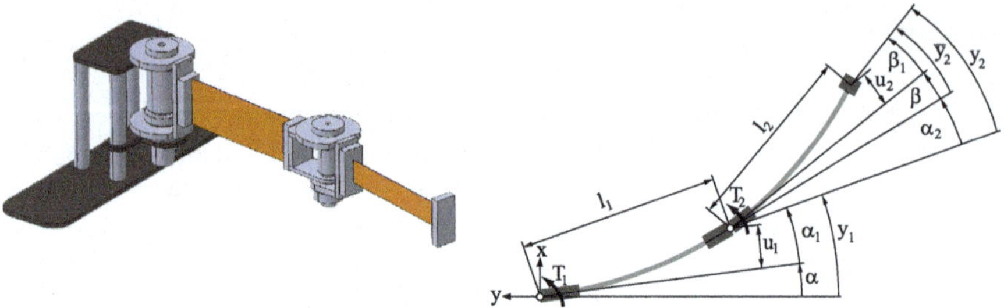

Fig. 1 Serial flexible manipulator

necessary which drives the system along the unstable manifold to a particular initial condition $\boldsymbol{q}_e(t_0)$, $\dot{\boldsymbol{q}}_e(t_0)$, while maintaining the constant output $\boldsymbol{y}_d = \boldsymbol{y}_d(t_0)$. Also a post-actuation phase $[t_f, t_{po}]$ is necessary to drive the internal dynamics along the stable manifold close to its resting position. The two-sided boundary value problem has to be solved numerically. This can be done by a finite difference method [17], e.g. using the Matlab solver *bvp5c*.

3 Determination of the Linearly Combined Output

The presented derivation of the inverse model depends on a linearly combined system output. In this section the derivation of the linearly combined output is demonstrated exemplarily for a serial flexible manipulator which moves in the horizontal plane. The manipulator consists of two elastic arms with rigid elements attached at their ends for mounting of the motors and end-effector mass, see Fig. 1. The total length of the first arm is denoted as l_1 and the rigid end parts have length l_{r11} and l_{r12}. The second arm has length l_2 and the rigid end parts have length l_{r21} and l_{r22}. The two motors produce the applied torques T_1 and T_2 which act on the joint angles $\boldsymbol{q}_r = [\alpha, \beta]^T$. The transverse elastic deformations of the two arms are described by the elastic coordinates $\boldsymbol{q}_e^1 \in \mathbb{R}^r$ and $\boldsymbol{q}_e^2 \in \mathbb{R}^s$. In the following the matrix $\boldsymbol{\Gamma}$ is derived from geometrical considerations and by optimization such that the outputs $\boldsymbol{y} = [y_1, y_2]^T$ yield a very good approximation of the end-effector position,

$$\boldsymbol{r}^{ef}(\boldsymbol{q}_r, \boldsymbol{q}_e) \approx \bar{\boldsymbol{r}}^{ef}(\boldsymbol{y}) = \begin{bmatrix} l_1 \sin(y_1) + l_2 \sin(y_1 + y_2) \\ -l_1 \cos(y_1) - l_2 \cos(y_1 + y_2) \end{bmatrix}. \tag{10}$$

Here y_1 and y_2 can be viewed as auxiliary angles, see Fig. 1. With this approximation the desired trajectories for the system output $\boldsymbol{y}_d$ can be computed by rigid body inverse kinematics from the desired trajectory $\boldsymbol{r}_d^{ef}$ of the end-effector point.

3.1 Choice of System Output Using Geometrical Considerations

Due to the body elasticity the tips of the arms are subjected to the displacements u_1 and u_2. These are perpendicular to the undeformed axes of the arm. Restricting to small elastic rotations, these elastic deformations are given by

$$\begin{aligned} u_1 &= \sum_{i=1}^{r} \Phi_i^1 q_{ei}^1 + l_{r12} \sum_{i=1}^{r} \Psi_i^1 q_{ei}^1, \\ u_2 &= \sum_{i=1}^{s} \Phi_i^2 q_{ei}^2 + l_{r21} \sum_{i=1}^{s} \Psi_i^2 q_{ei}^2. \end{aligned} \tag{11}$$

The ith elastic coordinate of the first arm is denoted by q_{ei}^1 and of the second arm by q_{ei}^2. The value of the ith displacement shape function at the end of the elastic parts of the first and second arm are denoted by Φ_i^1 and Φ_i^2, respectively. The values of the ith shape functions for the elastic rotation evaluated at the end of the elastic parts of arm one and two are denoted by Ψ_i^1 and Ψ_i^2, respectively. Thus, the first terms of Eq. (11) represent the transverse elastic deformation of the elastic parts of both bodies. Due to the elastic rotation of these elastic parts, the rigid end parts of the arms undergo an rotation with respect to their reference frame given by

$$\bar{\alpha}_2 = \sum_{i=1}^{r} \Psi_i^1 q_{ei}^1 \quad \text{and} \quad \bar{\beta}_2 = \sum_{i=1}^{s} \Psi_i^2 q_{ei}^2. \tag{12}$$

The influence of $\bar{\alpha}_2$ and $\bar{\beta}_2$ is represented by the second terms of Eq. (11). These deformations u_1, u_2 result in the approximate deformation angles α_1, β_1, see Fig. 1. For small displacements these two deformation angles α_1, β_1 can be determined and expressed as linear combinations of the elastic coordinates as

$$\alpha_1 \approx \frac{u_1}{l_1} = \sum_{i=1}^{r} \frac{\Phi_i^1 + l_{r12}\Psi_i^1}{l_1} q_{ei}^1 = \sum_{i=1}^{r} \Gamma_{1i} q_{ei}^1 \tag{13}$$

and

$$\beta_1 \approx \frac{u_2}{l_2} = \sum_{i=1}^{s} \frac{\Phi_i^2 + l_{r22}\Psi_i^2}{l_2} q_{ei}^2 = \sum_{i=1}^{s} \Gamma_{2(r+i)} q_{ei}^2. \tag{14}$$

From these two equations the linearly combined system outputs $y_1 = \alpha + \alpha_1$ and $\bar{y}_2 = \beta + \beta_1$ can be determined. For flexible manipulators without any rigid parts similar outputs are used for trajectory tracking by [11] and [18]. However, when using this output to approximate the end-effector point (10) of a multi-link flexible manipulator, the elastic rotation of the coordinate system attached to tip of arm 1 is neglected. Since in this coordinate system the motor angle β is described, the output $\bar{y}_2$ is not suitable for end-effector tracking, see [15]. Thus the system output y_2 has to be corrected by the additional angle α_2 given by

$$\alpha_2 = \bar{\alpha}_2 - \alpha_1 = \sum_{i=1}^{r} (\Psi_i^1 - \Gamma_{1i}) q_{ei}^1 = \sum_{i=1}^{r} \Gamma_{2i} q_{ei}^1. \tag{15}$$

From this follows the system output $y_2 = \bar{y}_2 + \alpha_2 = \beta + \beta_1 + \alpha_2$, which contains contributions of the elastic deformation of arm 1 and arm 2. The linearly combined system output which is suitable to approximate the end-effector point by Eq. (10) is then given by

$$\begin{aligned} \boldsymbol{y} &= \boldsymbol{q}_r + \boldsymbol{\Gamma}\boldsymbol{q}_e \\ &= \begin{bmatrix} \alpha \\ \beta \end{bmatrix} + \begin{bmatrix} \Gamma_{11} & \dots & \Gamma_{1r} & 0 & \dots & 0 \\ \Gamma_{21} & \dots & \Gamma_{2r} & \Gamma_{2(r+1)} & \dots & \Gamma_{2(r+s)} \end{bmatrix} \begin{bmatrix} q_{e1}^1 \\ \vdots \\ q_{er}^1 \\ q_{e1}^2 \\ \vdots \\ q_{es}^2 \end{bmatrix}. \end{aligned} \tag{16}$$

3.2 Design of System Output by Optimization

Output relocation is a method where a different system output $\hat{y}$ is chosen in order to achieve minimum phase property. In [11] output relocation for a flexible two arm manipulator is investigated. Thereby a linearly combined system output with block diagonal matrix $\boldsymbol{\Gamma}$ is used. It is shown that for the two outputs the entries of $\boldsymbol{\Gamma}$ can be scaled with a value between 0 and 1 to obtain minimum phase property. However, the influence of the elastic rotation (15) of the first body on the second system output is neglected and might result in large end-effector errors, see [15].

Therefore, an optimization based design procedure for a new system output $\hat{\boldsymbol{y}}$ is proposed here to obtain a minimum phase design of underactuated multibody systems and also a very good approximation of the end-effector point $\boldsymbol{r}^{ef}$. Thereby the new output $\hat{\boldsymbol{y}}$ is an artificial output, which does not represent a specific material point of the multibody system. For this task the linearly combined output is very convenient, since it provides an easy way for the parametrization of the design variables,

$$\hat{\boldsymbol{y}} = \boldsymbol{q}_r + \boldsymbol{\Gamma}(\boldsymbol{p})\boldsymbol{q}_e. \tag{17}$$

Here the design variables $\boldsymbol{p}$ are just the entries of the weighting matrix $\boldsymbol{\Gamma}$. Then, it follows with $\boldsymbol{y}_d = \boldsymbol{0}$, $\forall t$ from (9) that the zero dynamics, which depend only on the elastic coordinates $\boldsymbol{q}_e, \dot{\boldsymbol{q}}_e$ and the design variables $\boldsymbol{p}$, are given by

$$\left[\boldsymbol{M}_{ee}(\boldsymbol{q}_e) - \boldsymbol{M}_{re}^T(\boldsymbol{q}_r)\boldsymbol{\Gamma}(\boldsymbol{p})\right]\ddot{\boldsymbol{q}}_r = \boldsymbol{g}_r(\boldsymbol{q}_r, \dot{\boldsymbol{q}}_e) - \boldsymbol{k}_e(\boldsymbol{q}_e, \dot{\boldsymbol{q}}_e). \tag{18}$$

The linearization of the zero dynamics yield the system matrix $\boldsymbol{A}(\boldsymbol{p})$ which also depends on the design variables. The design goal is to achieve a stable zero dynamics by changing the system output, whereby the new system output should still yield a good approximation of the end-effector position $\boldsymbol{r}^{ef}$. Therefore, a two-step computation of the optimization criterion $f(\boldsymbol{p})$ is proposed, which should be minimized in the course of the optimization. The two steps of the optimization criterion computation are:

Step 1 Firstly, local asymptotic stability is checked. Thus, all eigenvalues of the linearized zero dynamics must be in the left half-plane,

$$\mathrm{Re}\big[\lambda\big(\boldsymbol{A}(\boldsymbol{p})\big)\big] < 0. \tag{19}$$

If at least one eigenvalue has a non-negative real part, a large default value for the optimization criterion $f(\boldsymbol{p})$ is returned. Otherwise, the linearized analysis shows asymptotic stability of the zero dynamics and it is proceeded with step 2.

Step 2 If all eigenvalues of the zero dynamics are in the left half-plane, the actual optimization criterion $f(\boldsymbol{p})$ will be calculated. Therefore a feedforward control for a test trajectory $\boldsymbol{y}_d$ is computed. From this inverse model the end-effector trajectory $\boldsymbol{r}^{ef}(\boldsymbol{q}_r, \boldsymbol{q}_e) \in \mathbb{R}^m$ and the effective deviation $e(t)$ from the desired trajectory can be determined as

$$e(t) = \sqrt{\sum_{i=1}^{m} e_i^2(t)} \quad \text{with } \boldsymbol{e}(t) = \boldsymbol{r}_d^{ef}(t) - \boldsymbol{r}^{ef}(\boldsymbol{q}_r, \boldsymbol{q}_e). \tag{20}$$

Then, the optimization criterion is chosen as the maximal effective end-effector deviation

$$f(\boldsymbol{p}) = \max_t e(t). \tag{21}$$

With the solution for the internal dynamics for the desired trajectory $\boldsymbol{y}_d$ the boundedness of the internal dynamics is verified. If unbounded states for this design occur the time integration fails and also a large default value is returned for the optimization criterion $f(\boldsymbol{p})$.

The optimization criterion is discontinuous due to the distinction between stable and unstable designs. Therefore, gradient based methods cannot be used and the stochastic particle swarm optimization algorithm is applied, see [13] for details on the used algorithm. The presented optimization based approach for designing a suitable system output yields a good tracking performance for a given desired output trajectory. If the optimized output $\boldsymbol{\Gamma}(\boldsymbol{p}^*)$ also yields a good end-effector approximation for a different trajectory has to be checked for each particular case.

4 Application to a Serial Manipulator

The feedforward control design is first demonstrated for a planar serial flexible manipulator as shown in Fig. 1. The first arm has length $l_1 = 351$ mm and consists of a first rigid part of length $l_{r11} = 63.5$ mm, an elastic part of length $l_{e1} = 209$ mm and a second rigid part of length $l_{r12} = 78.5$ mm. The elastic part has thickness 1.27 mm and height 76.2 mm. The second arm has length $l_2 = 287.5$ mm and consists of a first rigid part of length $l_{r21} = 62.5$ mm, an elastic part of length $l_e = 210$ mm and a rigid end-effector mass of length $l_{r22} = 15$ mm. The elastic part of the second arm

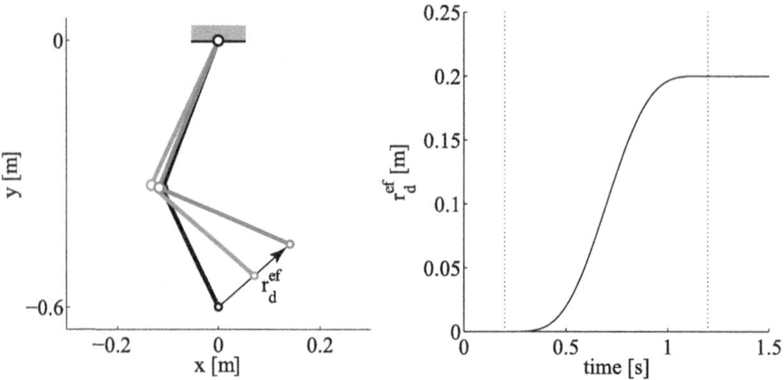

Fig. 2 Desired trajectory of the manipulator

has thickness 0.9 mm and height 38.1 mm. The rigid parts of the arms represent the joints, motor mounting and end-effector mass. All parts of the manipulator are made out of steel. For the description of the elastic deformation of the arms the first two bending eigenmodes are used as shape functions for each arm, i.e. $\boldsymbol{q}_e \in \mathbb{R}^4$. A tangent floating frame of reference is used. The end-effector of the manipulator should follow a straight test trajectory, whose path is shown in Fig. 2. Along the path, the trajectory is described by a polynomial of ninth order, such that the velocities and accelerations are zero at both the start and the end of the trajectory. The path length is 0.2 m and the time for following the trajectory is 1 s. The trajectory starts from rest at time 0.2 s and ends in rest at time 1.2 s.

In the following simulation results for this flexible manipulator are presented, see also [15] for more details. Thereby the computed feedforward control is tested by simulation, whereby it is combined with PID-feedback control for the joint trajectories $\boldsymbol{q}_r$. However, since the inverse model is exact, the PID-control has in these tests only to compensate numerical errors. In reality there are larger uncertainties and disturbances which the controller has to account for. The presented results should demonstrate the capacities of the feedforward control, and the maximal achievable accuracy in the ideal case. In the following the trajectory error in path direction e_{ip}, orthogonal to the path e_{op} and the absolute error $e_{abs} = \sqrt{e_{ip}^2 + e_{op}^2}$ are presented.

Firstly, a feedforward control based on a rigid system output is used, i.e. the elasticities are neglected yielding $\boldsymbol{\Gamma} = \boldsymbol{0}$. This is the so-called collocated output which yields a minimum phase system and classical inversion can be used. From the tracking error for this strategy, which is presented in Fig. 3, the strong influence of body flexibility is seen, yielding an unacceptable behavior and errors of several centimeters.

The inversion based feedforward control with linearly combined output is considered next. Using the elastic data of the manipulator and Eqs. (10)–(16) the matrix $\boldsymbol{\Gamma}$ is given by

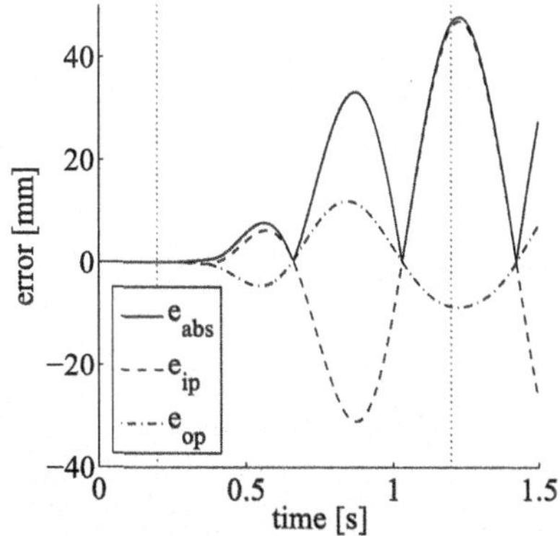

Fig. 3 Error of the end-effector trajectory using exact inversion with collocated output

$$\boldsymbol{\Gamma} = \begin{bmatrix} 25.779 & 47.552 & 0 & 0 \\ 13.500 & 88.897 & 32.316 & 39.470 \end{bmatrix}. \tag{22}$$

An analysis of the zero-dynamics of the flexible manipulator with this output shows, that the system is non-minimum phase. Thus, stable inversion is necessary and the Matlab boundary-value solver *bvp5c* is used. However, it turns out, that for the $\boldsymbol{\Gamma}$ values given by (22) no solution can be computed numerically using the *bvc5p*. But by slight variations of $\Gamma_{21} = 13.375$ and $\Gamma_{22} = 92.5$, a numerical solution of the boundary value problem is found. In Fig. 4 the end-effector trajectory error is shown for this feedforward control. A very high accuracy is achieved. The trajectory errors are in the magnitude of less than 0.1 mm. After reaching the final position at time 1.2 s only minor deviations of the end-effector point remain.

The optimization based system output design is applied next. For the optimization based design the non-zero entries of the weighting matrix $\boldsymbol{\Gamma}$ are used as design variables. Based on the values for $\boldsymbol{\Gamma}$ derived from geometrical considerations (22), the bounds for the design variables are defined. These are set such that a variation of the design variables of $+/-20\ \%$ around the geometrical case (22) is allowed. For the optimization 100 particles are used and yields the values

$$\boldsymbol{\Gamma} = \begin{bmatrix} 25.025 & 45.909 & 0 & 0 \\ 15.292 & 90.345 & 29.296 & 32.297 \end{bmatrix}. \tag{23}$$

The new system output yields a minimum phase system and feedforward control design by classical inversion can be applied. The simulation results are presented in Fig. 4 and show that with this approach high accuracy for the end-effector trajectory can be obtained. The maximal trajectory error is about 0.28 mm. Compared to the previous results the achievable accuracy for this minimum phase system is slightly worse than the output using geometrical considerations yielding a non-minimum phase system, see Fig. 4, but much better than using the collocated output, see Fig. 3. Comparing to results for this manipulator presented in [15], the optimization based output design yields also better results than the so-called quasi-static deformation compensation, which is an alternative approach for feedforward control design of flexible manipulators.

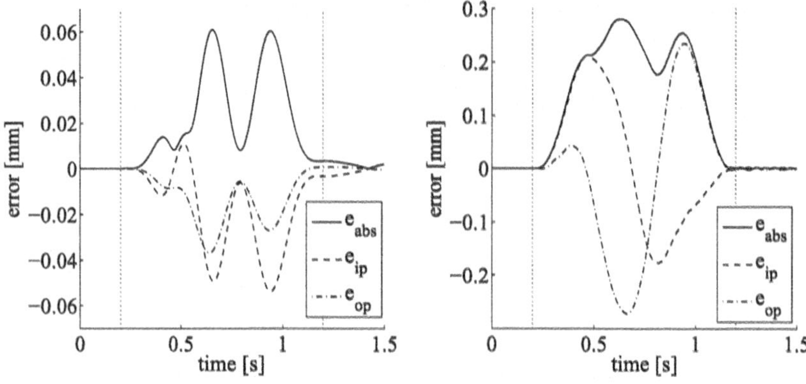

Fig. 4 Error of the end-effector trajectory using stable inversion with $\boldsymbol{\Gamma}$ from geometrical considerations (*left*) and optimization (*right*)

5 Extension to Parallel Manipulators

For multibody systems with kinematic loops and f_c degrees of freedom the description of the kinematics using a minimal set of generalized coordinates is in general not directly possible. Therefore, the kinematic loop is cut at a suitable joint, yielding a multibody system in tree structure with $f > f_c$ degrees of freedom. The equations of motion of the obtained open loop system are derived in analogy to systems with tree structure according to Eq. (1). By introducing n_c implicit algebraic loop closing constraint equations

$$\boldsymbol{c}(\boldsymbol{q}, t) = \boldsymbol{0}, \tag{24}$$

the equations of motion of the closed loop system can be formulated as a set of differential-algebraic equations (DAE)

$$\boldsymbol{M}(\boldsymbol{q})\ddot{\boldsymbol{q}} + \boldsymbol{k}(\boldsymbol{q}, \dot{\boldsymbol{q}}) = \boldsymbol{g}(\boldsymbol{q}, \dot{\boldsymbol{q}}) + \boldsymbol{B}(\boldsymbol{q})\boldsymbol{u} + \boldsymbol{C}^T(\boldsymbol{q})\boldsymbol{\lambda} \tag{25a}$$

$$\boldsymbol{c}(\boldsymbol{q}, t) = \boldsymbol{0} \tag{25b}$$

where the matrix $\boldsymbol{C}$ is the Jacobian matrix of the constraint equations $\boldsymbol{c}$ with respect to the generalized coordinates $\boldsymbol{q} \in \mathbb{R}^f$. The vector $\boldsymbol{\lambda} \in \mathbb{R}^{n_c}$ represents the generalized reaction forces in the cut joints. In order to derive the input-output normal-form of the system (25a)–(25b) according to Sect. 2, the set of differential-algebraic equations has to be transformed into a set of $f_c = f - n_c$ ordinary differential equations.

There are several ways of rephrasing differential-algebraic equations as purely differential equations. One way is partitioning the generalized coordinates $\boldsymbol{q} \in \mathbb{R}^f$ into a set of independent coordinates $\boldsymbol{q}_i \in \mathbb{R}^{f_c}$ and dependent coordinates $\boldsymbol{q}_d \in \mathbb{R}^{n_c}$ according to

$$\boldsymbol{q}^T = [\, \boldsymbol{q}_i^T \;\; \boldsymbol{q}_d^T \,] \tag{26}$$

and expressing the system dynamics in these independent coordinates, see e.g. [10]. The first step is to differentiate the constraint equations $\boldsymbol{c}$ twice with respect to the time t leading to the constraint equations on acceleration level

$$\ddot{\boldsymbol{c}}(\ddot{\boldsymbol{q}},\dot{\boldsymbol{q}},\boldsymbol{q},t)=\boldsymbol{C}(\boldsymbol{q})\ddot{\boldsymbol{q}}+\boldsymbol{c}''(\dot{\boldsymbol{q}},\boldsymbol{q},t)=\boldsymbol{0}, \tag{27}$$

with the vector $\boldsymbol{c}''$ representing the local accelerations due to the constraints. For linear independent constraint equations and a reasonable choice of dependent coordinates, the Jacobian matrix $\boldsymbol{C}$ can be split up in an independent and a dependent part leading to

$$\ddot{\boldsymbol{c}}(\ddot{\boldsymbol{q}},\dot{\boldsymbol{q}},\boldsymbol{q},t)=\boldsymbol{C}_d(\boldsymbol{q})\ddot{\boldsymbol{q}}_d+\boldsymbol{C}_i(\boldsymbol{q})\ddot{\boldsymbol{q}}_i+\boldsymbol{c}''(\dot{\boldsymbol{q}},\boldsymbol{q},t)=\boldsymbol{0}, \tag{28}$$

in which $\boldsymbol{C}_d \in \mathbb{R}^{n_c\times n_c}$ must be a regular matrix. Solving Eq. (28) for the dependent accelerations $\ddot{\boldsymbol{q}}_d$ and dropping the dependencies for better readability, the dependent accelerations $\ddot{\boldsymbol{q}}_d$ can be expressed as

$$\ddot{\boldsymbol{q}}_d=-\boldsymbol{C}_d^{-1}\left(\boldsymbol{C}_i\ddot{\boldsymbol{q}}_i+\boldsymbol{c}''\right). \tag{29}$$

Based on this relationship, the generalized accelerations can be written as

$$\ddot{\boldsymbol{q}}=\begin{bmatrix}\boldsymbol{I}\\-\boldsymbol{C}_d^{-1}\boldsymbol{C}_i\end{bmatrix}\ddot{\boldsymbol{q}}_i+\begin{bmatrix}\boldsymbol{0}\\-\boldsymbol{C}_d^{-1}\boldsymbol{c}''\end{bmatrix}=\boldsymbol{J}_c\ddot{\boldsymbol{q}}_i+\boldsymbol{b}''. \tag{30}$$

By inserting Eq. (30) into Eq. (25a) and by left-side multiplication with the transposed Jacobian matrix $\boldsymbol{J}_c$ the equations of motion in minimal coordinates of the multibody body system with kinematic loop are obtained,

$$\boldsymbol{J}_c^T\boldsymbol{M}(\boldsymbol{q})\left(\boldsymbol{J}_c\ddot{\boldsymbol{q}}_i+\boldsymbol{b}''\right)+\boldsymbol{J}_c^T\boldsymbol{k}(\boldsymbol{q},\dot{\boldsymbol{q}})=\boldsymbol{J}_c^T\boldsymbol{g}(\boldsymbol{q},\dot{\boldsymbol{q}})+\boldsymbol{J}_c^T\boldsymbol{B}(\boldsymbol{q})\boldsymbol{u}+\boldsymbol{J}_c^T\boldsymbol{C}^T(\boldsymbol{q})\boldsymbol{\lambda}. \tag{31}$$

Since the relation

$$\boldsymbol{J}_c^T\boldsymbol{C}^T=\begin{bmatrix}\boldsymbol{I} & -\boldsymbol{C}_i^T\boldsymbol{C}_d^{-T}\end{bmatrix}\begin{bmatrix}\boldsymbol{C}_i^T\\\boldsymbol{C}_d^T\end{bmatrix}=\boldsymbol{C}_i^T-\boldsymbol{C}_i^T\boldsymbol{C}_d^{-T}\boldsymbol{C}_d^T=\boldsymbol{0} \tag{32}$$

holds, the Lagrange multipliers $\boldsymbol{\lambda}$ vanish in Eq. (31) and the equations of motion can be displayed as

$$\overline{\boldsymbol{M}}(\boldsymbol{q})\ddot{\boldsymbol{q}}_i+\overline{\boldsymbol{k}}(\boldsymbol{q},\dot{\boldsymbol{q}})=\overline{\boldsymbol{g}}(\boldsymbol{q},\dot{\boldsymbol{q}})+\overline{\boldsymbol{B}}(\boldsymbol{q})\boldsymbol{u}. \tag{33}$$

These equations of motion in minimal form allow the partitioning of the independent accelerations $\ddot{\boldsymbol{q}}_i$ into the rigid accelerations $\ddot{\boldsymbol{q}}_r$ and the elastic accelerations $\ddot{\boldsymbol{q}}_e$ according to Eq. (2).

Based on this representation the equations of motion (33) can be transformed into the input-output normal-form, which permits the analysis of the internal dynamics and the application of the exact model inversion procedures discussed in the previous sections. This approach is applied to the flexible parallel manipulator shown in Fig. 5. Using for the coordinate partitioning the independent coordinates $\boldsymbol{q}_i=[\boldsymbol{q}_r,\boldsymbol{q}_e]$ with $\boldsymbol{q}_r=[s_1,\alpha]$, again a linearly combined system output can be used to approximate the end-effector position

$$\boldsymbol{r}^{ef}(\boldsymbol{q}_r,\boldsymbol{q}_e)\approx\boldsymbol{r}^{ef}(\boldsymbol{y})\quad\text{with } y_1=s_1,\ y_2=\alpha+\boldsymbol{\Gamma}\boldsymbol{q}_e. \tag{34}$$

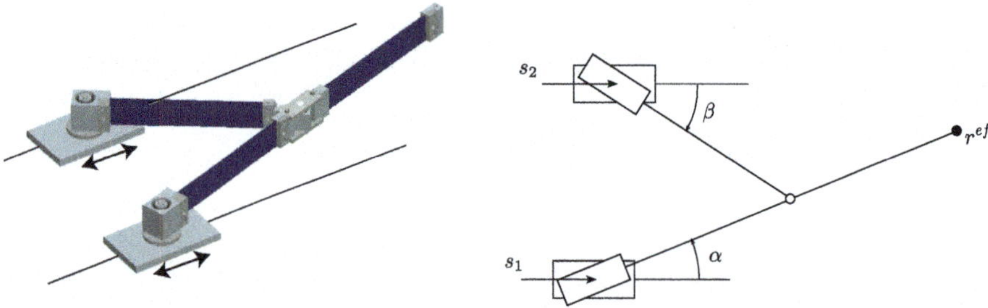

Fig. 5 Parallel flexible manipulator

The weighting matrix $\boldsymbol{\Gamma}$ can be derived similarly to Sect. 3. The transformation into input-output normal-form shows, that it is again a system of vector relative degree $\boldsymbol{r} = \{2, \ldots, 2\}$ and the elastic coordinates $\boldsymbol{q}_e$ describe the internal dynamics. Also in this case the weighting matrix $\boldsymbol{\Gamma}$ derived from geometric considerations yields a non-minimum phase system and requires a stable inversion. A simulation result is presented in Sect. 7.

6 Model Inversion Using Servo-Constraints and Projections

In this section an alternative approach for the solution of the exact model inversion problem is introduced, which is especially appealing for parallel flexible manipulators. Formulating the complete problem as a set of differential-algebraic equations according to

$$\boldsymbol{M}(\boldsymbol{q})\ddot{\boldsymbol{q}} + \boldsymbol{k}(\boldsymbol{q}, \dot{\boldsymbol{q}}) = \boldsymbol{g}(\boldsymbol{q}, \dot{\boldsymbol{q}}) + \boldsymbol{B}(\boldsymbol{q})\boldsymbol{u} + \boldsymbol{C}^T(\boldsymbol{q})\boldsymbol{\lambda} \tag{35a}$$

$$\boldsymbol{h}(\boldsymbol{q}, t) = \boldsymbol{y}(\boldsymbol{q}) - \boldsymbol{y}_d(t) = \boldsymbol{0} \tag{35b}$$

$$\boldsymbol{c}(\boldsymbol{q}, t) = \boldsymbol{0}, \tag{35c}$$

with the vector $\boldsymbol{c} \in \mathbb{R}^{n_c}$ representing the constraints due to the kinematic loops and the vector $\boldsymbol{h} \in \mathbb{R}^m$ describing the so-called servo-constraints [1, 14]. This unifies the handling of both types of constraints. Due to the fact that the constraint equations are on position level, it is not guaranteed that a numerical solution can be found easily. In order to ensure that a solution can be obtained, a formalism to reduce the index of the system without imposing a numerical drift is presented.

6.1 Differentiation Index

A very important characteristic of differential-algebraic equations is the differentiation index. It describes how a perturbation in the variables effects the solution of the DAE. The differentiation index is defined as the minimal number of differentiations of the constraint equations necessary to obtain ordinary differential equations for all variables. To ensure a better readability, the constraints in Eq. (35b) and Eq. (35c) are combined leading to the representation

$$\boldsymbol{M}(\boldsymbol{q})\ddot{\boldsymbol{q}} + \boldsymbol{k}(\boldsymbol{q},\dot{\boldsymbol{q}}) = \boldsymbol{g}(\boldsymbol{q},\dot{\boldsymbol{q}}) + \boldsymbol{\Theta}(\boldsymbol{q})\begin{bmatrix}\boldsymbol{u}\\ \boldsymbol{\lambda}\end{bmatrix} \tag{36a}$$

$$\boldsymbol{\xi}(\boldsymbol{q},t) = \boldsymbol{0} \tag{36b}$$

with the matrix $\boldsymbol{\Theta} = [\boldsymbol{B}\ \boldsymbol{C}^T] \in \mathbb{R}^{f\times(m+n_c)}$ and the vector $\boldsymbol{\xi}^T = [\boldsymbol{h}^T\ \boldsymbol{c}^T]^T \in \mathbb{R}^{(m+n_c)}$ of the constraints. Firstly, the constraint equations are differentiated with respect to the time

$$\begin{aligned}\frac{d\boldsymbol{\xi}}{dt} &= \frac{\partial\boldsymbol{\xi}}{\partial\boldsymbol{q}}\dot{\boldsymbol{q}} + \frac{\partial\boldsymbol{\xi}}{\partial\boldsymbol{\lambda}}\dot{\boldsymbol{\lambda}} + \frac{\partial\boldsymbol{\xi}}{\partial\boldsymbol{u}}\dot{\boldsymbol{u}} + \frac{\partial\boldsymbol{\xi}}{\partial t} = \boldsymbol{0}\\ \dot{\boldsymbol{\xi}} &= \begin{bmatrix}\boldsymbol{H}\\ \boldsymbol{C}\end{bmatrix}\dot{\boldsymbol{q}} + \boldsymbol{\xi}' = \boldsymbol{\Xi}\dot{\boldsymbol{q}} + \boldsymbol{\xi}' = \boldsymbol{0},\end{aligned} \tag{37}$$

in which the matrix $\boldsymbol{\Xi} = [\boldsymbol{H}^T, \boldsymbol{C}^T]^T \in \mathbb{R}^{(m+n_c)\times f_t}$ summarized the Jacobian matrices of the servo-constraints $\boldsymbol{H}$ and geometric constraints $\boldsymbol{C}$. As the constraint equations do neither depend on the reaction forces $\boldsymbol{\lambda}$ nor on the system inputs $\boldsymbol{u}$, the Jacobian matrix of the unified constraints $\boldsymbol{\xi}$ with respect to $\boldsymbol{\lambda}$ and $\boldsymbol{u}$ vanish. Since Eq. (37) also does not depend on $\boldsymbol{\lambda}$ or $\boldsymbol{u}$, the second partial differentiation can be displayed as

$$\ddot{\boldsymbol{\xi}} = \boldsymbol{\Xi}\ddot{\boldsymbol{q}} + \boldsymbol{\xi}'' = \boldsymbol{0}. \tag{38}$$

By solving Eq. (36a) for the generalized accelerations $\ddot{\boldsymbol{q}}$ and inserting them into Eq. (38), the constraint equations on velocity level can be presented as

$$\ddot{\boldsymbol{\xi}} = \boldsymbol{\Xi}(\boldsymbol{q})\boldsymbol{M}(\boldsymbol{q})^{-1}\left(\boldsymbol{g}(\boldsymbol{q},\dot{\boldsymbol{q}}) - \boldsymbol{k}(\boldsymbol{q},\dot{\boldsymbol{q}}) + \boldsymbol{\Theta}(\boldsymbol{q})\begin{bmatrix}\boldsymbol{u}\\ \boldsymbol{\lambda}\end{bmatrix}\right) + \boldsymbol{\xi}'' = \boldsymbol{0}. \tag{39}$$

The final differentiation yields to the formulation

$$\begin{aligned}\dddot{\boldsymbol{\xi}} &= \frac{\partial\ddot{\boldsymbol{\xi}}}{\partial\boldsymbol{q}}\dot{\boldsymbol{q}} + \frac{\partial\ddot{\boldsymbol{\xi}}}{\partial\dot{\boldsymbol{q}}}\ddot{\boldsymbol{q}} + \frac{\partial\ddot{\boldsymbol{\xi}}}{\partial\boldsymbol{u}}\dot{\boldsymbol{u}} + \frac{\partial\ddot{\boldsymbol{\xi}}}{\partial\boldsymbol{\lambda}}\dot{\boldsymbol{\lambda}} + \frac{\partial\ddot{\boldsymbol{\xi}}}{\partial t}\\ &= \frac{\partial\ddot{\boldsymbol{\xi}}}{\partial\boldsymbol{q}}\dot{\boldsymbol{q}} + \frac{\partial\ddot{\boldsymbol{\xi}}}{\partial\dot{\boldsymbol{q}}}\ddot{\boldsymbol{q}} + \boldsymbol{\Xi}(\boldsymbol{q})\boldsymbol{M}(\boldsymbol{q})^{-1}\boldsymbol{\Theta}(\boldsymbol{q})\begin{bmatrix}\dot{\boldsymbol{u}}\\ \dot{\boldsymbol{\lambda}}\end{bmatrix} + \frac{\partial\ddot{\boldsymbol{\xi}}}{\partial t} = \boldsymbol{0}\end{aligned} \tag{40}$$

providing a set of differential equations for the generalized reaction forces $\boldsymbol{\lambda}$ and the system inputs $\boldsymbol{u}$, if the matrix $\boldsymbol{R} = \boldsymbol{\Xi}\boldsymbol{M}^{-1}\boldsymbol{\Theta}$ is regular. In this case, the system has a differentiation index of 3. If the matrix $\boldsymbol{R}$ is not invertible, the differentiation index is accordingly higher. For the special case, that the system has only independent geometric constraints, $\boldsymbol{R}$ reduces to $\boldsymbol{R} = \boldsymbol{C}\boldsymbol{M}^{-1}\boldsymbol{C}^T$ which is positive definite and therewith invertible. If only servo-constraints occur it is $\boldsymbol{R} = \boldsymbol{H}\boldsymbol{M}^{-1}\boldsymbol{B}$. This is exactly the decoupling matrix occurring in Eq. (4) of the symbolic coordinate transformation into input-output normal-form. Thus, if $\boldsymbol{R}$ is invertible the system has differentiation index 3 and vector relative degree $\boldsymbol{r} = \{2,\ldots,2\}$. In [2] it is discussed that the differentiation index is one higher than the relative degree, if the internal dynamics are not affected by a constraint. In the following it is assumed, that servo-constraints and geometric constraints occur and that the matrix $\boldsymbol{R}$ has full rank, providing differentiation index 3.

6.2 Projections

In order to reduce the index of the system, the equations of motion (36a) and the constraint equations on acceleration level (39) are taken into account. The first step is eliminating the generalized reaction forces $\boldsymbol{\lambda}$ and the system inputs $\boldsymbol{u}$ with a projection. For classical DAEs this method is sometimes referred as the null space method, see [7]. For servo-constraint problems, the corresponding projection matrices are obtained with two QR decompositions according to

$$\boldsymbol{\Xi}^T = \boldsymbol{Q}_r \boldsymbol{R}_r = [\, \boldsymbol{Q}_{r,1} \quad \boldsymbol{J}_r \,] \begin{bmatrix} \boldsymbol{R}_{r,1} \\ \boldsymbol{0} \end{bmatrix} = \boldsymbol{Q}_{r,1} \boldsymbol{R}_{r,1}, \tag{41a}$$

$$\boldsymbol{\Theta} = \boldsymbol{Q}_l \boldsymbol{R}_l = [\, \boldsymbol{Q}_{l,1} \quad \boldsymbol{J}_l \,] \begin{bmatrix} \boldsymbol{R}_{l,1} \\ \boldsymbol{0} \end{bmatrix} = \boldsymbol{Q}_{l,1} \boldsymbol{R}_{l,1}. \tag{41b}$$

The matrix $\boldsymbol{\Xi}^T$ is split into an orthogonal $\boldsymbol{Q}_r \in \mathbb{R}^{f \times f}$ and a triangular matrix $\boldsymbol{R}_r \in \mathbb{R}^{f \times (m+n_c)}$, whose $f - (m + n_c)$ lower rows are zero. Therefore, the matrix $\boldsymbol{\Xi}^T$ can be displayed as the product of the first $(m + n_c)$ columns of $\boldsymbol{Q}_r$, called $\boldsymbol{Q}_{r,1}$, and the first $(m + n_c)$ rows of $\boldsymbol{R}_r$, called $\boldsymbol{R}_{r,1}$. With this relationship, the matrix $\boldsymbol{\Xi}$ can be written as

$$\boldsymbol{\Xi} = \boldsymbol{R}_{r,1}^T \boldsymbol{Q}_{r,1}^T. \tag{42}$$

Due to the fact that the matrix $\boldsymbol{Q}_r$ is orthogonal, the matrix product

$$\boldsymbol{\Xi} \boldsymbol{J}_r = \boldsymbol{R}_{r,1}^T \underbrace{\boldsymbol{Q}_{r,1}^T \boldsymbol{J}_r}_{=\boldsymbol{0}} \tag{43}$$

vanishes. Therefore, the columns of the matrix $\boldsymbol{J}_r$ span the null space of the matrix $\boldsymbol{\Xi}$ and the columns of the matrix $\boldsymbol{Q}_{r,1}$ span the row space of the matrix $\boldsymbol{\Xi}$, respectively. Since the dimensions of $\boldsymbol{\Xi}^T$ and $\boldsymbol{\Theta}$ match, an analog procedure shows that the columns of the matrix $\boldsymbol{J}_l$ span the left null space of the matrix $\boldsymbol{\Theta}$ and the columns of the matrix $\boldsymbol{Q}_{l,1}$ span the column space of the matrix $\boldsymbol{\Theta}$.

The properties of the matrix $\boldsymbol{Q}_r$ are used to introduce a new set of generalized accelerations $\ddot{z}$ according to

$$\ddot{q} = \boldsymbol{Q}_r \ddot{z} = \boldsymbol{J}_r \ddot{z}_i + \boldsymbol{Q}_{r,1} \ddot{z}_d, \tag{44}$$

with the independent accelerations $\ddot{z}_i \in \mathbb{R}^{f-(m+n_c)}$ and the dependent accelerations $\ddot{z}_d \in \mathbb{R}^{(m+n_c)}$. Substituting the generalized accelerations in Eq. (38) with Eq. (44) results in

$$\ddot{\xi} = \Xi(J_r\ddot{z}_i + Q_{r,1}\ddot{z}_d) + \xi'' = R_{r,1}^T \underbrace{Q_{r,1}^T J_r}_{=0} \ddot{z}_i + R_{r,1}^T \underbrace{Q_{r,1}^T Q_{r,1}}_{=I} \ddot{z}_d + \xi'' = 0. \tag{45}$$

Solving Eq. (45) for the dependent accelerations $\ddot{z}_d$ leads to

$$\ddot{z}_d = -R_{r,1}^{-T}\xi''. \tag{46}$$

Therefore, using Eq. (44), the generalized coordinates can be expressed as

$$\ddot{q} = J_r\ddot{z}_i - Q_{r,1}R_{r,1}^{-T}\xi''. \tag{47}$$

The next step is to eliminate the reaction forces $\boldsymbol{\lambda}$ and system inputs $\boldsymbol{u}$ and to express the equations of motion in terms of the independent accelerations $\ddot{z}_i$. By substituting the generalized accelerations $\ddot{q}$ in Eq. (36a) with Eq. (47) and multiplying the arising equations with the transposed Jacobian matrix J_l from the left yields to

$$J_l^T M(q)(J_r\ddot{z}_i - Q_{r,1}R_{r,1}^{-T}\xi'') + J_l^T k(q,\dot{q}) = J_l^T g(q,\dot{q}) + \underbrace{J_l^T \Theta(q)}_{=0}\begin{bmatrix} u \\ \lambda \end{bmatrix}. \tag{48}$$

This leads to the equations of motion in the new, independent coordinates z_i. These equations describe the internal dynamics of the servo-constraint problem given in Eqs. (35a)–(35c). This corresponds to the internal dynamics given by Eq. (9) using symbolic coordinate transformation. In order to solve the initial value problem, these equations have to be transformed back to the original set of coordinates q. Solving Eq. (48) for independent accelerations $\ddot{z}_i$ according to

$$\ddot{z}_i = (J_l^T M(q) J_r)^{-1} J_l^T (g(q,\dot{q}) - k(q,\dot{q}) + M(q)Q_{r,1}R_{r,1}^{-T}\xi'') \tag{49}$$

and inserting the independent accelerations $\ddot{z}_i$ into Eq. (47) leads to

$$\ddot{q} = J_r(J_l^T M(q) J_r)^{-1} J_l^T (g - k + M(q)Q_{r,1}R_{r,1}^{-T}\xi'') - Q_{r,1}R_{r,1}^{-T}\xi''. \tag{50}$$

In order to derive a state space representation, the same procedure is done for the velocities. Similar to Eq. (47), the independent velocities $\dot{z}_i$ can be expressed as

$$\dot{q} = J_r\dot{z}_i - Q_{r,1}R_{r,1}^{-T}\xi'. \tag{51}$$

Multiplying Eq. (51) with the transposed Jacobian matrix J_l from the left and solving for $\dot{z}_i$ results in

$$\dot{z}_i = (J_l^T J_r)^{-1} J_l^T (\dot{q} + Q_{r,1}R_{r,1}^{-T}\xi'). \tag{52}$$

Inserting this equation into Eq. (51) results in the representation necessary for state space representation

$$\dot{q} = J_r(J_l^T J_r)^{-1} J_l^T (\dot{q} + Q_r R_r^{-T}\xi') - Q_r R_r^{-T}\xi', \tag{53}$$

in which the velocities do not violate the constraint equations. One of the benefits of this representation is, that the equations of motion no longer dependent on the reaction forces nor on the system inputs and at the same time comply with the constraints. But the back transformation also introduced $2(m + f_c)$ zero eigenvalues leading to a numerical drift in the solution [3]. The same effect occurs, when the Lagrange equations of the first kind are applied to constraint mechanical systems.

A suitable way of avoiding the drift is to reformulate the system given by Eq. (50) and Eq. (53) as an index 1-DAE by only considering $2(f - (m + n_c))$ differential equations, which describe the internal dynamics, and taking the remaining $2(m + f_c)$ coordinates as algebraic equations into account. The generalized coordinates are split according to $\boldsymbol{q}^T = [\boldsymbol{q}_i^T \ \boldsymbol{q}_d^T]$ with $\boldsymbol{q}_i$ being the coordinates describing the internal dynamics and $\boldsymbol{q}_d$ being the coordinates describing the driven dynamics. From the analysis in Sect. 5 it is known that the internal dynamics can be described by the elastic coordinates and thus it is $\boldsymbol{q}_d = \boldsymbol{q}_e$. Then, the equations of motion of the servo-constraint problem in state space representation can be expressed as

$$\begin{bmatrix} \boldsymbol{I} & \boldsymbol{0} & \boldsymbol{0} & \boldsymbol{0} \\ \boldsymbol{0} & \boldsymbol{I} & \boldsymbol{0} & \boldsymbol{0} \\ \boldsymbol{0} & \boldsymbol{0} & \boldsymbol{0} & \boldsymbol{0} \\ \boldsymbol{0} & \boldsymbol{0} & \boldsymbol{0} & \boldsymbol{0} \end{bmatrix} \begin{bmatrix} \dot{\boldsymbol{q}}_i \\ \ddot{\boldsymbol{q}}_i \\ \boldsymbol{q}_d \\ \dot{\boldsymbol{q}}_d \end{bmatrix} = \begin{bmatrix} \boldsymbol{P}(\boldsymbol{J}_r(\boldsymbol{J}_l^T \boldsymbol{J}_r)^{-1} \boldsymbol{J}_l^T (\dot{\boldsymbol{q}} + \boldsymbol{Q}_{r,1} \boldsymbol{R}_{r,1}^{-T} \boldsymbol{\xi}') - \boldsymbol{Q}_{r,1} \boldsymbol{R}_{r,1}^{-T} \boldsymbol{\xi}') \\ \boldsymbol{P}(\hat{\boldsymbol{M}}(\boldsymbol{g} - \boldsymbol{k} + \boldsymbol{M}(\boldsymbol{q}) \boldsymbol{Q}_{r,1} \boldsymbol{R}_{r,1}^{-T} \boldsymbol{\xi}'') - \boldsymbol{Q}_{r,1} \boldsymbol{R}_{r,1}^{-T} \boldsymbol{\xi}'') \\ \boldsymbol{\xi}(\boldsymbol{q}, t) \\ \dot{\boldsymbol{\xi}}(\dot{\boldsymbol{q}}, \boldsymbol{q}, t) \end{bmatrix}, \tag{54}$$

in which the matrix $\boldsymbol{P}$ represents the Jacobian matrix of the vector $\boldsymbol{q}_i$ with respect to the generalized coordinates $\boldsymbol{q}$ and the matrix $\hat{\boldsymbol{M}}$ equals $\boldsymbol{J}_r(\boldsymbol{J}_l^T \boldsymbol{M}(\boldsymbol{q}) \boldsymbol{J}_r)^{-1} \boldsymbol{J}_l^T$. The numerical integration of these equations as an initial value problem does not provoke a drift in the solution.

In analogy to Sect. 2.2 the desired system inputs $\boldsymbol{u}_d$ together with the reaction forces $\boldsymbol{\lambda}$, which both are eliminated in Eq. (54), can be derived by solving Eq. (36a) for the generalized accelerations and substituting them with Eq. (47) according to

$$\boldsymbol{J}_r \ddot{\boldsymbol{z}}_i - \boldsymbol{Q}_{r,1} \boldsymbol{R}_{r,1}^{-T} \boldsymbol{\xi}'' = \boldsymbol{M}(\boldsymbol{q})^{-1} \big(\boldsymbol{g}(\boldsymbol{q}, \dot{\boldsymbol{q}}) - \boldsymbol{k}(\boldsymbol{q}, \dot{\boldsymbol{q}})\big) + \boldsymbol{M}(\boldsymbol{q})^{-1} \boldsymbol{\Theta}(\boldsymbol{q}) \begin{bmatrix} \boldsymbol{u} \\ \boldsymbol{\lambda} \end{bmatrix}. \tag{55}$$

Multiplying this equation from the left with the transposed matrix $\boldsymbol{Q}_{r_1}$ yields, in case of an index 3 problem, a regular matrix $\boldsymbol{Q}_{r,1}^T \boldsymbol{M}^{-1} \boldsymbol{\Theta}$. Thus solving for the desired reaction forces and system inputs leads to

$$\begin{bmatrix} \boldsymbol{u}_d \\ \boldsymbol{\lambda} \end{bmatrix} = -(\boldsymbol{Q}_{r,1}^T \boldsymbol{M}^{-1} \boldsymbol{\Theta})^{-1} \big(\boldsymbol{R}_{r,1}^{-T} \boldsymbol{\xi}'' + \boldsymbol{Q}_{r,1}^T \boldsymbol{M}^{-1} (\boldsymbol{g} - \boldsymbol{k}) - \underbrace{\boldsymbol{Q}_{r,1}^T \boldsymbol{J}_r \ddot{\boldsymbol{z}}_i}_{=\boldsymbol{0}}\big). \tag{56}$$

6.3 Solution of the Two-Point Boundary Value Problem

In order to solve the two-point boundary value problem for this approach, some additional considerations have to be taken into account. Firstly, the question arises

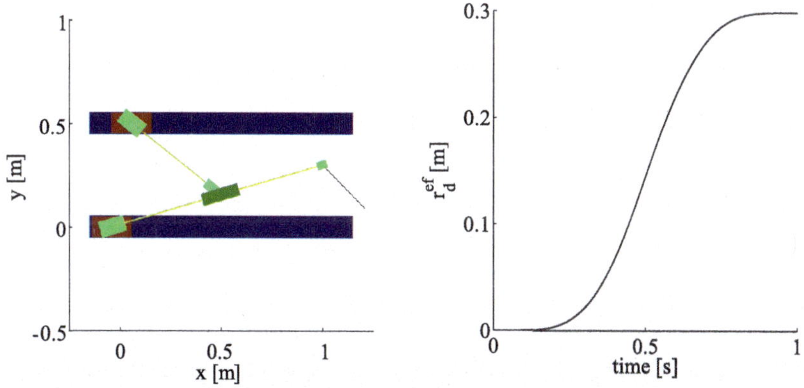

Fig. 6 Desired trajectory of the parallel manipulator

which representation is supposed to be used to solve the boundary value problem. There are three qualified candidates providing different advantages and drawbacks. The first candidate is the representation describing solely the internal dynamics of the model described by the first two rows of Eq. (54). By using this representation the formulation of the boundary conditions follows the description in Sect. 2, but it is necessary to compute the coordinates of the driven dynamics with a root search at each time step.

If a corresponding solver is available, the differential-algebraic representation according to Eq. (54) can be used to obtain a bounded solution. The third possibility is using the projected dynamics according to Eqs. (50) and (53). In this case, the numerical overhead is considerably reduced. The last two candidates have in common that the formulation of the boundary conditions has to be modified. In addition to the already mentioned boundary conditions based on the eigenspaces of the internal dynamics, it is necessary to use a combination of the constraint equations on position and velocity level at the time t_0 or t_f as well. A reasonable choice is ensuring that the solution fulfills the constraint equations on position level at t_0 and t_f.

7 Application to Parallel Manipulators

The model inversion formalisms using coordinate transformation presented in Sects. 2 and 5 and servo-constraints presented in Sect. 6 are applied to the parallel flexible manipulator shown in Fig. 5. The trajectory presented in Fig. 6 is used. The system consists of a long and a short arm each mounted on a car. The long arm is composed of three rigid parts connected with two elastic links and the short arm is composed of two rigid parts connected with an elastic link. The three identical elastic links, which are made out of steel, have length $l_e = 400$ mm, height $h_e = 80$ mm and depth $d_e = 2$ mm. The overall length of the long arm is $l_l = 1081$ mm, whereas the length of the short arm is $l_s = 560$ mm. The end of the short arm is connected

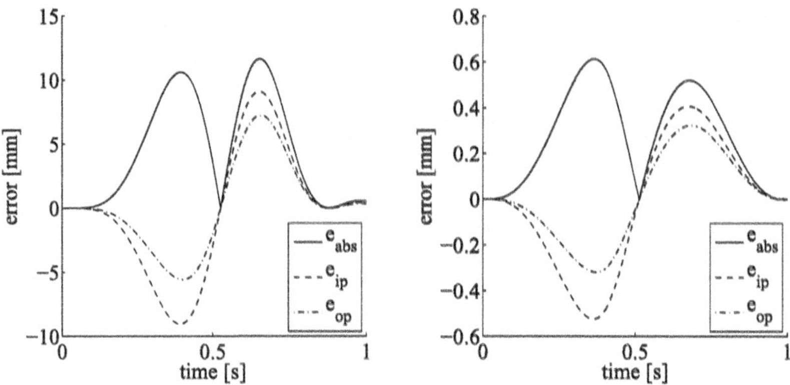

Fig. 7 Error of the end-effector trajectory with rigid output (*left*) and using optimized weights and classical inversion (*right*)

to the middle of the long arm by a revolute joint. The long arm is modeled as one elastic body consisting of two beam elements which are connected with rigid bodies. A model order reduction of the long arm based on proper orthogonal decomposition, see [5], results in a reduced elastic body with twelve shape functions to describe the elastic deformations. A similar procedure leads to a reduced model of the short arm with six shape functions. Due to the revolute joint the elastic deformations of the long arm are described in a secant floating frame of reference, whereas the deformations of the short arm are described in a tangent floating frame of reference. Therefore, the kinematics of the system with cut kinematic loop can be described with four rigid coordinates $\boldsymbol{q}_r = [s_1\ s_2\ \alpha\ \beta]^T$ and 18 elastic coordinates $\boldsymbol{q}_e$. In analogy to the serial manipulator, the end-effector of the long arm is supposed to follow a straight test trajectory, see Fig. 6. The end-effector point is the system output and can be approximated using the system output

$$y = \begin{bmatrix} s_1 \\ 0 \end{bmatrix} + \begin{bmatrix} \cos(\alpha) & -\sin(\alpha) \\ \sin(\alpha) & \cos(\alpha) \end{bmatrix} \left(\begin{bmatrix} l_{long} \\ 0 \end{bmatrix} + \begin{bmatrix} 0 \\ \sum_{i=1}^{f_e} w_i \Phi_i q_{ei} \end{bmatrix} \right), \tag{57}$$

where Φ_i is the ith shape function evaluated at the end-effector point. For $w_i = 1$ the exact end-effector position is obtained. In this case the system is non-minimum phase, and an output relocation can be performed for obtaining a minimum phase system. In this case the weights w_i are used as the design parameters for the optimization as presented in Sect. 3.2. For the servo-constraint approach the system output (57) is used, while for the coordinate transformation approach the linearly combined system output (34) is used.

Four different cases are studied. First of all, the servo-constraint approach is used and the weights w_i are all set to zero leading to an output omitting the elastic deformations, corresponding to a rigid output. In this case, the internal dynamics are stable and the forward integration of Eq. (54) can be used. The resulting error of the end-effector trajectory tracking is presented in Fig. 7, which shows a very large deviation of approximately 12 mm.

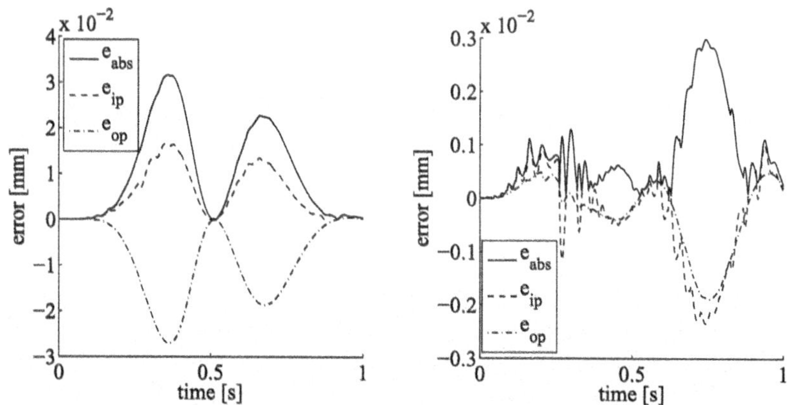

Fig. 8 Error of the end-effector trajectory using coordinate transformation (*left*) and servo constraints (*right*)

In order to improve the achievable accuracy of the trajectory tracking problem, an optimization of the weights is performed. Therefore, the algorithm presented in Sect. 3.2 is applied to the system dynamics described in Eq. (54). In this case, the design parameters $\boldsymbol{p}$ are the weights $\boldsymbol{w}$, which are varied from -1 to 1. The optimized output yields to a minimum phase system with a minimal trajectory tracking error. Then, for feedforward control design the servo-constraint approach is used in combination with forward integration. Figure 8 shows, that the error obtained is about 0.6 mm and thus less than a tenth of the error obtained with the rigid output.

Next, the stable inversion is applied to the system with end-effector point as system output, which has an unbounded internal dynamics. First, the coordinate transformation approach is used with linearly combined output (34), to approximate the end-effector point. The boundary value problem is solved with the Matlab solver *bvp5c*. Figure 8 shows the obtained error of the end-effector trajectory, which is around 0.03 mm. While the linearly combined system output is reproduced nearly exactly, this small tracking error of the end-effector point originates form its approximation by the linearly combined output.

Finally, the servo-constraint approach is used with exact output (57). Unlike the previous computations using the servo-constraint approach, only the differential part of Eq. (54) is considered, because the boundary value problem solver *bvp5c* does not support differential-algebraic equations. The error of the solution obtained by the boundary value problem is presented in Fig. 8. The maximal discrepancy is about 0.003 mm. Unlike the other cases, this error represents solely the solver tolerance. In summary, both stable inversion based approaches yield nearly exact reproduction of the end-effector trajectory.

8 Summary

The derivation of feedforward control designs for serial and flexible manipulators were presented. Firstly, exact inverse model based on concepts from differential geometric control theory were used and applied to serial and parallel flexible manipulators. It was shown, that the stability properties of the internal dynamics determines the complexity of the feedforward control design. By output optimization stable internal dynamics can be obtained, while keeping the end-effector tracking error small. In addition an alternative approach for feedforward control based on servo-constraints was presented and applied to a parallel flexible manipulator. By using numerical projection into the unconstrained subspace the description of the internal dynamics is obtained, while its differentiation index is reduced. Then, for the solution the same concepts as in the first feedforward control approach can be used. Both approaches provide powerful tools to design accurate feedforward control for flexible manipulators

Acknowledgements The authors would like to thank the German Research Foundation (DFG) for financial support of this work within the Cluster of Excellence in Simulation Technology (EXC 310/1) at the University of Stuttgart.

References

1. Blajer, W., Kolodziejczyk, K.: A geometric approach to solving problems of control constrains: theory and a DAE framework. Multibody Syst. Dyn. **11**, 343–364 (2004)
2. Campbell, S.: High-index differential algebraic equations. Mech. Based Des. Struct. Mach. **23**(2), 199–222 (1995)
3. Campbell, S., Leimkuhler, B.: Differentiation of constraints in differential-algebraic equations. Mech. Based Des. Struct. Mach. **19**(1), 19–39 (1991)
4. Devasia, S., Chen, D., Paden, B.: Nonlinear inversion-based output tracking. IEEE Trans. Autom. Control **41**(7), 930–942 (1996)
5. Fehr, J., Lehner, M., Eberhard, P.: On the automated generation of reduced order models in flexible multibody dynamics. In: Proceedings in Applied Mathematics and Mechanics (2007)
6. Hirschorn, R.: Invertibility of multivariable nonlinear control systems. IEEE Trans. Autom. Control **24**, 855–865 (1979)
7. Ider, S., Amirouche, F.: Coordinate reduction in the dynamics of constrained multibody systems—a new approach. J. Appl. Mech. **55**(4), 899–904 (1988)
8. Isidori, A.: Nonlinear Control Systems. Springer, London (1995)
9. Kurz, T., Eberhard, P., Henninger, C., Schiehlen, W.: From Neweul to Neweul-M^2: symbolical equations of motion for multibody system analysis and synthesis. Multibody Syst. Dyn. **24**, 25–41 (2010)
10. Leister, G., Bestle, D.: Symbolic-numerical solution of multibody systems with closed loops. Veh. Syst. Dyn. **21**(1), 129–142 (1992)
11. Moallem, M., Patel, R., Khorasani, K.: Flexible-Link Robot Manipulators: Control Techniques and Structural Design. Springer, London (2000)
12. Sastry, S.: Nonlinear Systems: Analysis, Stability and Control. Springer, New York (1999)
13. Sedlaczek, K., Eberhard, P.: Using augmented Lagrangian particle swarm optimization for constrained problems in engineering. Struct. Multidiscip. Optim. **32**, 277–286 (2006)
14. Seifried, R.: Two approaches for feedforward control and optimal design of underactuated multibody systems. Multibody Syst. Dyn. **27**(1), 75–93 (2012)

15. Seifried, R., Held, A., Dietmann, F.: Analysis of feed-forward control design approaches for flexible multibody systems. J. Syst. Des. Dyn. (Special Issue ACMD 2010) **5**(3), 429–440 (2011)
16. Shabana, A.: Dynamics of Multibody Systems. Cambridge University Press, Cambridge (2005)
17. Taylor, D., Li, S.: Stable inversion of continuous-time nonlinear systems by finite-difference methods. IEEE Trans. Autom. Control **47**(3), 537–542 (2002)
18. Zhao, H., Chen, D.: Tip trajectory tracking for multilink flexible manipulators using stable inversion. J. Guid. Control Dyn. **21**(2), 314–320 (1998)

A 3D Shear Deformable Finite Element Based on the Absolute Nodal Coordinate Formulation

Karin Nachbagauer, Peter Gruber, and Johannes Gerstmayr

Abstract The absolute nodal coordinate formulation (ANCF) has been developed for the modeling of large deformation beams in two or three dimensions. The absence of rotational degrees of freedom is the main conceptual difference between the ANCF and classical nonlinear beam finite elements that can be found in literature. In the present approach, an ANCF beam finite element is presented, in which the orientation of the cross section is parameterized by means of slope vectors. Based on these slope vectors, a thickness as well as a shear deformation of the cross section is included. The proposed finite beam element is investigated by an eigenfrequency analysis of a simply supported beam. The high frequencies of thickness modes are of the same magnitude as the shear mode frequencies. Therefore, the thickness modes do not significantly influence the performance of the finite element in dynamical simulations. The lateral buckling of a cantilevered right-angle frame under an end load is investigated in order to show a large deformation example in statics, as well as a dynamic application. A comparison to results provided in the literature reveals that the present element shows accuracy and high order convergence.

1 Introduction

The absolute nodal coordinate formulation (ANCF) has been developed by Shabana [21, 22] for the modeling of large deformation structural problems in two or three dimensions. In contrast to the classical large rotation vector formulation [12], the

K. Nachbagauer (✉)
Institute of Technical Mechanics, Johannes Kepler University Linz, Altenbergerstraße 69, 4040 Linz, Austria
e-mail: karin.nachbagauer@jku.at

P. Gruber · J. Gerstmayr
Linz Center of Mechatronics GmbH, Altenbergerstraße 69, 4040 Linz, Austria

P. Gruber
e-mail: peter.gruber@lcm.at

J. Gerstmayr
e-mail: johannes.gerstmayr@lcm.at

J.-C. Samin, P. Fisette (eds.), *Multibody Dynamics*,
Computational Methods in Applied Sciences 28,
DOI 10.1007/978-94-007-5404-1_4, © Springer Science+Business Media Dordrecht 2013

orientation of an ANCF element is not defined via rotational parameters, but in terms of slope vectors. The big advantage of ANCF elements is a constant mass matrix with respect to the generalized coordinates, which is advantageous in numerical procedures.

In the present chapter, a three-dimensional ANCF beam finite element is proposed, which is similar to a three-dimensional approach by Yakoub and Shabana [28]. The latter authors developed a shear-deformable beam element which provides a constant mass matrix and a geometrical description free of singularities. In case the Poisson effect is not neglected, the standard ANCF approach in [28] does not converge due to locking effects, see [19] for more details.

The elastic forces are investigated using a structural mechanics based formulation based on Reissner's nonlinear rod theory, as well as a continuum mechanics based formulation for a St. Venant Kirchhoff material. The importance of the *continuum mechanics* based formulation is emphasized in Irschik and Gerstmayr [9]. Since any suitable constitutive law can be utilized in this approach, a wide range of applications can be covered. Additionally, it has to be pointed out that the continuum mechanics based approach is in accordance with fully three-dimensional computations with solid finite elements, see [9]. The continuum mechanics based approach holds for simple cross sections only and has been investigated for rectangular cross sections so far. In contrast to the continuum mechanics based formulation, the *structural mechanics* based approach can be used for arbitrary cross sections, since standard beam stiffness parameters, e.g. bending stiffnesses, for stress resultants of the cross section are used, which are well-defined and available easily in engineering books. The structural mechanics based formulation can be interpreted as elastic line approach. In contrast to the continuum mechanics based approach, in which large strain considerations are possible, the structural mechanics based approach is limited to small strain considerations for large deformations.

In the original three-dimensional continuum mechanics based ANCF beam finite element by Yakoub and Shabana [28], the strain energy is based on a St. Venant Kirchhoff material law using a linear relation between the Green strain tensor and the second Piola-Kirchhoff stress tensor. Since this formulation suffers from Poisson locking, an enhanced locking-free approach is presented in this chapter.

In Reissner's nonlinear rod theory the strain energy is based on generalized strain measures for axial extension, bending, shear and torsion, which can be directly related to the generalized coordinates of the ANCF element, see [11] for details of the two-dimensional element, which is also depicted in Fig. 1(b). Moreover, the special choice of degrees of freedom in the present approach allows for a cross section deformation of the proposed elements, which is of particular interest e.g. in industrial applications, such as rolling mills. In Sugiyama et al. [25, 26] the Green strain tensor is used to define the beam kinematic properties and the generalized strains; however the equilibrium equations are formed on the continuum level. To be more precise, in [26] Green-Lagrange strains are utilized to define the curvature, while in the present approach only the cross section frame and its derivatives are used.

A vast number of different interpolation strategies has been discussed in literature, e.g., displacements and rotations are used as primal interpolated variables in

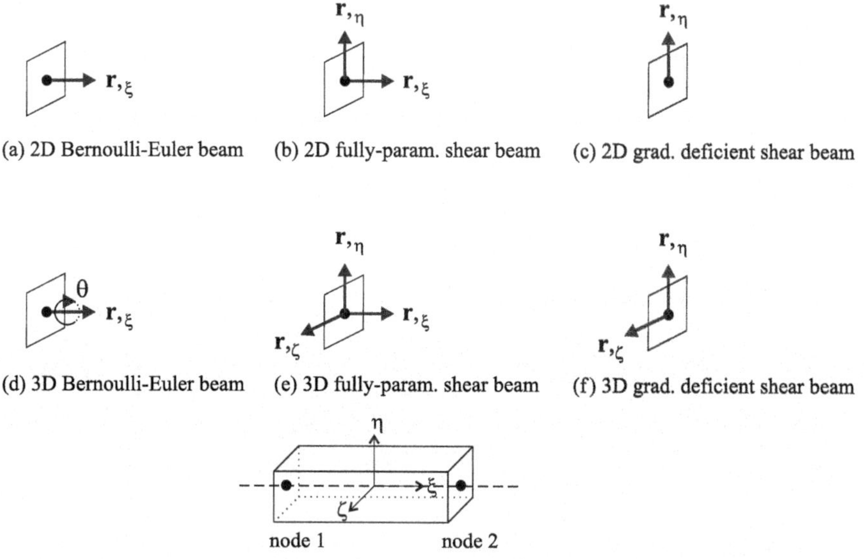

Fig. 1 Overview of different types of elements and their nodal coordinates

Simo and Vu-Quoc [24], displacements, slopes and rotations in Dmitrochenko [4] or strains in [5, 29]. In contrast to the formulations mentioned above, the present approach is based on the interpolation of displacements and slopes, as in Shabana [21]. For the interpolation of the displacements and the slopes, shape functions are chosen quadratic in axial direction and linear in the transversal directions. In so-called "fully-parameterized" elements, the nodal coordinates are based on three slope vectors (Fig. 1(e)), representing the position gradient, see e.g. [10, 25]. Here, the proposed formulation is based on two slope vectors only (Fig. 1(f)) and therefore belongs to the so-called "gradient deficient" elements. Figure 1(c) shows the two-dimensional analogue approach, which has been already presented in [18]. The same nodal coordinates are chosen in Kerkkänen et al. [13] for a two-noded ANCF element and in García-Vallejo et al. [6] a three-noded analogue is presented. In case of a Bernoulli-Euler beam element, in which the cross sections remain rigid and orthogonal to the beam axis, the position and the axial slope vector are used as nodal coordinates, e.g. in [7], see Fig. 1(a) for the two-dimensional case and Fig. 1(d) for the three-dimensional case, in which an additional angle parameter has to be added to include torsion, see [17].

The content of this chapter is arranged as follows. A detailed geometric description of the proposed ANCF beam finite element and the choice of degrees of freedom can be found in Sect. 2. Section 3 deals with the different definitions of the strain energy for the ANCF beam element. The proposed element is investigated for an eigenfrequency analysis of a simply-supported beam in Sect. 4.1. A comparison of the computed eigenfrequencies to results in [11] and to analytical values show good agreement. Since high thickness modes appear, time integrators should

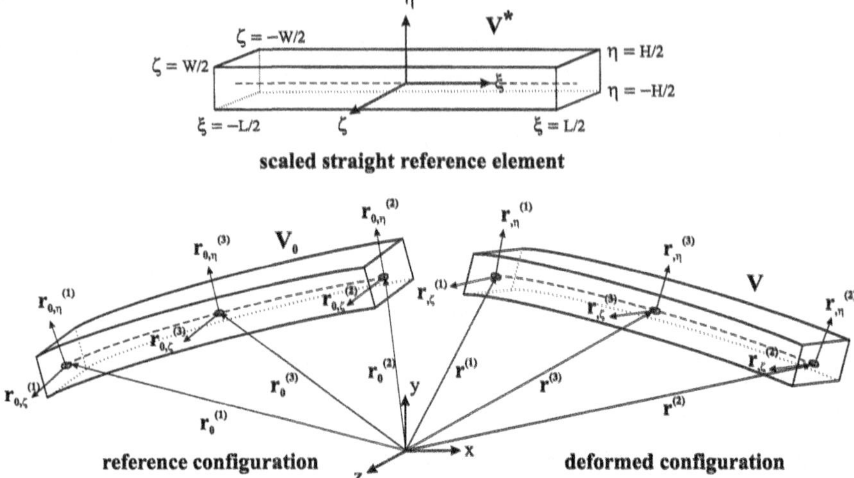

Fig. 2 Different configurations of the finite beam element: scaled straight reference element with volume V^* in the beam-fixed coordinate system (ξ, η, ζ), the undeformed beam in reference configuration with volume V_0 and the beam in deformed configuration with volume V, both depicted in the global coordinate system (x, y, z)

be applied to stiff dynamic problems. In addition, the possible mode shapes of the simply-supported beam are displayed, see Sect. 4.2. Further, the lateral buckling of a cantilevered right-angle frame under an end load is investigated in Sect. 4.3. This example produces a complete three-dimensional response and therefore is often used also for dynamic investigations in literature, e.g. in [1, 24].

Several static problems have been already shown in [16]. A comparison to results provided in the literature, to analytical solutions, and to the solution found by commercial finite element software shows accuracy and high order convergence in statics. Dynamic applications will be published in [14].

2 Setup of the ANCF Beam Element

In the present chapter, a spatial ANCF beam finite element is proposed. The analogue beam finite element in the planar case has been presented in [18]. There, the degrees of freedom are the displacement vector and the transversal derivative of the displacement. This choice is different from "fully-parameterized" elements [10, 25], in which the position gradients are used as degrees of freedom. See again Fig. 1 for the different choices of nodal coordinates.

The geometry of the proposed spatial finite beam element is described by means of the position and the two transversal slope vectors, which are defined on a scaled and straight reference element, see Fig. 2. For a sketch of the position vectors, slope vectors and the location of the three nodes at the end points and at the midpoint of the beam axis for the undeformed beam in reference configuration and for the beam in deformed configuration see again Fig. 2. The degrees of freedom are chosen as the nodal displacements and change of slope vectors. The displacement vector is denoted by $\mathbf{u}$ and the directional derivatives of the displacements are $\mathbf{u}_{,\eta} = \frac{\partial \mathbf{u}}{\partial \eta}$ and $\mathbf{u}_{,\zeta} = \frac{\partial \mathbf{u}}{\partial \zeta}$. While $\mathbf{u}$ are nodal coordinates, $\mathbf{u}_{,\eta}$ and $\mathbf{u}_{,\zeta}$ are denoted as generalized coordinates. Hence, the vector of degrees of freedom follows as

$$\mathbf{q} = \begin{bmatrix} \mathbf{u}^{(1)T} & \mathbf{u}_{,\eta}^{(1)T} & \mathbf{u}_{,\zeta}^{(1)T} & \mathbf{u}^{(2)T} & \mathbf{u}_{,\eta}^{(2)T} & \mathbf{u}_{,\zeta}^{(2)T} & \mathbf{u}^{(3)T} & \mathbf{u}_{,\eta}^{(3)T} & \mathbf{u}_{,\zeta}^{(3)T} \end{bmatrix}^T. \quad (1)$$

Since nine degrees of freedom are specified in each node, a three-noded beam element has 27 degrees of freedom. The shape functions are chosen similar to those given in [18] and are defined on a scaled and straight reference element in coordinate system (ξ, η, ζ), see Fig. 2,

$$\begin{aligned}
S_1(\xi,\eta,\zeta) &= -\frac{2}{L^2}\xi\left(\frac{L}{2} - \xi\right), \\
S_2(\xi,\eta,\zeta) &= \eta S_1, \qquad S_3(\xi,\eta,\zeta) = \zeta S_1, \\
S_4(\xi,\eta,\zeta) &= +\frac{2}{L^2}\xi\left(\frac{L}{2} + \xi\right), \\
S_5(\xi,\eta,\zeta) &= \eta S_4, \qquad S_6(\xi,\eta,\zeta) = \zeta S_4, \\
S_7(\xi,\eta,\zeta) &= -\frac{4}{L^2}\left(\xi - \frac{L}{2}\right)\left(\xi + \frac{L}{2}\right), \\
S_8(\xi,\eta,\zeta) &= \eta S_7, \qquad S_9(\xi,\eta,\zeta) = \zeta S_7.
\end{aligned} \quad (2)$$

The displacement vector $\mathbf{u}(\xi, \eta, \zeta)$ follows from

$$\mathbf{r}(\xi,\eta,\zeta) = \mathbf{r}_0(\xi,\eta,\zeta) + \mathbf{u}(\xi,\eta,\zeta) = \mathbf{S}(\mathbf{q}_0 + \mathbf{q}), \quad (3)$$

in which $\mathbf{q}_0$ is the generalized coordinate vector of the element in reference configuration and $\mathbf{S}$ represents the shape function matrix

$$\mathbf{S} = [\, S_1\mathbf{I} \quad \ldots \quad S_9\mathbf{I}\,], \quad (4)$$

in which $\mathbf{I}$ is the 3-by-3 identity matrix.

3 Virtual Work of Elastic Forces

In the first instance, the virtual work of elastic forces is derived from a continuum mechanics based formulation, using the Green-Lagrange strain tensor and the second Piola-Kirchhoff stress tensor according to [28]. This formulation is described in Sect. 3.1 and referred to as *standard continuum mechanics based formulation*. The main problem of this formulation is the Poisson locking effect. To avoid this locking phenomenon, the strain energy is modified in Sect. 3.2 according to the well known selective reduced integration. To be more precise, the elasticity tensor is split into two parts. The first part, related to bending, axial deformation, shear and cross section deformation, does not take into account the Poisson ratio ν, while the second part covers the Poisson effect constant over cross section. This splitting leads to a more efficient formulation regarding computational aspects. This improved formulation is called *enhanced continuum mechanics based formulation* and abbreviated by CMF. Note that the Poisson locking also occurs in the original fully parameterized element of Yakoub and Shabana [28]. There are alternatives to the proposed selective reduced integration, see e.g. Sugiyama and Suda [26] or Schwab [20], however, those alternatives have not been rigorously tested concerning static and dynamic linear and nonlinear examples. In Sect. 3.3, the virtual work of elastic forces is derived from a *structural mechanics based formulation* based on generalized strains, which is defined according to Simo and Vu-Quoc [23] and can be interpreted as elastic line approach. The presented formulation is enhanced in order to include cross sectional deformation and is abbreviated by SMF in the following.

3.1 Standard Continuum Mechanics Based Formulation

According to Yakoub and Shabana [28], the virtual work of the elastic forces can be derived from the nonlinear continuum mechanics considerations, using the relation between the Green-Lagrange strain tensor and the second Piola-Kirchhoff stress tensor. For more details on the derivation, see Bonet and Wood [3]. The Green-Lagrange strain tensor $\mathbf{E}$ reads

$$\mathbf{E} = \frac{1}{2}(\mathbf{C} - \mathbf{I}), \tag{5}$$

with the right Cauchy-Green tensor $\mathbf{C}$, which is defined as

$$\mathbf{C} = \mathbf{F}^T\mathbf{F}. \tag{6}$$

The deformation gradient $\mathbf{F}$ is described via the derivatives of the position as follows

$$\mathbf{F} = \frac{\partial \mathbf{r}}{\partial \mathbf{r}_0} = \frac{\partial \mathbf{r}}{\partial \boldsymbol{\xi}}\frac{\partial \boldsymbol{\xi}}{\partial \mathbf{r}_0} = \begin{bmatrix} \frac{\partial r_1}{\partial \xi} & \frac{\partial r_1}{\partial \eta} & \frac{\partial r_1}{\partial \zeta} \\ \frac{\partial r_2}{\partial \xi} & \frac{\partial r_2}{\partial \eta} & \frac{\partial r_2}{\partial \zeta} \\ \frac{\partial r_3}{\partial \xi} & \frac{\partial r_3}{\partial \eta} & \frac{\partial r_3}{\partial \zeta} \end{bmatrix} \begin{bmatrix} \frac{\partial r_{01}}{\partial \xi} & \frac{\partial r_{01}}{\partial \eta} & \frac{\partial r_{01}}{\partial \zeta} \\ \frac{\partial r_{02}}{\partial \xi} & \frac{\partial r_{02}}{\partial \eta} & \frac{\partial r_{02}}{\partial \zeta} \\ \frac{\partial r_{03}}{\partial \xi} & \frac{\partial r_{03}}{\partial \eta} & \frac{\partial r_{03}}{\partial \zeta} \end{bmatrix}^{-1}, \tag{7}$$

in which the vector $\boldsymbol{\xi} = (\xi, \eta, \zeta)$ denotes the coordinates of the scaled straight reference element in unit configuration, see Fig. 2. The transformation between the scaled straight element and the possibly distorted element in reference configuration is expressed in the element Jacobian $\mathbf{J} = \frac{\partial \mathbf{r}_0}{\partial \boldsymbol{\xi}}$. In case of a straight and undistorted reference configuration, the element Jacobian simplifies to $\mathbf{J} = \mathbf{I}$. For the Green-Lagrange strain tensor $\mathbf{E}$ and the second Piola-Kirchhoff stress tensor $\mathbf{S}$, the engineering strain vector $\boldsymbol{\varepsilon}$ and stress vector $\boldsymbol{\sigma}$ reads as follows

$$\boldsymbol{\varepsilon} = [\, E_{xx} \quad E_{yy} \quad E_{zz} \quad 2E_{yz} \quad 2E_{xz} \quad 2E_{xy} \,]^T, \tag{8}$$

$$\boldsymbol{\sigma} = [\, S_{xx} \quad S_{yy} \quad S_{zz} \quad S_{yz} \quad S_{xz} \quad S_{xy} \,]^T. \tag{9}$$

The stress-strain relation,

$$\boldsymbol{\sigma} = \mathbf{D}\boldsymbol{\varepsilon}, \tag{10}$$

is written in terms of the elasticity matrix $\mathbf{D}$ defined by

$$\mathbf{D} = \frac{E\nu}{(1+\nu)(1-2\nu)} \begin{bmatrix} \frac{1-\nu}{\nu} & 1 & 1 & 0 & 0 & 0 \\ 1 & \frac{1-\nu}{\nu} & 1 & 0 & 0 & 0 \\ 1 & 1 & \frac{1-\nu}{\nu} & 0 & 0 & 0 \\ 0 & 0 & 0 & \frac{1-2\nu}{2\nu} & 0 & 0 \\ 0 & 0 & 0 & 0 & \frac{1-2\nu}{2\nu} & 0 \\ 0 & 0 & 0 & 0 & 0 & \frac{1-2\nu}{2\nu} \end{bmatrix}, \tag{11}$$

in which E denotes Young's modulus and ν is Poisson's ratio. Then, the virtual work of elastic forces is written as follows

$$\begin{aligned} U_{\text{stand.}}^{\text{CMF}} &= \frac{1}{2}\int_{V_0} \boldsymbol{\varepsilon}^T \boldsymbol{\sigma}\, dV_0 \\ &= \frac{1}{2}\int_{V_0} \boldsymbol{\varepsilon}^T \mathbf{D}\boldsymbol{\varepsilon}\, dV_0, \end{aligned} \tag{12}$$

which holds for a rectangular cross section with

$$dV_0 = \det(\mathbf{J})\, dV^*, \tag{13}$$

$$dV^* = d\xi\, d\eta\, d\zeta. \tag{14}$$

In Eq. (12), the quantities are defined in the global coordinate system (x, y, z). In order to compare continuum mechanics to structural mechanics based formulations regarding also shear strain considerations, shear correction factors are required in the elasticity matrix. The function of the shear correction factors is the minimization of the error between the computed constant and the real parabolic shear stress distribution for transverse shear, see [27]. Since the initiation of the shear correction factors into the elasticity matrix is dependent on the orientation of the cross section of the beam, a local frame basis $(\mathbf{e}_1^0, \mathbf{e}_2^0, \mathbf{e}_3^0)$ is introduced. The basis vectors $\mathbf{e}_1^0$, $\mathbf{e}_2^0$ and $\mathbf{e}_3^0$ are defined by means of the slope vectors $\mathbf{r}_{0,\eta}$ and $\mathbf{r}_{0,\zeta}$, which describe derivatives of a reference position $\mathbf{r}_0$ in a possible pre-deformed element,

$$\mathbf{e}_1^0 = \frac{\bar{\mathbf{e}}_1^0}{\|\bar{\mathbf{e}}_1^0\|}, \qquad \bar{\mathbf{e}}_1^0 = \mathbf{r}_{0,\eta} \times \mathbf{r}_{0,\zeta}, \tag{15}$$

$$\mathbf{e}_3^0 = \frac{\bar{\mathbf{e}}_3^0}{\|\bar{\mathbf{e}}_3^0\|}, \qquad \bar{\mathbf{e}}_3^0 = \mathbf{r}_{0,\zeta}, \tag{16}$$

$$\mathbf{e}_2^0 = \frac{\bar{\mathbf{e}}_2^0}{\|\bar{\mathbf{e}}_2^0\|}, \qquad \bar{\mathbf{e}}_2^0 = \bar{\mathbf{e}}_3^0 \times \bar{\mathbf{e}}_1^0 = \mathbf{r}_{0,\zeta} \times (\mathbf{r}_{0,\eta} \times \mathbf{r}_{0,\zeta}). \tag{17}$$

In Eqs. (15)–(17) it is assumed that the slope vectors $\mathbf{r}_{0,\eta}$ and $\mathbf{r}_{0,\zeta}$ are perpendicular in the reference configuration, which is no restriction to generality in relation to

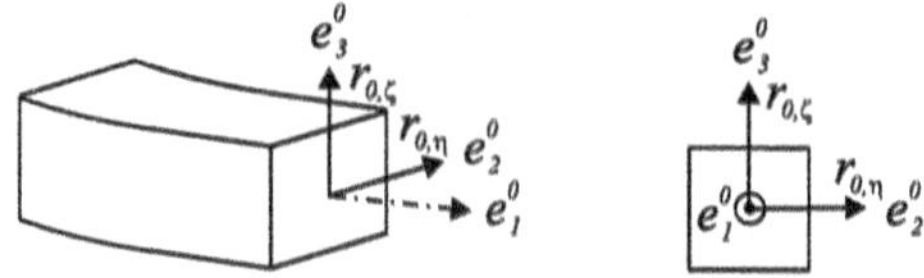

Fig. 3 Perspective and front view of the cross section with local frame basis vectors: $\mathbf{e}_3^0$ is the unit vector in direction of $\mathbf{r}_{0,\zeta}$, $\mathbf{e}_1^0$ is the unit vector perpendicular to $\mathbf{r}_{0,\eta}$ and $\mathbf{r}_{0,\zeta}$ and $\mathbf{e}_2^0$ is perpendicular to $\mathbf{e}_1^0$ and $\mathbf{e}_3^0$

classical beam theory in which the cross section is undeformed. In most cases, it can be assumed that $\mathbf{r}_{0,\eta}$ and $\mathbf{r}_{0,\zeta}$ have unit length. For a sketch of the local basis vectors, see Fig. 3. A tensor $\mathbf{A}_0$, which describes the transformation of the global coordinate system (x, y, z) to the local frame basis $(\mathbf{e}_1^0, \mathbf{e}_2^0, \mathbf{e}_3^0)$ can be defined now as

$$\mathbf{A}_0 = \left[\mathbf{e}_1^0 \mid \mathbf{e}_2^0 \mid \mathbf{e}_3^0\right]. \tag{18}$$

Using the transformation tensor $\mathbf{A}_0$, an arbitrary position $\mathbf{r}_0$ in Eq. (3) given in the global coordinate system can be easily transformed to a representation in the local frame basis, indicated by symbol $*$, as

$$\mathbf{r}_0^* = \mathbf{A}_0^T \mathbf{r}_0. \tag{19}$$

It has to be mentioned here that the transformation tensor has to be constant on the cross section, but changes with the axial coordinate. Note, the transformation takes into account the whole cross section and not only the beam axis.

Now, the elasticity matrix including shear correction factors, given in the local frame basis, is written as

$$\mathbf{D}^* = \frac{E\nu}{(1+\nu)(1-2\nu)} \begin{bmatrix} \frac{1-\nu}{\nu} & 1 & 1 & 0 & 0 & 0 \\ 1 & \frac{1-\nu}{\nu} & 1 & 0 & 0 & 0 \\ 1 & 1 & \frac{1-\nu}{\nu} & 0 & 0 & 0 \\ 0 & 0 & 0 & \frac{1-2\nu}{2\nu} & 0 & 0 \\ 0 & 0 & 0 & 0 & \frac{1-2\nu}{2\nu}k_2 & 0 \\ 0 & 0 & 0 & 0 & 0 & \frac{1-2\nu}{2\nu}k_3 \end{bmatrix}, \tag{20}$$

with the shear correction factors k_2 and k_3. Note, that a correction factor for torsional shear deformation cannot be accounted for.

The entries of matrix $\mathbf{D}^*$ have to be transformed from the local basis frame to the global coordinate system using the transformation tensor in Eq. (18). It has to be mentioned that in case the beam is oriented along the x-axis, matrix $\mathbf{D}^*$ holds in the local frame basis as well as in the global coordinate system and therefore need not be transformed. Otherwise, the transformed elasticity matrix given now in the global coordinate system is used to compute Eq. (12), or alternatively, all other quantities in Eq. (12) have to be transformed to the local frame basis. The transformation from $\mathbf{D}^*$ to the according elasticity matrix given in the global coordinate system is technically complex, since the independent components of the 6×6 matrix $\mathbf{D}^*$ have to be reorganized. Hence, the transformations are applied to $\mathbf{E}$ and $\mathbf{S}$ in order to define the quantities in the principle of virtual work in the same coordinate system.

The variational formulation of the elastic forces can be written as

$$\delta U^{\mathrm{CMF}}_{\mathrm{stand.}} = \int_V \mathbf{S}\delta\mathbf{E}\,dV. \tag{21}$$

To compute the principle of virtual work in Eq. (21), the following steps have to be performed. The Green-Lagrange strain tensor $\mathbf{E}$ is transformed to its according tensor in the beam-fixed coordinate system $\mathbf{E}^*$ by using the relation

$$\mathbf{E}^* = \mathbf{A}_0^T\mathbf{E}\mathbf{A}_0. \tag{22}$$

The six independent components of $\mathbf{E}^*$ are entered in the engineering strain vector $\boldsymbol{\varepsilon}^*$, i.e. $(\mathbf{E}^*)^{3\times 3} \rightarrow (\boldsymbol{\varepsilon}^*)^6$, see Eq. (8). Now the engineering stress vector $\boldsymbol{\sigma}^*$ can be computed by the stress-strain relation using $\mathbf{D}^*$

$$\boldsymbol{\sigma}^* = \mathbf{D}^*\boldsymbol{\varepsilon}^*. \tag{23}$$

The components of the engineering stress vector $\boldsymbol{\sigma}^*$ are filled in the second Piola-Kirchhoff stress tensor $\mathbf{S}^*$, i.e. $(\boldsymbol{\sigma}^*)^6 \rightarrow (\mathbf{S}^*)^{3\times 3}$, see Eq. (9). Note, that $\mathbf{S}^*$ can be written in the absolute coordinate system by the transformation

$$\mathbf{S} = \mathbf{A}_0\mathbf{S}^*\mathbf{A}_0^T. \tag{24}$$

Finally, Eq. (21) can be solved.

3.2 Enhanced Continuum Mechanics Based Formulation—CMF

The standard continuum mechanics formulation in Sect. 3.1 suffers from Poisson locking in case of bending, since the Poisson ratio ν couples axial strains E_{xx} and the transverse normal strains E_{yy} and E_{zz} in the stress-strain relation in Eq. (10). The locking is caused by the fact that in case of bending, the rectangular cross section would deform into a trapezoidal shape because of the Poisson effect. As this deformation mode is not included in the considered ANCF shape functions, this leads to an overly stiff behavior. In the two-dimensional case, Gerstmayr et al. [11] suggested to split the elasticity matrix $\mathbf{D}^*$ into two parts,

$$\mathbf{D}^* = \mathbf{D}^{0*} + \mathbf{D}^{\nu*}, \tag{25}$$

in which $\mathbf{D}^{0*}$ does not include the Poisson ratio ν, while $\mathbf{D}^{\nu*}$ involves the Poisson effect only. To be more precise, $\mathbf{D}^{0*}$ includes the Young's modulus E and the shear modulus G in the diagonal;

$$\mathbf{D}^{0*} = \mathrm{diag}(E, E, E, G, Gk_2, Gk_3), \tag{26}$$

in which the shear correction factors k_2 and k_3 again account for the distribution of the shear stress along the cross section. The matrix $\mathbf{D}^{\nu *}$ includes the Poisson effect in the axial and transverse deformation as follows

$$\mathbf{D}^{\nu *} = \frac{E\nu}{(1+\nu)(1-2\nu)} \begin{bmatrix} 2\nu & 1 & 1 & 0 & 0 & 0 \\ 1 & 2\nu & 1 & 0 & 0 & 0 \\ 1 & 1 & 2\nu & 0 & 0 & 0 \\ 0 & 0 & 0 & 0 & 0 & 0 \\ 0 & 0 & 0 & 0 & 0 & 0 \\ 0 & 0 & 0 & 0 & 0 & 0 \end{bmatrix}. \tag{27}$$

Hence, the strain energy in (12) is split into two parts according to Eq. (25) and follows to

$$\begin{aligned} U_{\text{enh.}}^{\text{CMF}} &= \frac{1}{2}\int_{-W/2}^{W/2}\int_{-H/2}^{H/2}\int_{-L/2}^{L/2} \boldsymbol{\varepsilon}^T \mathbf{D}^0 \boldsymbol{\varepsilon} \det(\mathbf{J})\, d\xi\, d\eta\, d\zeta \\ &\quad + \frac{1}{2} HW \int_{-L/2}^{L/2} \boldsymbol{\varepsilon}^T \mathbf{D}^\nu \boldsymbol{\varepsilon} \det(\mathbf{J})\, d\xi, \end{aligned} \tag{28}$$

considering the Poisson effect only at the beam axis $\eta = 0$ and $\zeta = 0$. This technique can be interpreted as a selective reduced integration scheme. This enhanced formulation in Eq. (28) eliminates the Poisson locking effect and is referred to as CMF.

It has to be mentioned that again the components of the elasticity matrices $\mathbf{D}^{0*}$ and $\mathbf{D}^{\nu *}$ in Eq. (25) are given in the local frame basis.

3.3 Structural Mechanics Based Formulation—SMF

The elastic forces can also be defined by a structural mechanics based formulation, following the approach of Simo [23]. There, the strain energy is written in terms of generalized strains, namely the axial strain Γ_1, the shear strains Γ_2 and Γ_3, the torsional strain κ_1 and the bending strains κ_2 and κ_3:

$$U_{\text{Simo}}^{\text{SMF}} = \frac{1}{2}\int_{-L/2}^{L/2} \left(EA\Gamma_1^2 + GAk_2\Gamma_2^2 + GAk_3\Gamma_3^2 + GJk_t\kappa_1^2 + EI_2\kappa_2^2 + EI_3\kappa_3^2\right) d\xi. \tag{29}$$

It has to be mentioned that the principle axes of the cross section are in η- and ζ-direction, such that no coupling of κ_2 and κ_3 occurs. The variational form of Eq. (29) reads

$$\begin{aligned} \delta U_{\text{Simo}}^{\text{SMF}} &= \int_{-L/2}^{L/2} (EA\Gamma_1\delta\Gamma_1 + GAk_2\Gamma_2\delta\Gamma_2 + GAk_3\Gamma_3\delta\Gamma_3 \\ &\quad + GJk_t\kappa_1\delta\kappa_1 + EI_2\kappa_2\delta\kappa_2 + EI_3\kappa_3\delta\kappa_3)\, d\xi. \end{aligned} \tag{30}$$

Equation (29) uses the following abbreviations for the beam stiffness parameters. E represents Young's modulus and A is the area of the cross section,

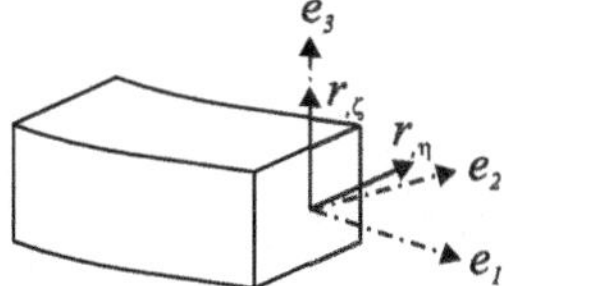

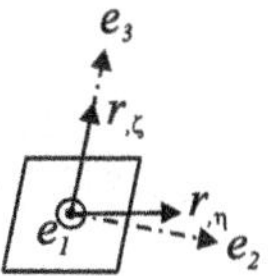

Fig. 4 Perspective and front view of the cross section with cross section frame basis vectors $\mathbf{e}_1$, $\mathbf{e}_2$ and $\mathbf{e}_3$, which are represented by the slope vectors $\mathbf{r},_\eta$ and $\mathbf{r},_\zeta$ only

$G = E/2(1+\nu)$ is the shear modulus, EI_2 is the bending stiffness with respect to the y-axis and EI_3 is the bending stiffness with respect to the z-axis, while $J = I_2 + I_3$ is the torsional moment of inertia. Hence, GJ represents the torsional stiffness combined with the torsional correction factor k_t, EA is the axial stiffness and GA is the shear stiffness with the shear correction factors k_2 and k_3. A similar approach is also given in a recent work [19], in which correction factors are not included explicitly. In Sugiyama et al. [25, 26] the virtual work of elastic forces follows from Simo [23] too, but in contrast to the proposed approach, in which the curvature follows the material measure of curvature, there a geometrical curvature is used. A comparison of results computed with the proposed formulation to analytical solutions considering arbitrary digits shows correctness of the proposed formulation, see also [8]. Equation (29) can be rewritten in matrix notation as follows:

$$U_{\text{Simo}}^{\text{SMF}} = \frac{1}{2}\int_{-L/2}^{L/2} \left(\boldsymbol{\Gamma}^T \operatorname{diag}(EA, GAk_2, GAk_3)\boldsymbol{\Gamma} + \boldsymbol{\kappa}^T \operatorname{diag}(GJk_t, EI_2, EI_3)\boldsymbol{\kappa}\right) d\xi. \tag{31}$$

3.3.1 Cross Section Frame

For the definition of the generalized strains in the succeeding section, a cross section frame is constructed based on a rotation tensor $\mathbf{A}$, which is related to $\mathbf{A}_0$ in Eq. (18). The rotation tensor $\mathbf{A}$ is represented by the slope vectors $\mathbf{r},_\eta$ and $\mathbf{r},_\zeta$ only. The rotation tensor is written in a simple way, but it has to be mentioned that it is not symmetric regarding η and ζ, since the slope vectors $\mathbf{r},_\eta$ and $\mathbf{r},_\zeta$ are almost perpendicular only, see Fig. 4. The local basis vectors $\mathbf{e}_1$, $\mathbf{e}_2$, and $\mathbf{e}_3$ describe the rotation tensor $\mathbf{A}$ by means of

$$\mathbf{A} = [\mathbf{e}_1 \mid \mathbf{e}_2 \mid \mathbf{e}_3]. \tag{32}$$

In this case, the basis vectors $\mathbf{e}_1$, $\mathbf{e}_2$ and $\mathbf{e}_3$ are defined as

$$\mathbf{e}_1 = \frac{\bar{\mathbf{e}}_1}{\|\bar{\mathbf{e}}_1\|}, \qquad \bar{\mathbf{e}}_1 = \mathbf{r},_\eta \times \mathbf{r},_\zeta, \tag{33}$$

$$\mathbf{e}_3 = \frac{\bar{\mathbf{e}}_3}{\|\bar{\mathbf{e}}_3\|}, \qquad \bar{\mathbf{e}}_3 = \mathbf{r},_\zeta, \tag{34}$$

$$\mathbf{e}_2 = \frac{\bar{\mathbf{e}}_2}{\|\bar{\mathbf{e}}_2\|}, \qquad \bar{\mathbf{e}}_2 = \bar{\mathbf{e}}_3 \times \bar{\mathbf{e}}_1 = \mathbf{r},_\zeta \times (\mathbf{r},_\eta \times \mathbf{r},_\zeta), \tag{35}$$

which simplifies the representation of the variation of the generalized strains, see the following derivations.

3.3.2 Definition of Generalized Strains

The axial strain Γ_1 is defined as

$$\Gamma_1 = \mathbf{e}_1^T \mathbf{r}' - 1. \tag{36}$$

Throughout the following, primes indicate the derivative with respect to the local axial coordinate ξ, $\mathbf{r}' = \frac{\partial \mathbf{r}}{\partial \xi}$. The shear strains Γ_2 and Γ_3 then read

$$\Gamma_2 = \mathbf{e}_2^T \mathbf{r}' \quad \text{and} \quad \Gamma_3 = \mathbf{e}_3^T \mathbf{r}'. \tag{37}$$

For the implementation of Eq. (30), the variations of $\boldsymbol{\Gamma}$ and $\boldsymbol{\kappa}$ have to be computed. The variation for the axial strain Γ_1 follows as

$$\Gamma_1 = \mathbf{e}_1^T \mathbf{r}' - 1 = \frac{\bar{\mathbf{e}}_1^T}{\|\bar{\mathbf{e}}_1\|} \mathbf{r}' - 1 \tag{38}$$

$$\delta \Gamma_1 = \delta \mathbf{e}_1^T \mathbf{r}' + \mathbf{e}_1^T \delta \mathbf{r}' \tag{39}$$

$$= \left(\frac{-(\bar{\mathbf{e}}_1^T \delta \bar{\mathbf{e}}_1)\mathbf{e}_1}{\|\bar{\mathbf{e}}_1\|^3} + \frac{\delta \bar{\mathbf{e}}_1}{\|\bar{\mathbf{e}}_1\|} \right)^T \mathbf{r}_\xi + \frac{\bar{\mathbf{e}}_1^T}{\|\bar{\mathbf{e}}_1\|} \mathbf{S}_\xi \delta \mathbf{q}, \tag{40}$$

with

$$\bar{\mathbf{e}}_1^T := \mathbf{r}_\eta \times \mathbf{r}_\zeta \tag{41}$$

$$\delta \bar{\mathbf{e}}_1^T = \delta \mathbf{r}_\eta \times \mathbf{r}_\zeta + \mathbf{r}_\eta \times \delta \mathbf{r}_\zeta \tag{42}$$

$$= (\mathbf{S}_\eta \delta \mathbf{q}) \times \mathbf{r}_\zeta + \mathbf{r}_\eta \times (\mathbf{S}_\zeta \delta \mathbf{q}). \tag{43}$$

The variations for the bending strains Γ_2 and Γ_3 are computed analogously. For the definition of the torsion κ_1 and the bending strains κ_2 and κ_3 again the rotation tensor is used. The vector of twist and curvature $\mathbf{k}$ is defined by the columns $\mathbf{e}_i$ of the rotation tensor $\mathbf{A}$ as follows:

$$\mathbf{k} = \frac{1}{2} \sum_{i=1}^{3} \mathbf{e}_i \times \mathbf{e}_i', \tag{44}$$

with $\mathbf{e}_i' = \mathbf{k} \times \mathbf{e}_i$ and $k_i = \mathbf{e}_i^T \mathbf{k}$, in which the index i defines the i-th component of the considered vector. The torsion κ_1 and the bending strains κ_2 and κ_3 are the entries of vector

$$\boldsymbol{\kappa} = \sum_{i=1}^{3} \kappa_i \mathbf{e}_i = \sum_{i=1}^{3} k_i \mathbf{e}_i. \tag{45}$$

Notice, the components of $\mathbf{k}$ are considered in the local basis $\mathbf{e}_i$, i.e., there holds

$$\kappa_i = \left(\mathbf{A}^T \boldsymbol{\kappa} \right)_i. \tag{46}$$

For the definition of the vector of twist and curvature, the derivative with respect to the beam axis coordinate ξ is used in the present approach, while in Sugiyama et al. [25, 26] the derivative with respect to the arc length s in actual configuration is used.

Note, that the vector $\mathbf{k}$ in Eq. (44) can be rewritten as follows

$$\mathbf{k} = \frac{1}{2}\sum_{i=1}^{3} \mathbf{e}_i \times \mathbf{e}_i' = \frac{1}{2}\sum_{i=1}^{3}\left(\frac{\bar{\mathbf{e}}_i}{\|\bar{\mathbf{e}}_i\|} \times \frac{-(\bar{\mathbf{e}}_i^T\bar{\mathbf{e}}_i')\bar{\mathbf{e}}_i + \bar{\mathbf{e}}_i'\|\bar{\mathbf{e}}_i\|^2}{\|\bar{\mathbf{e}}_i\|^3}\right)$$
$$= \frac{1}{2}\sum_{i=1}^{3}\left(\frac{1}{\|\bar{\mathbf{e}}_i\|^4}(\bar{\mathbf{e}}_i \times \bar{\mathbf{e}}_i')\|\bar{\mathbf{e}}_i\|^2\right) = \frac{1}{2}\sum_{i=1}^{3}\left(\frac{1}{\|\bar{\mathbf{e}}_i\|^2}(\bar{\mathbf{e}}_i \times \bar{\mathbf{e}}_i')\right), \tag{47}$$

with $\bar{\mathbf{e}}_i$ defined in Eqs. (33)–(35). For the implementation, the variation $\delta\boldsymbol{\kappa}$ is necessary:

$$\delta\boldsymbol{\kappa} = \sum_{i=1}^{3}(\delta k_i \mathbf{e}_i + k_i \delta\mathbf{e}_i), \tag{48}$$

using the formula for the variation of $\delta\mathbf{k}$:

$$\delta\mathbf{k} = \sum_{i=1}^{3}\left(\frac{1}{2}\delta\left(\frac{1}{\|\bar{\mathbf{e}}_i\|^2}\right)(\bar{\mathbf{e}}_i \times \bar{\mathbf{e}}_i') + \frac{1}{2}\frac{1}{\|\bar{\mathbf{e}}_i\|^2}\delta(\bar{\mathbf{e}}_i \times \bar{\mathbf{e}}_i')\right) \tag{49}$$
$$= \sum_{i=1}^{3}\left(\frac{1}{2}\left(-(\bar{\mathbf{e}}_i^T\bar{\mathbf{e}}_i)^{-2}2\bar{\mathbf{e}}_i\delta\bar{\mathbf{e}}_i\right)(\bar{\mathbf{e}}_i \times \bar{\mathbf{e}}_i') + \frac{1}{2}\frac{1}{\|\bar{\mathbf{e}}_i\|^2}(\delta\bar{\mathbf{e}}_i \times \bar{\mathbf{e}}_i' + \bar{\mathbf{e}}_i \times \delta\bar{\mathbf{e}}_i')\right) \tag{50}$$
$$= \sum_{i=1}^{3}\left(-\frac{\bar{\mathbf{e}}_i\delta\bar{\mathbf{e}}_i}{\|\bar{\mathbf{e}}_i\|^4}(\bar{\mathbf{e}}_i \times \bar{\mathbf{e}}_i') + \frac{1}{2\|\bar{\mathbf{e}}_i\|^2}(\delta\bar{\mathbf{e}}_i \times \bar{\mathbf{e}}_i' + \bar{\mathbf{e}}_i \times \delta\bar{\mathbf{e}}_i')\right). \tag{51}$$

Pre-curvature can be easily implemented in the theory if $\boldsymbol{\kappa}$ is enhanced by an according term as follows:

$$\boldsymbol{\kappa} = \sum_{i=1}^{3}\kappa_i\mathbf{e}_i = \sum_{i=1}^{3}(k_i - k_{0i})\mathbf{e}_i, \tag{52}$$

in which the torsional strain κ_1 and bending strains κ_2, κ_3 are expressed as the difference of the components of the vector of twist and curvature in deformed and undeformed state with

$$\mathbf{k}_0 = \frac{1}{2}\sum_{i=1}^{3}\mathbf{e}_{0i} \times \mathbf{e}_{0i}' = \sum_{i=1}^{3}k_{0i}\mathbf{e}_i. \tag{53}$$

If the beam is initially straight, the term $\mathbf{k}_0$ as well as $\delta(\mathbf{A}\mathbf{k}_0)$ vanish and $\delta\boldsymbol{\kappa}$ turns into $\delta\boldsymbol{\kappa} = \delta\mathbf{k}$.

3.3.3 Definition of the Cross Section Deformation Energy

The cross section deformation is included by design of the shape functions in the ANCF, while it is not included in classical beam finite elements. In Betsch and Steinmann [2], a straightforward approach using constraints is presented. Here, in order to avoid the locking phenomenon, no Poisson coupling is involved, following the idea of the continuum mechanics based approach in Sect. 3.2. This approach can be interpreted as penalty formulation, see Gerstmayr et al. [11] for the two-dimensional case. It has to be mentioned here that the eigenfrequencies corresponding to the thickness modes are limited to twice size of shear mode frequencies, see again [11]. In the three-dimensional case, the virtual work related to the cross section deformation strain energy U^{CSD} can be consistently derived from the continuum mechanics based approach in Eq. (12) when neglecting the Poisson effect

$$U^{\mathrm{CSD}} = \frac{1}{2}\int_{-L/2}^{L/2} \left(EA\left(E_{\eta\eta}^2 + E_{\zeta\zeta}^2\right) + 2GAE_{\eta\zeta}^2\right) d\xi. \tag{54}$$

Note, that in difference to the two-dimensional case, the cross section may undergo not only a thickness, but also a shear deformation. The respective components of the Green-Lagrange strain tensor $\mathbf{E} = \frac{1}{2}(\mathbf{F}^T\mathbf{F} - \mathbf{I})$ without consideration of the element transformation read as follows

$$E_{\eta\eta} = \frac{1}{2}\left(\mathbf{r}_{,\eta}^T\mathbf{r}_{,\eta} - 1\right), \qquad E_{\zeta\zeta} = \frac{1}{2}\left(\mathbf{r}_{,\zeta}^T\mathbf{r}_{,\zeta} - 1\right), \qquad E_{\eta\zeta} = \frac{1}{2}\mathbf{r}_{,\eta}^T\mathbf{r}_{,\zeta}. \tag{55}$$

The quantities in Eq. (55) only depend on the respective length of the slope vectors $|\mathbf{r}_{,\eta}|$ and $|\mathbf{r}_{,\zeta}|$, whereas $\boldsymbol{\Gamma}$ and $\boldsymbol{\kappa}$ are expressed in terms of the normalized vectors $\mathbf{e}_1$, $\mathbf{e}_2$, $\mathbf{e}_3$ and the direction of the beam axis $\mathbf{r}'$. Hence, the total strain energy U^{SMF} can be defined straightforward as the sum of Eq. (29) and Eq. (54) as

$$U^{\mathrm{SMF}} = U^{\mathrm{SMF}}_{\mathrm{Simo}} + U^{\mathrm{CSD}}, \tag{56}$$

and is referred to as SMF. The variation of Eq. (54) is computed by

$$\delta U^{\mathrm{CSD}} = \int_{-L/2}^{L/2} \left(EAE_{\eta\eta}\delta E_{\eta\eta} + EAE_{\zeta\zeta}\delta E_{\zeta\zeta} + 2GAE_{\eta\zeta}\delta E_{\eta\zeta}\right) d\xi. \tag{57}$$

4 Numerical Examples

The proposed ANCF element has been implemented in the framework of the multibody and finite element research code HOTINT.[1]

The investigated formulations, namely the standard and the enhanced continuum mechanics formulation, as well as the structural mechanics based formulation, described in Sect. 3, have been implemented. Regarding the numerical applications, only the enhanced continuum mechanics based and the structural based formulation are investigated. The enhanced continuum mechanics formulation given by Eq. (28) is abbreviated by CMF, while SMF indicates the usage of the structural mechanics formulation given by Eq. (56). In order to validate the proposed ANCF element and to illustrate the performance, the proposed element has been already tested numerically by using e.g. small and large deformation static problems. A comparison

[1] http://tmech.mechatronik.uni-linz.ac.at/staff/gerstmayr/hotint.html.

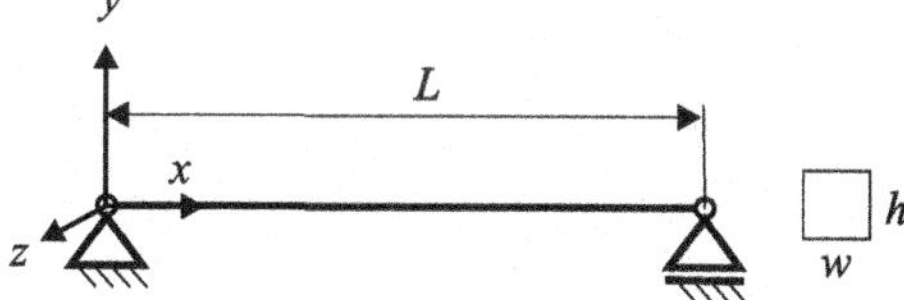

Fig. 5 Simply supported beam investigated for eigenfrequency analysis

to results in literature and to analytical solutions shows good accuracy and high convergence order of the proposed element. Detailed results of such investigations have been already shown in [15]. As an extension of the extensive tests in [15], in the present chapter, an eigenfrequency analysis of a simply supported beam is presented. Results are compared to given eigenfrequencies in literature and to analytical solutions. Additionally, the corresponding mode shapes are presented. Furthermore, a lateral buckling test of a cantilevered right-angle frame under an end load is investigated. This is a standard but challenging test for large deformation beam finite elements and has not been investigated with the ANCF so far. The obtained numerical results are compared to results given in literature [1].

4.1 Eigenfrequency Analysis

Similar to [10], the eigenfrequencies of an unstressed, simply supported beam are computed, which is of length $L = 2$ m, height $h = 0.4$ m and width $w = 0.4$ m. The density is chosen as $\rho = 7850$ kg/m^3, Young's modulus is $E = 1 \cdot 10^9$ N/m^2 with a Poisson ratio of $\nu = 0.3$. For a sketch of the problem setup, see Fig. 5. To realize the bearing on the left hand side of the simply supported beam the displacements of the axis in x-, y- and z-direction are set to zero, for the slide bearing on the right only the displacement in y- and z-direction are fixed. The computed values are compared in Table 1 to theoretical values denoted by *Timoshenko (anal.)* taken from Tables 4 and 5 in [10]. In addition to [10] we provide the analytical value for the torsional frequency w_t

$$w_t = \pi \sqrt{\frac{G k_t}{\rho L^2}}, \tag{58}$$

which gives 319.350 rad/s in this example problem. Due to the fact that cross-section deformation influences the torsional eigenfrequency (cf. Eq. (54)), this term had to be penalized (by a factor 1000) in the SMF. Otherwise, the converged value for the torsional eigenfrequency reads 316.732 rad/s. Similarly, in order to obtain the analytical values regarding axial and thickness deformation, the Poisson ratio was set $\nu = 0$ for the CMF. If not doing so (i.e., $\nu = 0.3$), one obtains the converged values 280.177 rad/s for the first axial eigenfrequency, 837.026 rad/s for the second axial eigenfrequency, and 2710.97 rad/s for the first thickness eigenfrequency.

Table 1 Eigenfrequencies for an unstressed simply supported beam. Values are given in rad/s

# Elements	1st bend.	1st axial	1st tors.	2nd bend.
CMF element				
1	105.149	281.373	353.399	–
32	95.6340	280.321	319.346	332.235
SMF element				
1	105.148	281.373	352.130	–
32	95.6341	280.321	319.347	332.236
Timoshenko (anal.)	95.6340	280.321	319.350	332.235
# Elements	**2nd axial**	**1st shear**	**2nd shear**	**1st thick.**
CMF element				
1	1012.36	1766.99	1896.70	3090.98
32	840.962	1766.99	1878.87	3090.98
SMF element				
1	1012.36	1766.99	1896.70	3090.98
32	840.962	1766.99	1878.87	3090.98
Timoshenko (anal.)	840.962	1766.99	1878.87	3090.98

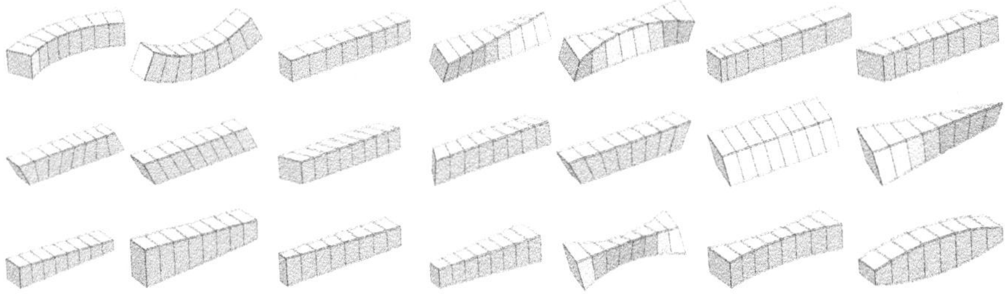

Fig. 6 Representation of the possible mode shapes for a simply supported beam

4.2 Eigenmodes

In this computation, a single ANCF beam element is used to display the possible mode shapes of the presented element. The proposed element consists of nine degrees of freedom per node and therefore one element is represented by 27 degrees of freedom. Since five degrees of freedom are eliminated by the boundary constraints, 22 eigenvalues can be computed. See Fig. 6 for a visualization of the computed mode shapes. It has to be mentioned, that one rigid body mode is generated, which is not displayed in this figure.

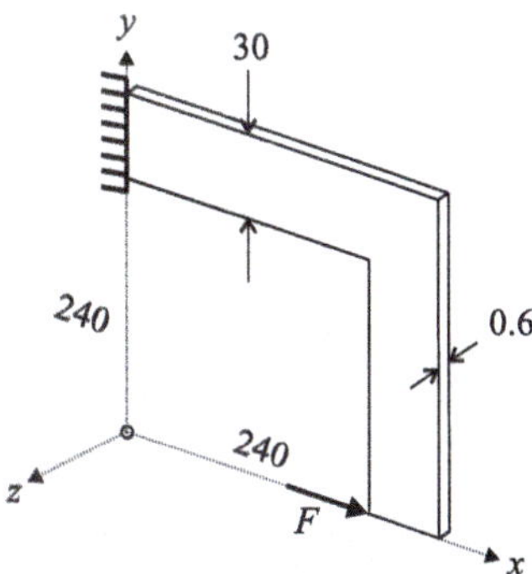

Fig. 7 Geometry of the cantilevered right-angle frame under in-plane end load F

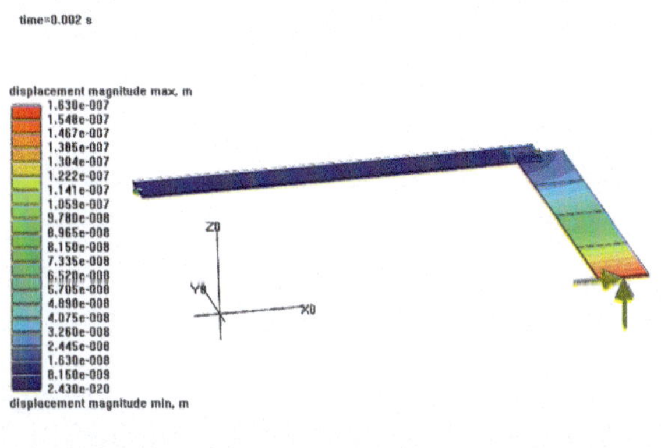

Fig. 8 Visualization of the cantilevered right-angle frame under in-plane end load and a small disturbance load normal to the plane of the frame in HOTINT

4.3 Lateral Buckling of a Right-Angle Frame Under an End Load

This example considers the buckling behavior of a right-angle frame under an in-plane end load and a small disturbance load at the free end normal to the plane of the frame. See Fig. 7 for a sketch of the problem setup. According to Argyris et al. [1], the geometry of the two connected beams is chosen as follows: length $L = 0.24$ m, height $h = 0.03$ m and width $w = 0.0006$ m. The extreme slenderness of the cross section has to be regarded. The Young's modulus is chosen as $E = 7.124 \cdot 10^{10}$ N/m^2 with a Poisson ratio $\nu = 0.31$. The free end is loaded with a linearly increasing transverse force F in order to obtain the critical load given in [1] as $F_{\mathrm{cr}} = 1.088$ N for beam elements and $F_{\mathrm{cr}} = 1.1453$ N using triangular plate elements. A small transversal disturbance load acts normal to the plane of the frame and is set to $F_{\mathrm{dist}} = 0.0001$ N, see Fig. 8 for the load case. In Simo and Vu-Quoc [24], the critical load can be found as $F_{\mathrm{cr}} = 1.09$ N. There, the projection of final deformed shape onto the x–z plane is shown for an applied load of magnitude $F = 1.485$ N. In the present test, the main in-plane load is set to $F = 1.485$ N and the transversal disturbance

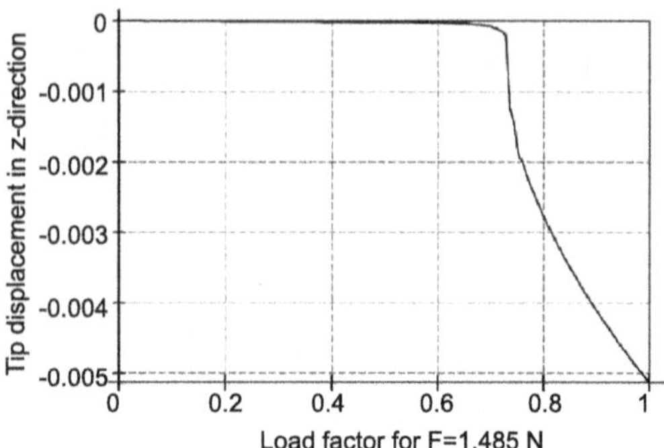

Fig. 9 A jump in the tip displacement in z-direction at an applied load $0.734F$ shows that the correct buckling load is obtained

load F_{dist} is applied. Figure 9 shows the tip displacement in z-direction versus the applied load factor λ which represents the applied load with $\lambda \cdot F$. The critical load is found to be $0.734 \cdot F$, which yields $F_{\mathrm{cr}} = 1.0899$ N.

5 Conclusion

In this chapter, a three-dimensional beam finite element including axial, bending, shear, and torsional deformation is presented. The formulation of the proposed element is based on the absolute nodal coordinate formulation (ANCF), which is used to describe large deformations in multibody dynamics problems. In contrast to the standard large rotation vector formulation, the rotational degrees of freedom are not employed in the ANCF and therefore does not necessarily suffer from singularities emerging from angle parameterizations. However, cross section deformation is possible. Different approaches for the deformation energy are presented. A continuum mechanics based formulation for the elastic forces is discussed and enhanced in order to avoid Poisson locking. Additionally, the elastic forces are defined by a structural mechanics based approach, which includes a term accounting for cross section deformation, which is not considered in Reissner's classical theory. Previously, the investigation of several static problems have already shown high convergence order and accuracy of the proposed element. Here, an eigenfrequency test and a lateral buckling test of a right-angle frame are presented, the computations imply good agreement to results given in literature and it can be now judged that specially designed ANCF finite elements are suitable for large deformation three-dimensional structural problems. The extension to warping is still an open problem and will be investigated in future.

Acknowledgements K. Nachbagauer and P. Gruber acknowledge support from the Austrian Science Funds (FWF): I337-N18, J. Gerstmayr from the K2-Comet Austrian Center of Competence in Mechatronics (ACCM).

References

1. Argyris, J.H., Balmer, H., Doltsinis, J.St., Dunne, P.C., Haase, M., Kleiber, M., Malejannakis, G.A., Mlejenek, J.P., Muller, M., Scharpf, D.W.: Finite element method—the natural approach. Comput. Methods Appl. Mech. Eng. **17–18**, 1–106 (1979)
2. Betsch, P., Steinmann, P.: Frame-indifferent beam finite elements based upon the geometrically exact beam theory. Int. J. Numer. Methods Eng. **54**, 1775–1788 (2002)
3. Bonet, J., Wood, R.D.: Nonlinear Continuum Mechanics for Finite Element Analysis. Cambridge University Press, Cambridge (1997)
4. Dmitrochenko, O.: A new finite element of thin spatial beam in absolute nodal coordinate formulation. In: Proceedings of Multibody Dynamics, ECCOMAS Thematic Conference, Madrid, Spain (2005)
5. Gams, M., Planinc, I., Saje, M.: The strain-based beam finite elements in multibody dynamics. J. Sound Vib. **305**, 194–210 (2007)
6. García-Vallejo, D., Mikkola, A.M., Escalona, J.L.: A new locking-free shear deformable finite element based on absolute nodal coordinates. Nonlinear Dyn. **50**, 249–264 (2007)
7. Gerstmayr, J., Irschik, H.: On the appropriateness of the absolute nodal coordinate formulation for nonlinear problems of rods. In: Proceedings of the 4th European Conference on Structural Control (4ECSC), vol. 1 (2008)
8. Gerstmayr, J., Irschik, H.: On the correct representation of bending and axial deformation in the absolute nodal coordinate formulation with an elastic line approach. J. Sound Vib. **318**, 461–487 (2008)
9. Gerstmayr, J., Irschik, H.: A continuum-mechanics interpretation of Reissner's non-linear shear-deformable beam theory. Math. Comput. Model. Dyn. Syst. **17**, 19–29 (2011)
10. Gerstmayr, J., Matikainen, M.K.: Analysis of stress and strain in the absolute nodal coordinate formulation. Mech. Based Des. Struct. Mach. **34**, 409–430 (2006)
11. Gerstmayr, J., Matikainen, M.K., Mikkola, A.M.: A geometrically exact beam element based on the absolute nodal coordinate formulation. Multibody Syst. Dyn. **20**, 359–384 (2008)
12. Ibrahimbegović, A.: On finite element implementation of geometrically nonlinear Reissner's beam theory: three-dimensional curved beam elements. Comput. Methods Appl. Mech. Eng. **122**, 11–26 (1995)
13. Kerkkänen, K.S., Sopanen, J.T., Mikkola, A.M.: A linear beam finite element based on the absolute nodal coordinate formulation. J. Mech. Des. **127**, 621–630 (2005)
14. Nachbagauer, K., Gerstmayr, J.: Nonlinear dynamic analysis of three-dimensional shear deformable ANCF beam finite elements. In: Proceedings of the 2nd Joint International Conference on Multibody System Dynamics, Stuttgart, Germany, May 29–June 1, 2012
15. Nachbagauer, K., Gruber, P., Gerstmayr, J.: A 3D shear deformable finite element based on the absolute nodal coordinate formulation. In: Samin, J.C., Fisette, P. (eds.) Proceedings of the Multibody Dynamics 2011, ECCOMAS Thematic Conference, Brussels, Belgium, 4–7 July 2011
16. Nachbagauer, K., Gruber, P., Gerstmayr, J.: Structural and continuum mechanics approaches for a 3D shear deformable ANCF beam finite element: application to static and linearized dynamic examples. J. Comput. Nonlinear Dyn. **8**, 021004 (2013)
17. Nachbagauer, K., Gruber, P., Vetyukov, Y., Gerstmayr, J.: A spatial thin beam finite element based on the absolute nodal coordinate formulation without singularities. In: Proceedings of the ASME 2011 International Design Engineering Technical Conferences and Computers and Information in Engineering Conference IDETC/CIE 2011, Washington, DC, USA, Paper No. DETC2011/MSNDC-47732 (2011)
18. Nachbagauer, K., Pechstein, A., Irschik, H., Gerstmayr, J.: A new locking-free formulation for planar, shear deformable, linear and quadratic beam finite elements based on the absolute nodal coordinate formulation. Multibody Syst. Dyn. **26**(3), 245–263 (2011)
19. Romero, I.: A comparison of finite elements for nonlinear beams: the absolute nodal coordinate and geometrically exact formulations. Multibody Syst. Dyn. **20**, 51–68 (2008)

20. Schwab, A.L., Meijaard, J.P.: Comparison of three-dimensional flexible beam elements for dynamic analysis: classical finite element formulation and absolute nodal coordinate formulation. Journal of Computational and Nonlinear Dynamics **5**, 011010 (2010)
21. Shabana, A.A.: Definition of the slopes and the finite element absolute nodal coordinate formulation. Multibody Syst. Dyn. **1**(3), 339–348 (1997)
22. Shabana, A.A.: Dynamics of Multibody Systems, 3rd edn. Cambridge University Press, New York (2005)
23. Simo, J.C.: A finite strain beam formulation. The three-dimensional dynamic problem. Part I. Comput. Methods Appl. Mech. Eng. **49**, 55–70 (1985)
24. Simo, J.C., Vu-Quoc, L.: A three-dimensional finite-strain rod model. Part II: Computational aspects. Comput. Methods Appl. Mech. Eng. **58**, 79–116 (1986)
25. Sugiyama, H., Gerstmayr, J., Shabana, A.A.: Deformation modes in the finite element absolute nodal coordinate formulation. J. Sound Vib. **298**, 1129–1149 (2006)
26. Sugiyama, H., Suda, Y.: A curved beam element in the analysis of flexible multi-body systems using the absolute nodal coordinates. Proc. IMechE, Part K: J. Multi-body Dyn. **221**, 219–231 (2007)
27. Timoshenko, S., Goodier, J.N.: Theory of Elasticity. McGraw-Hill, New York (1951)
28. Yakoub, R.Y., Shabana, A.A.: Three dimensional absolute nodal coordinate formulation for beam elements. J. Mech. Des. **123**, 606–621 (2001)
29. Zupan, D., Saje, M.: Finite-element formulation of geometrically exact three-dimensional beam theories based on interpolation of strain measures. Comput. Methods Appl. Mech. Eng. **192**, 5209–5248 (2003)

A Variational Approach to Multirate Integration for Constrained Systems

Sigrid Leyendecker and Sina Ober-Blöbaum

Abstract The simulation of systems with dynamics on strongly varying time scales is quite challenging and demanding with regard to possible numerical methods. A rather naive approach is to use the smallest necessary time step to guarantee a stable integration of the fast frequencies. However, this typically leads to unacceptable computational loads. Alternatively, multirate methods integrate the slow part of the system with a relatively large step size while the fast part is integrated with a small time step. In this work, a multirate integrator for constrained dynamical systems is derived in closed form via a discrete variational principle on a time grid consisting of macro and micro time nodes. Being based on a discrete version of Hamilton's principle, the resulting variational multirate integrator is a symplectic and momentum preserving integration scheme and also exhibits good energy behaviour. Depending on the discrete approximations for the Lagrangian function, one obtains different integrators, e.g. purely implicit or purely explicit schemes, or methods that treat the fast and slow parts in different ways. The performance of the multirate integrator is demonstrated by means of several examples.

1 Introduction

Mechanical systems with dynamics on varying time scales, in particular those including highly oscillatory motion, impose challenging questions for numerical integration schemes. Tiny step sizes are required to guarantee a stable integration of the fast frequencies. However, for the simulation of the slow dynamics, integration

S. Leyendecker (✉)
Chair of Applied Dynamics, University of Erlangen-Nuremberg, Konrad-Zuse-Str. 3/5, 91052 Erlangen, Germany
e-mail: sigrid.leyendecker@ltd.uni-erlangen.de

S. Ober-Blöbaum
Computational Dynamics and Optimal Control, University of Paderborn, Warburger Str. 100, 33098 Paderborn, Germany
e-mail: sinaob@math.uni-paderborn.de

J.-C. Samin, P. Fisette (eds.), *Multibody Dynamics*, Computational Methods in Applied Sciences 28,
DOI 10.1007/978-94-007-5404-1_5,

with a larger time step is accurate enough. Here, small time steps increase integration times unnecessarily, especially for costly function evaluations. Typical examples of systems exhibiting dynamics on different time scales can be found in astrophysics, where depending on the distances between planets, the resulting gravitational forces can be extremely strong or weak leading to different time scales for a flight trajectory through space, or in molecular dynamics, where locally extremely high frequencies superpose global folding processes. In multibody dynamics, such systems occur e.g. in combustion engines with chain drives or in vehicle dynamics, or generally in systems being composed of rigid and elastic parts with varying and in particular with high stiffness. In this chapter, variational integrators are constructed for the efficient and structure preserving simulation of such systems.

1.1 Variational Integrators

The key feature of variational integrators is that they are based on a discrete variational formulation of the underlying system, e.g. a discrete version of Hamilton's principle for conservative mechanical systems. More concretely, the time stepping schemes are derived from a discrete variational principle based on a discrete action function that approximates the continuous one. This is opposed to the standard derivation of integration methods that start with a continuous equation of motion and replace the continuous quantities, in particular the derivatives with respect to time, by discrete approximations. The variational theory of discrete mechanics provides a theoretical framework that parallels continuous variational dynamics. Discrete analogues to the Euler-Lagrange equations, Noether's theorem, and the Legendre transform are derived from a discrete Lagrangian by performing similar steps as in the continuous theory. The resulting time stepping schemes are structure preserving, i.e. they are symplectic-momentum conserving and exhibit good energy behaviour, meaning that no artificial dissipation is present and the energy error stays bounded over longterm simulations. There exist many works on symplectic integrators like [12, 13, 15, 16, 18, 25, 27] to mention just a few. A detailed introduction and a survey on the history and literature on the variational view of discrete mechanics is given in [24]. Choosing different variational formulations (e.g. Hamilton, Lagrange-d'Alembert, Hamilton-Pontryagin, etc.), variational integrators have been developed for classical conservative mechanical systems (for an overview see [19, 20]), forced [13] and controlled [26] systems, constrained systems (holonomic [21, 22] and nonholonomic systems [14]), nonsmooth systems [7], stochastic systems [5], and multiscale systems [30]. In this chapter, we focus on holonomically constrained systems in the framework of discrete variational mechanics for which the constraints are enforced using Lagrange multipliers. Thus, a discrete version of the index 3 DAEs is solved.

1.2 Integration Methods for Multirate Systems

For systems comprising fast and slow dynamics, different integration methods have been developed to save computational work while preserving the accuracy of the simulation. Here, the methods distinguish with respect to the simulation goals and the structure of the underlying system (for an overview of numerical methods for oscillatory, multiscale Hamiltonian systems see e.g. [6]).

Considering systems with a slow potential that is expensive to evaluate while the fast potential is cheap to evaluate, splitting methods have been developed to accurately capture the slow dynamics without resolving the fast one. One possibility to achieve this is via implicit-explicit methods that treat the fast potential implicitly and the slow one explicitly as e.g. the so called impulse method (see e.g. [12, 17]). This method can also be interpreted as a particular variational splitting method found in the literature under the name IMEX [28]. To refine the resolution for the fast dynamics associated with the fast potential, smaller time steps can be used to perform its implicit time integration. If a structure preserving integrator is used, the composition ensures that its properties are inherited. The explicit treatment of the expensive potential certainly decreases computational costs, however, here, the fast integration is performed for all variables, also the slow degrees of freedom.

Another alternative for the efficient simulation of multirate systems is averaging. Here one is not interested in resolving the fast dynamics, but considers it to be sufficient to feed an average of the fast dynamics into the slow equations of motion. HMM (heterogeneous multiscale methods [32]) aim to link models at different scales and provides a general framework for designing and analysing very heterogeneous, multiscale or even multiphysics problems. Relying on a top-down strategy, the missing information is filled in an incomplete model on the macro scale by estimating what happens on the micro scale through averaging. Thereby, one avoids the isolated pointwise evaluation of oscillatory functions, instead relies on averaged quantities. FLAVORS (flow averaging integrators [30]) are another example of averaging methods. These integrators are formulated using variational methods and the average of the flow is performed via a splitting and resynchronisation technique.

The separation of the unknowns into fast and slow degrees of freedom enables to resolve the fast dynamics in an efficient way if different time grids are used for different parts of the system. These multirate integration methods (for an overview see e.g. [10–12] and references therein) integrate the slow part of the system with a relatively large step size while the fast part is integrated with a small time step. Thereby, main challenges are the identification of fast and slow parts (e.g. either by separating the system's energy or by defining disjunct sets of degrees of freedom), the synchronisation of their different dynamics and in particular the treatment of mixed parts as they often appear when fast and slow dynamics are coupled either via potentials or by constraints. Furthermore, resonance phenomena impose restrictions on the combination of large and small time steps. Hence, another challenge is the stability analysis of multirate stepping schemes as done for linear problems e.g. in [3, 8]. Similar to our approach, the latter work is based on a variational derivation, however the resulting multirate schemes are different from those presented here. There are many examples in the literature where multirate schemes based on backward differentiation formulas (BDF) or Runge-Kutta methods are applied e.g. to electric circuit systems [11, 29, 31], but also to mechanical problems [2]. However, most of them do not focus on the preservation of the underlying system's structure. The aim of this chapter is to develop a structure preserving multirate integrator based on variational mechanics.

1.3 Contribution and Outline

Sections 2 and 3 give an overview of Lagrangian dynamics. Basic definitions and properties such as energy conservation, symplecticity and Noether's theorem are revisited in the continuous and the discrete setting, respectively. In particular, the variational formulation for constrained variational mechanics including the Lagrange multiplier theorem is presented. The variational framework provides the basis for the derivation of the variational multirate integrator described in Sect. 4. The multirate integrator is derived in closed form via a discrete variational principle on a time grid consisting of macro and micro time nodes and thus falls into the class of structure preserving integrators which generally exhibit very good longterm stability. The use of different quadrature rules in the approximation of the appearing integrals and its influence on the degree of coupling in the resulting system of discrete equations of motion, the number of necessary function evaluations and the possibility to treat fast and slow parts in an implicit or an explicit way, respectively, is discussed. The performance of the variational multirate integrator is demonstrated by means of a standard benchmark problem for multirate integration, the Fermi-Pasta-Ulam problem, and for an example from constrained multibody dynamics in Sect. 5.

2 Lagrangian Dynamics

Basic definitions and properties of Lagrangian dynamics like the conservation of energy, symplecticity and Noether's theorem are recalled in Sect. 2.1, before Lagrangian dynamics subject to scleronomic, holonomic constraints is considered in Sect. 2.2. All notation has been introduced in [23, 24], where a large part of the theory presented here can be found.

2.1 Lagrangian Dynamics—Definitions and Properties

Consider an n-dimensional mechanical system in a configuration manifold $Q \subseteq \mathbb{R}^n$ with configuration vector $q(t) \in Q$ and velocity vector $\dot{q}(t) \in T_{q(t)}Q$ in the tangent space, where t denotes the time variable in the bounded interval $[t_0, t_N] \subset \mathbb{R}$. Let, the Lagrangian $L: TQ \to \mathbb{R}$ of the mechanical system consists of the difference of the kinetic energy $T(\dot{q})$ and a potential $U(q)$. Let $\mathscr{C}(Q) = \mathscr{C}([t_0, t_N], Q, q_0, q_N)$ denote the space of smooth curves $q : [t_0, t_N] \to Q$ satisfying $q(t_0) = q_0$ and $q(t_N) = q_N$, where $q_0, q_N \in Q$ are fixed endpoints. For $q \in \mathscr{C}(Q)$, the action integral is defined as

$$\mathfrak{S}(q) = \int_{t_0}^{t_N} L(q, \dot{q})\, dt$$

Requiring that the first variation of this action vanishes, i.e. $\delta\mathfrak{S} = 0$, Hamilton's principle of stationary action yields the Euler-Lagrange equations of motion of a conservative mechanical system

$$\frac{\partial L(q,\dot{q})}{\partial q} - \frac{d}{dt}\left(\frac{\partial L(q,\dot{q})}{\partial \dot{q}}\right) = 0 \tag{1}$$

2.1.1 Energy Conservation

The total energy $E : TQ \to \mathbb{R}$ of a Lagrangian L is given by

$$E(q,\dot{q}) = \dot{q} \cdot \frac{\partial L}{\partial \dot{q}} - L(q,\dot{q})$$

It is conserved along a solution of the Euler-Lagrange equations (1). More generally, solutions of (1) can be identified with the Lagrangian flow $F_L : [t_0, t_N] \times TQ \to TQ$ that takes a given initial state $(q(t_0), \dot{q}(t_0)) \in TQ$ forward in time to the actual state at $t \in [t_0, t_N]$ via $F_L^t : TQ \to TQ$ with $F_L^t : (q(t_0), \dot{q}(t_0)) \mapsto (q(t), \dot{q}(t))$. In other words, $E \circ F_L^t = E$ for all $t \in [t_0, t_N]$, i.e. the total energy is conserved along the Lagrangian flow.

2.1.2 Symplecticity

For hyperregular Lagrangians, the Lagrangian two form $\Omega_L : T(TQ) \times T(TQ) \to \mathbb{R}$ (being a two form means that Ω_L is a skew symmetric bilinear form on TQ) is symplectic, i.e. it is a closed, weakly nondegenerate two form. A coordinate expression of the Lagrangian symplectic form is given by

$$\Omega_L(q,\dot{q}) = \frac{\partial^2 L}{\partial q^i \partial \dot{q}^j} dq^i \wedge dq^j + \frac{\partial^2 L}{\partial \dot{q}^i \partial \dot{q}^j} d\dot{q}^i \wedge dq^j$$

where Einstein's summation convention is used. An important property of the Lagrangian flow is that it is symplectic in the sense that it preserves the Lagrangian symplectic form, i.e.

$$(F_L^t)^*(\Omega_L) = \Omega_L$$

where $(F_L^t)^*(\Omega_L)$ denotes the pull back of Ω_L. As a consequence of symplecticity, the volume in state space is conserved.

2.1.3 Noether's Theorem

Another key property of the Lagrangian flow is its behaviour with respect to the action of a Lie group G (with Lie algebra $\mathfrak{g}$). The Lagrangian is said to be G-invariant, if the Lie group acts on the configuration via $\phi : G \times Q \to Q$ and on the velocity via the tangent lift $\phi^{TQ} : G \times TQ \to TQ$ and $L \circ \phi_g^{TQ} = L$ holds for all $g \in G$ with $\phi_g^{TQ}(q, \dot{q}) = \phi^{TQ}(g, (q, \dot{q}))$. In this case, the group is said to be a symmetry of the Lagrangian, leading to a momentum map $J_L : TQ \to \mathfrak{g}^*$ that is preserved along the Lagrangian flow, so that $J_L \circ F_L^t = J_L$ for all times $t \in [t_0, t_N]$.

Example 1 Some classical examples of symmetries are the invariance of the Lagrangian with respect to translation and rotation, leading to the conservation of total linear momentum and total angular momentum, respectively.

2.2 Constrained Lagrangian Dynamics

Now, let the motion be constrained by the vector valued function of holonomic, scleronomic constraints requiring $g(q) = 0 \in \mathbb{R}^m$. It is assumed that $0 \in \mathbb{R}^m$ is a regular value of the constraints, such that

$$C = g^{-1}(0) = \left\{ q \,\middle|\, q \in Q, g(q) = 0 \right\} \subset Q$$

is an $(n - m)$-dimensional submanifold, called constraint manifold. Just as C can be embedded in Q via $i : C \to Q$, its $2(n - m)$-dimensional tangent bundle

$$TC = \left\{ (q, \dot{q}) \,\middle|\, (q, \dot{q}) \in T_q Q, g(q) = 0, G(q) \cdot \dot{q} = 0 \right\} \subset TQ \tag{2}$$

can be embedded in TQ in a natural way by tangent lift $Ti : TC \to TQ$. Here and in the sequel $G(q) = Dg(q)$ denotes the $m \times n$ Jacobian of the constraints. Note that according to (2), admissible velocities are constrained to the null space of the constraint Jacobian.

A Lagrangian $L : TQ \to \mathbb{R}$ can be restricted to $L^C = L_{|TC} : TC \to \mathbb{R}$. To investigate the relation of the dynamics of L^C on TC and the dynamics of L on TQ, the following notation is used. Let $q_0, q_N \in C$ be fixed endpoints and consider $\mathscr{C}(Q) = \mathscr{C}([t_0, t_N], Q, q_0, q_N)$ and the corresponding space of curves in C denoted by $\mathscr{C}(C) = \mathscr{C}([t_0, t_N], C, q_0, q_N)$. Furthermore, set $\mathscr{C}(\mathbb{R}^m) = \mathscr{C}([t_0, t_N], \mathbb{R}^m)$ to be the space of curves $\lambda : [t_0, t_N] \to \mathbb{R}^m$ with no boundary conditions.

Theorem 1 *Suppose that* 0 *is a regular value of the scleronomic holonomic constraints* $g : Q \to \mathbb{R}^m$ *and set* $C = g^{-1}(0) \subset Q$. *Let* $L : TQ \to \mathbb{R}$ *be a Lagrangian and* $L^C = L_{|TC}$ *its restriction to* TC. *Then the following statements are equivalent:*

(i) $q \in \mathscr{C}(C)$ *extremises the action integral* $\mathfrak{S}^C(q) = \int_{t_0}^{t_N} L^C(q, \dot{q})\, dt$ *and hence solves the Euler-Lagrange equations for* L^C.

(ii) $q \in \mathscr{C}(Q)$ *and* $\lambda \in \mathscr{C}(\mathbb{R}^m)$ *satisfy the constrained Euler-Lagrange equations*

$$\frac{\partial L(q,\dot{q})}{\partial q} - \frac{d}{dt}\left(\frac{\partial L(q,\dot{q})}{\partial \dot{q}}\right) - G^T(q)\cdot\lambda = 0 \qquad (3)$$
$$g(q) = 0$$

(iii) $(q,\lambda) \in \mathscr{C}(Q \times \mathbb{R}^m)$ *extremise*

$$\bar{\mathfrak{S}}(q,\lambda) = \int_{t_0}^{t_N} L(q,\dot{q}) - g^T(q)\cdot\lambda\, dt \qquad (4)$$

and hence, solve the Euler-Lagrange equations for the augmented Lagrangian $\bar{L} : T(Q \times \mathbb{R}^m) \to \mathbb{R}$ *defined by* $\bar{L}(q,\lambda,\dot{q},\dot{\lambda}) = L(q,\dot{q}) - g^T(q)\cdot\lambda$.

The proof given in [24] makes use of the Lagrange multiplier theorem (see e.g. [1]). The term $-G^T(q)\cdot\lambda \in (TC)^{\perp}$ in $(3)_1$ represents the constraint forces that prevent the system from deviation of the constraint manifold. As can be seen, the constrained system on TC is a standard Lagrangian systems and so it has the usual conservation properties. In particular, the constrained Lagrangian system $L^C : TC \to \mathbb{R}$ has a flow map that preserves the symplectic two form $\Omega_{L^C} = (Ti)^*\Omega_L$. Furthermore, Noether's theorem holds for both, the unconstrained as well as the constrained case. Thus, if the lifted group action leaves L^C on TC invariant, the same momentum map is preserved.

3 Discrete Variational Dynamics

The variational theory of discrete mechanics provides a theoretical framework that parallels continuous variational dynamics. Discrete analogues to the Euler-Lagrange equations, the symplectic structure and Noether's theorem are derived from a discrete Lagrangian by performing similar steps as in the continuous theory.

3.1 Discrete Variational Dynamics—Definitions and Properties

Corresponding to TQ, the discrete state space is defined by $Q \times Q$ which is locally isomorphic to TQ. For a discrete time grid $\{t_0, t_0+\Delta t, \ldots, t_0+N\Delta t = t_N\}$ with $N \in \mathbb{N}$ and constant step size $\Delta t \in \mathbb{R}$, let $\mathscr{C}_d(Q) = \mathscr{C}(\{t_0, t_0+\Delta t, \ldots, t_0+N\Delta t = t_N\}, Q, q_0, q_N)$ denote the space of discrete trajectories $q_d : \{t_0, t_0+\Delta t, \ldots, t_0+N\Delta t = t_N\} \to Q$ satisfying $q_d(t_0) = q_0$ and $q_d(t_N) = q_N$ for given $q_0, q_N \in Q$. A continuous trajectory $q : [t_0, t_N] \to Q$ is replaced by a discrete trajectory $q_d = \{q_k\}_{k=0}^N$. Here, $q_k = q_d(t_0 + k\Delta t)$ is viewed as an approximation to $q(t_0 + k\Delta t)$.

According to the key idea of variational integrators, the variational principle is discretised rather than the resulting equations of motion. The action integral is approximated in a time interval $[t_k, t_{k+1}]$ using the discrete Lagrangian $L_d : Q \times Q \to \mathbb{R}$ via

$$L_d(q_k, q_{k+1}) \approx \int_{t_k}^{t_{k+1}} L(q, \dot{q})\, dt \tag{5}$$

The quadrature used to approximate the integral in (5) determines the actual time stepping scheme (6) and in particular its order of accuracy. For $q_d \in \mathscr{C}_d(Q)$, variation of the discrete action sum

$$\mathfrak{S}_d(q_d) = \sum_{k=0}^{N-1} L_d(q_k, q_{k+1})$$

reads

$$\begin{aligned}\delta\mathfrak{S}_d = \delta q_0^T \cdot D_1 L_d(q_0, q_1) + \sum_{k=1}^{N-1} \delta q_k^T \cdot \big(D_2 L_d(q_{k-1}, q_k) + D_1 L_d(q_k, q_{k+1})\big)\\ + \delta q_N^T \cdot D_2 L_d(q_{N-1}, q_N)\end{aligned}$$

Requiring its stationarity for all $\{\delta q_k\}_{k=1}^{N-1}$ and $\delta q_0 = \delta q_N = 0$ yields the discrete (unconstrained) Euler-Lagrange equations

$$D_1 L_d(q_k, q_{k+1}) + D_2 L_d(q_{k-1}, q_k) = 0 \quad \text{for } k = 1, \ldots, N-1 \tag{6}$$

For a given initial configuration $q_0 = q(t_0) \in Q$ and initial velocity $\dot{q}(t_0) \in T_{q_0}Q$ with corresponding initial conjugate momentum

$$p_0 = p(t_0) = \frac{\partial L(q(t_0), \dot{q}(t_0))}{\partial \dot{q}} \in T_{q_0}^* Q$$

the first discrete configuration can be computed by solving

$$p_0 = -D_1 L_d(q_0, q_1)$$

Then for two given subsequent configurations, (6) can be used to integrate forward in time. See [23, 24] for the theory on (discrete) Legendre transforms.

3.1.1 Symplecticity

The discrete object corresponding to the Lagrangian flow is the discrete Lagrangian map $F_{L_d} : Q \times Q \to Q \times Q$ with $F_{L_d} : (q_{k-1}, q_k) \mapsto (q_k, q_{k+1})$ according to (6). One can show that the discrete Lagrangian map inherits the properties we summarised for the continuous Lagrangian flow. That means the discrete Lagrangian symplectic form with coordinate expression

$$\Omega_{L_d}(q_0, q_1) = \frac{\partial^2 L_d}{\partial q_0^i \partial q_1^j} dq_0^i \wedge dq_1^j$$

is preserved under the discrete Lagrangian map, i.e.

$$(F_{L_d})^*(\Omega_{L_d}) = \Omega_{L_d}$$

and we say that F_{L_d} is symplectic.

3.1.2 Energy Behaviour

Due to the symplecticity of the discrete Lagrangian map, backward error analysis can be used to prove that no energy is dissipated numerically, see [12]. As a consequence, the total energy oscillates (with a small amplitude) close to the real value and no energy is gained or lost artificially along the discrete trajectory q_d. Thus, the integration runs very stable, even when relatively large time steps are used. We speak about good longterm energy behaviour.

On the other hand, for exactly energy conserving time stepping schemes, the energy is conserved up to numerical accuracy, see e.g. [4] and many references therein. It is well known [9] that numerical integrators based on constant time steps cannot be symplectic and exactly energy conserving at the same time.

3.1.3 Discrete Noether's Theorem

Consider a given discrete Lagrangian system $L_d : Q \times Q \to \mathbb{R}$ which is invariant under the lift $\phi^{Q\times Q} : G \times (Q \times Q) \to Q \times Q$ of the action $\phi : G \times Q \to Q$, i.e. $L_d \circ \phi_g^{Q\times Q} = L_d$ for all $g \in G$. Then the corresponding discrete Lagrangian momentum map $J_{L_d} : Q \times Q \to \mathfrak{g}^*$ is a conserved quantity of the discrete Lagrangian map, such that $J_{L_d} \circ F_{L_d} = J_{L_d}$. Note that discrete momentum maps are conserved exactly, i.e. up to the numerical accuracy to which the (often nonlinear) discrete equations of motion are solved.

Example 2 As in the continuous case, most common classical examples are the conservation of total linear momentum and total angular momentum, when the discrete Lagrangian is invariant with respect to translation and rotation, respectively. In general, a value of the momentum map can be computed from the initial data, and this value is exactly preserved along the discrete trajectory.

3.2 Constrained Discrete Variational Dynamics

Let $q_0, q_N \in C$ be fixed end points. Consider $\mathscr{C}_d(Q) = \mathscr{C}(\{t_0, t_0 + \Delta t, \dots, t_0 + N\Delta t = t_N\}, Q, q_0, q_N)$ and let $\mathscr{C}_d(C)$ denote the corresponding set of discrete trajectories in C. Furthermore, let $\mathscr{C}_d(\mathbb{R}^m) = \mathscr{C}(\{t_0, t_0 + \Delta t, \dots, t_0 + N\Delta t = t_N\}, \mathbb{R}^m)$ be the set of maps $\lambda_d : \{t_0, t_0 + \Delta t, \dots, t_0 + N\Delta t = t_N\} \to \mathbb{R}^m$ with no boundary conditions. Then, $\lambda_d = \{\lambda_k\}_{k=0}^{N-1}$ with $\lambda_k = \lambda_d(t_k)$ approximates the Lagrange multiplier $\lambda(t_k)$ at $t_k = t_0 + k\Delta t$.

To include scleronomic holonomic constraints in the discrete variational principle, the integral over $[t_k, t_{k+1}]$ of the scalar product of the constraints and the corresponding Lagrange multiplier in (4) is approximated by the trapezoidal rule

$$\frac{1}{2} g_d^T(q_k) \cdot \lambda_k + \frac{1}{2} g_d^T(q_{k+1}) \cdot \lambda_{k+1} \approx \int_{t_k}^{t_{k+1}} g^T(q) \cdot \lambda \, dt$$

whereby $g_d^T(q_k) = \Delta t g^T(q_k)$ is used and let $G_d^T(q_k) = D g_d^T(q_k)$.

Analogue to Theorem 1, the relation between the constrained discrete Lagrangian system on $Q \times Q$ and that corresponding to a discrete Lagrangian restricted to $C \times C$ is stated in the following theorem which has again been taken from [24].

Theorem 2 *Suppose that* 0 *is a regular value of the scleronomic holonomic constraints* $g : Q \to \mathbb{R}^m$ *and set* $C = g^{-1}(0) \subset Q$. *Let* $L_d : Q \times Q \to \mathbb{R}$ *be a discrete Lagrangian and* $L_d^C = L_{d|C\times C}$ *its restriction to* $C \times C$. *Then the following statements are equivalent*:

(i) $q_d = \{q_k\}_{k=0}^N \in \mathcal{C}_d(C)$ *extremises the discrete action* $\mathfrak{S}_d^C = \mathfrak{S}_{d|C\times C}$ *and hence solves the discrete Euler-Lagrange equations for* L_d^C.

(ii) $\{q_k\}_{k=0}^N \in \mathcal{C}_d(Q)$ *and* $\{\lambda_k\}_{k=1}^{N-1} \in \mathcal{C}_d(\mathbb{R}^m)$ *satisfy the constrained discrete Euler-Lagrange equations*

$$D_1 L_d(q_k, q_{k+1}) + D_2 L_d(q_{k-1}, q_k) - G_d^T(q_k) \cdot \lambda_k = 0$$
$$g(q_{k+1}) = 0$$

(iii) $(q_d, \lambda_d) \in \mathcal{C}_d(Q \times \mathbb{R}^m)$ *extremise* $\bar{\mathfrak{S}}_d(q_d, \lambda_d) = \mathfrak{S}_d(q_d) - \langle \lambda_d, g_d(q_d) \rangle$ *and hence, solve the Euler-Lagrange equations for the augmented Lagrangian* $\bar{L}_d : (Q \times \mathbb{R}^m) \times (Q \times \mathbb{R}^m) \to \mathbb{R}$ *defined by*

$$\bar{L}_d(q_k, \lambda_k, q_{k+1}, \lambda_{k+1}) = L_d(q_k, q_{k+1}) - \frac{1}{2} g_d^T(q_k) \cdot \lambda_k - \frac{1}{2} g_d^T(q_{k+1}) \cdot \lambda_{k+1}$$

As in the continuous case, the structure preservation properties remain untouched by the presence of constraints, the constrained discrete Lagrangian map preserves the standard discrete symplectic form on $C \times C$ and the same discrete momentum maps that are preserved in the discrete unconstrained case (again, the preserved value can be computed as the continuous momentum maps at the initial condition).

4 Variational Multirate Integrator

Having reviewed Lagrangian dynamics for constrained systems in the time continuous case in Sect. 2 and in the discrete setting in Sect. 3, we now focus on constrained systems with dynamics on different time scales.

4.1 Slow and Fast Potential and Constraints

Let the fact that the Lagrangian contains slow and fast dynamics be characterised by the possibility to additively split the potential energy $U(q) = V(q) + W(q)$ into a slow potential V and a fast potential W. Then, the constrained Euler-Lagrange equations of motion on a time interval $[t_0, t_N] \subset \mathbb{R}$

$$\begin{aligned} \frac{\partial V}{\partial q} + \frac{\partial W}{\partial q} - \frac{d}{dt}\frac{\partial T}{\partial \dot{q}} - \left(\frac{\partial g}{\partial q}\right)^T \cdot \lambda = 0 \\ g(q) = 0 \end{aligned} \tag{7}$$

can be derived via Hamilton's principle requiring stationarity of the action. See Sect. 5 for examples of such additively split potentials.

4.2 Slow and Fast Variables

We further assume that the n-dimensional configuration variable q can be divided into n^s slow variables $q^s \in Q^s$ and n^f fast variables $q^f \in Q^f$ such that $Q^s \times Q^f = Q$ and $q = (q^s, q^f)$ with $n^s + n^f = n$. Let the fast potential depend of the fast degrees of freedom only, i.e. $W = W(q^f)$ while the slow potential $V = V(q)$ depends on the complete configuration variable as does the constraint function $g = g(q)$. With these assumptions, the Euler-Lagrange equations (7) take the form

$$\begin{aligned} \frac{\partial V}{\partial q^s} - \frac{d}{dt}\frac{\partial T}{\partial \dot{q}^s} - \left(\frac{\partial g}{\partial q^s}\right)^T \cdot \lambda = 0 \\ \frac{\partial V}{\partial q^f} + \frac{\partial W}{\partial q^f} - \frac{d}{dt}\frac{\partial T}{\partial \dot{q}^f} - \left(\frac{\partial g}{\partial q^f}\right)^T \cdot \lambda = 0 \\ g(q) = 0 \end{aligned}$$

Remark 1 If in addition, the slow potential depends on the slow variables only and on top of that the kinetic energy does not contain any entries coupling $\dot{q}^s$ and $\dot{q}^f$, then the system is completely decoupled and simulation can be performed independently in parallel, without any exchange of information. This case is trivial and we focus on the scenario described above. Note that the inclusion of additional potentials or constraint functions depending on the fast or the slow variable only is straightforward.

4.3 Discrete Variational Principle on Macro and Micro Grid

Rather than choosing one time grid for the approximation as for standard variational integrators, for the multirate integrator, two different time grids are introduced, see Fig. 1. With the time steps ΔT and Δt (where $\Delta T \geq \Delta t$), a macro time grid $\{t_k = k\Delta T \mid k = 0, \ldots, N\}$ and a micro time grid $\{t_k^m = k\Delta T + m\Delta t \mid k = 0, \ldots, N-1, m = 0, \ldots, p\}$ are defined. Note that except for the boundary nodes t_0, t_N, two micro time nodes coincide with a macro time node, i.e. $t_{k-1}^p = t_k^0 = t_k$

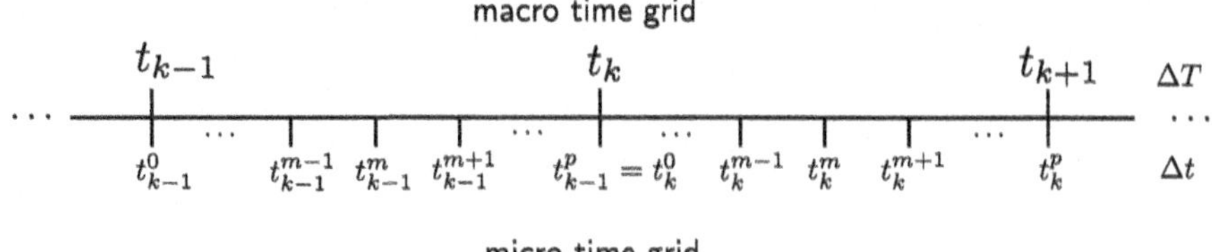

Fig. 1 Macro and micro time grid

for $k = 1, \ldots, N-1$, and $\Delta T = p\Delta t$, see Fig. 1. The macro time grid provides the domain for the discrete macro trajectory of the slow variables

$$q_d^s = \{q_k^s\}_{k=0}^N \quad \text{with } q_k^s \approx q^s(t_k)$$

and the discrete micro trajectory of the fast variables lives on the micro grid

$$q_d^f = \{q_k^f\}_{k=0}^{N-1} = \left\{\{q_k^{f,m}\}_{m=0}^p\right\}_{k=0}^{N-1} \quad \text{with } q_k^{f,m} \approx q^f(t_k^m)$$

Since the constraints depend on the complete configuration variables, the Lagrange multipliers cannot be separated in a fast and a slow part and must be computed on the fine time grid. Thus, the discrete trajectory of Lagrange multipliers takes the form

$$\lambda_d = \{\lambda_k\}_{k=0}^{N-1} = \left\{\{\lambda_k^m\}_{m=0}^p\right\}_{k=0}^{N-1} \quad \text{with } \lambda_k^m \approx \lambda(t_k^m)$$

Note that $t_{k-1}^p = t_k^0$ and therefore also $q_{k-1}^{f,p} = q_k^{f,0}$ and $\lambda_{k-1}^p = \lambda_k^0$ hold.

As an approximation to $\bar{\mathfrak{S}}$ in (4), the augmented discrete action is defined as

$$\bar{\mathfrak{S}}_d(q_d^s, q_d^f, \lambda_d) = \sum_{k=0}^{N-1} \left[L_d(q_k^s, q_{k+1}^s, q_k^f) - h_d(q_k^s, q_{k+1}^s, q_k^f, \lambda_k)\right] \tag{8}$$

The discrete Lagrangian $L_d = T_d - V_d - W_d$ approximates $\int_{t_k}^{t_{k+1}} L(q, \dot{q})\,dt$ and reads

$$L_d(q_k^s, q_{k+1}^s, q_k^f) = T_d(q_k^s, q_{k+1}^s, q_k^f) - V_d(q_k^s, q_{k+1}^s, q_k^f) - W_d(q_k^f) \tag{9}$$

while h_d is approximating $\int_{t_k}^{t_{k+1}} g(q)^T \cdot \lambda\,dt$. Omitting the arguments of L_d and h_d, stationarity of the discrete action

$$\delta\bar{\mathfrak{S}}_d = \sum_{k=0}^{N-1} \Bigg\{ D_{q_k^s}(L_d + h_d) \cdot \delta q_k^s + D_{q_{k+1}^s}(L_d + h_d) \cdot \delta q_{k+1}^s + \sum_{m=0}^{p} \left[D_{q_k^{f,m}}(L_d + h_d) \cdot \delta q_k^{f,m} + D_{\lambda_k^m} h_d \cdot \delta\lambda_k^m\right] \Bigg\} = 0$$

with independent variations δq_k^s for $k = 0, \ldots, N$ and $\delta q_k^{f,m}, \delta\lambda_k^m$ for $k = 0, \ldots, N-1$ and $m = 0, \ldots, p$ yields the discrete Euler-Lagrange equations. Let $k = 0$

and assume that an initial configuration $(q_0^s, q_0^{f,0})$ being consistent with the constraints, i.e. $g((q_0^s, q_0^{f,0})) = 0$, and an initial conjugate momentum $(p_0^s, p_0^{f,0})$ is given. Then for $k = 0$, the unknowns $q_1^s, q_0^{f,1}, \ldots, q_0^{f,p}$ and $\lambda_0^0, \ldots, \lambda_0^{p-1}$ are determined by solving the following set of equations for $m = 1, \ldots, p-1$.

$$
\begin{aligned}
&(IC)^s && D_{q_0^s}\left(L_d(q_0^s, q_1^s, q_0^f) + h_d(q_0^s, q_1^s, q_0^f, \lambda_0)\right) = -p_0^s \\
&(IC)^f && D_{q_0^{f,0}}\left(L_d(q_0^s, q_1^s, q_0^f) + h_d(q_0^s, q_1^s, q_0^f, \lambda_0)\right) = -p_0^{f,0} \\
& && D_{\lambda_0^1} h_d(q_0^s, q_1^s, q_0^f, \lambda_0) = 0 \\
&(DEL)_0^{f,m} && D_{q_0^{f,m}}\left(L_d(q_0^s, q_1^s, q_0^f) + h_d(q_0^s, q_1^s, q_0^f, \lambda_0)\right) = 0 \\
& && D_{\lambda_0^{m+1}} h_d(q_0^s, q_1^s, q_0^f, \lambda_0) \\
& && \quad + \delta_{m,p-1} D_{\lambda_1^0} h_d(q_1^s, q_2^s, q_1^f, \lambda_1) = 0
\end{aligned} \tag{10}
$$

These equations can be considered as initial conditions, since they determine the unknowns in the first macro time interval from given initial data. Note that variation with respect to λ_0^0 is unnecessary, since the initial configuration does fulfil the constraints a priori. Analog to the variational integrators for constrained systems on a single time grid as described in Sect. 3.2 (see also e.g. [21, 24]), here variation with respect to λ_0^m yields the condition $g((q_0^s, q_0^{f,m})) = 0$. Therefore, the last condition $g((q_1^s, q_0^{f,p})) = 0$ is composed by contributions from variation with respect to the multipliers λ_0^p and λ_1^0 (which are equal) which is ensured using the Dirac delta in the last equation. To proceed further in time for $k = 1, \ldots, N-1$ (assuming that $q_{k-1}^s, q_k^s, q_{k-1}^{f,0}, \ldots, q_{k-1}^{f,p}$ are given), solving the following discrete Euler-Lagrange equations for $m = 1, \ldots, p-1$ determines $q_{k+1}^s, q_k^{f,1}, \ldots, q_k^{f,p}$ and $\lambda_k^0, \ldots, \lambda_k^{p-1}$.

$$
\begin{aligned}
&(DEL)_k^s \\
&\quad D_{q_k^s}\left(L_d(q_k^s, q_{k+1}^s, q_k^f) + L_d(q_{k-1}^s, q_k^s, q_{k-1}^f)\right. \\
&\quad\quad \left. + h_d(q_k^s, q_{k+1}^s, q_k^f, \lambda_k) + h_d(q_{k-1}^s, q_k^s, q_{k-1}^f, \lambda_{k-1})\right) = 0 \\
&(DEL)_k^{f,0} \\
&\quad D_{q_k^{f,0}}\left(L_d(q_k^s, q_{k+1}^s, q_k^f) + L_d(q_{k-1}^s, q_k^s, q_{k-1}^f)\right. \\
&\quad\quad \left. + h_d(q_k^s, q_{k+1}^s, q_k^f, \lambda_k) + h_d(q_{k-1}^s, q_k^s, q_{k-1}^f, \lambda_{k-1})\right) = 0 \\
&\quad D_{\lambda_k^1} h_d(q_k^s, q_{k+1}^s, q_k^f, \lambda_k) = 0 \\
&(DEL)_k^{f,m} \\
&\quad D_{q_k^{f,m}}\left(L_d(q_k^s, q_{k+1}^s, q_k^f) + h_d(q_k^s, q_{k+1}^s, q_k^f, \lambda_k)\right) = 0 \\
&\quad D_{\lambda_k^{m+1}} h_d(q_k^s, q_{k+1}^s, q_k^f, \lambda_k) \\
&\quad\quad + \delta_{m,p-1}(1 - \delta_{k,N-1}) D_{\lambda_{k+1}^0} h_d(q_{k+1}^s, q_{k+2}^s, q_{k+1}^f, \lambda_{k+1}) = 0
\end{aligned} \tag{11}
$$

Again, at the macro nodes the constraint equations include an additional term which is added using the Dirac delta. Note however, that this term does not exist at the very end node t_N.

Remark 2 Due to the variational derivation of the multirate integrator, we can state that a discrete symplectic form is preserved along the discrete solution trajectory. Furthermore, if the discrete Lagrangian is invariant under a group action on the macro grid, then the corresponding momentum map is preserved at the macro time nodes $\{t_k\}_{k=0}^N$.

4.4 Discrete Action—Influence of Quadrature

The quadrature rules in use for the discrete Lagrangian (9) and the discrete constraint term in (8) determine the degree of coupling between the discrete equations (10) and (11), respectively. This can range from a fully implicit scheme over variants being explicit in the macro and implicit in the micro quantities to fully explicit schemes. We consider Lagrangians of the form $L(q,\dot{q}) = T(\dot{q}) - V(q) - W(q^f)$.

4.4.1 Kinetic Energy

Assume that the kinetic energy can be decomposed in a contribution from the fast and the slow variables, i.e.

$$T(\dot{q}) = \frac{1}{2}\dot{q}^T \cdot M \cdot \dot{q} = \frac{1}{2}(\dot{q}^s)^T \cdot M^s \cdot \dot{q}^s + \frac{1}{2}(\dot{q}^f)^T \cdot M^f \cdot \dot{q}^f$$

where M^s and M^f are the mass matrices for the slow and fast variables, respectively, yielding the total mass matrix as $M = \operatorname{diag}(M^s, M^f)$. In the sequel the velocities $\dot{q}^s$ and $\dot{q}^f$ are approximated using backward difference operators on the macro and micro grid. Then the discrete kinetic energy is defined on the time interval $[t_k, t_{k+1}]$ as

$$\begin{aligned} T_d = {} & \frac{\Delta T}{2}\left(\frac{q_{k+1}^s - q_k^s}{\Delta T}\right)^T \cdot M^s \cdot \left(\frac{q_{k+1}^s - q_k^s}{\Delta T}\right) \\ & + \sum_{m=0}^{p-1} \frac{\Delta t}{2}\left(\frac{q_k^{f,m+1} - q_k^{f,m}}{\Delta t}\right)^T \cdot M^f \cdot \left(\frac{q_k^{f,m+1} - q_k^{f,m}}{\Delta t}\right) \end{aligned}$$

4.4.2 Constraints

The discrete function h_d approximates the integral $\int_{t_k}^{t_{k+1}} g(q)^T \cdot \lambda \, dt$. Similar to the approximation on a standard time grid as described in Sect. 3.2 (see also [21]), a trapezoidal rule is used here.

$$\begin{aligned} (q_k^s, q_{k+1}^s, q_k^f, \lambda_k) = \sum_{m=0}^{p-1} \Bigg[& \frac{1}{2} g_d^T\left(q_k^s, q_{k+1}^s, q_k^{f,m}\right) \cdot \lambda_k^m \\ & + \frac{1}{2} g_d^T\left(q_k^s, q_{k+1}^s, q_k^{f,m+1}\right) \cdot \lambda_k^{m+1} \Bigg] \end{aligned}$$

For the discrete constraint function g_d, an intuitive example is the following.

Example 3 The slow variables q^s can be linearly interpolated between q_k^s and q_{k+1}^s on the micro time grid as

$$q_k^{s,m} = \frac{1}{p}\big((p-m)q_k^s + mq_{k+1}^s\big) \quad \text{for } m = 0, \ldots, p \tag{12}$$

Then, the discrete constraint function reads

$$g_d\big(q_k^s, q_{k+1}^s, q_k^{f,m}\big) = \Delta t g\big((q_k^{s,m}, q_k^{f,m})\big)$$

4.4.3 Potential Energy

When standard variational integrators are used for problems with very stiff potentials, their discrete counterparts are often based on midpoint evaluations of the continuous potentials such that the corresponding integration scheme is implicit. On the other hand, softer potentials can be approximated by evaluations of the continuous potential on the left or right node yielding explicit schemes (at least as long as there are no constraints present), which are of course much cheaper regarding the computational costs. For the multirate integrator, a large variety of combinations is possible.

Example 4 (Implicit fast and explicit slow forces) Let's first consider a special case where the dynamics is not subject to any constraints. Then, choosing an affine combination as approximation in the slow potential that involves only macro nodes

$$V_d\big(q_k^s, q_{k+1}^s, q_k^f\big) = \Delta T\big(\alpha V\big((q_k^s, q_k^{f,0})\big) + (1-\alpha)V\big((q_{k+1}^s, q_k^{f,p})\big)\big) \tag{13}$$

with $0 \le \alpha \le 1$ and a micro node based midpoint rule in the fast potential

$$W_d\big(q_k^f\big) = \sum_{m=0}^{p-1} \Delta t W\left(\frac{q_k^{f,m} + q_k^{f,m+1}}{2}\right) \tag{14}$$

leads to discrete conservative forces in the discrete Euler-Lagrange equations which are explicit for the slow potential and implicit for the fast one. Thus, only few evaluations of the gradient of V are necessary which is advantageous when the slow potential's evaluation is very costly compared to the fast one. The resulting scheme can be interpreted as a variational splitting method which is symmetric and symplectic, since it is a symmetric composition of symmetric and symplectic methods. When this method is formulated with $\alpha = \frac{1}{2}$ on only one time grid with a constant time step (i.e. $\Delta t = \Delta T$ and $p = 1$) and without splitting the configuration variable into fast and slow variables, one obtains the IMEX method in [28] which is an example of an impulse method, see [12] and references therein.

Example 5 (Fully implicit scheme) In this example, the slow variables are interpolated according to (12) and then midpoints are inserted into the slow potential

$$V_d(q_k^s, q_{k+1}^s, q_k^f) = \sum_{m=0}^{p-1} \Delta t V\left(\left(\frac{q_k^{s,m} + q_k^{s,m+1}}{2}, \frac{q_k^{f,m} + q_k^{f,m+1}}{2}\right)\right) \quad (15)$$

and into the fast potential as in (14). As a result, the discrete Euler-Lagrange equations are fully coupled and have to be solved simultaneously using an iteration method. Another quadrature yielding a fully implicit scheme for $0 \le \alpha \le 1$ is given by

$$V_d(q_k^s, q_{k+1}^s, q_k^f) = \sum_{m=0}^{p-1} \Delta t\left(\alpha V\left(\left(q_k^{s,m}, q_k^{f,m}\right)\right) + (1-\alpha) V\left(\left(q_k^{s,m+1}, q_k^{f,m+1}\right)\right)\right) \quad (16)$$

Example 6 (Fully explicit scheme) In the absence of constraints, using the affine combination of the slow potential evaluated at the macro nodes in (13) and the affine combination of micro node evaluations of the fast potential

$$W_d(q_k^f) = \sum_{m=0}^{p-1} \Delta t\left(\alpha W\left(q_k^{f,m}\right) + (1-\alpha) W\left(q_k^{f,m+1}\right)\right)$$

with $0 \le \alpha \le 1$ leads to discrete Euler-Lagrange equations (11) that can subsequently be solved without iteration, i.e. first $q_k^{f,1}$ is obtained from $(\text{DEL})_k^{f,0}$, then $(\text{DEL})_k^{f,m}$ yields $q_k^{f,m+1}$ for $m = 1, \dots, p-1$. At any time, q_{k+1}^s can be computed from $(\text{DEL})_k^s$. For $\alpha = \frac{1}{2}$, this choice of quadrature leads to the scheme in [8] for the special case that a synchronised time grid is used there.

More general multirate schemes are obtained for different choices and combinations of quadrature. Depending on the complexity of the evaluation of the potential functions and their gradients, the computational costs of the overall simulation is heavily influenced by the choice of quadrature.

5 Numerical Examples

5.1 Fermi-Pasta-Ulam Problem

The performance of the presented multirate approach is first demonstrated by means of the Fermi-Pasta-Ulam (FPU) problem (see e.g. [12]). Consider $2l$ unit point masses that are chained together by soft and stiff springs as shown in Fig. 2. With an appropriate choice of the coordinates it is possible to separate the slow and the fast variables of the multirate system. The slow variables q_i^s, $i = 1, \dots, l$, correspond to the location of i-th stiff spring's centre, while the length of i-th stiff spring is a

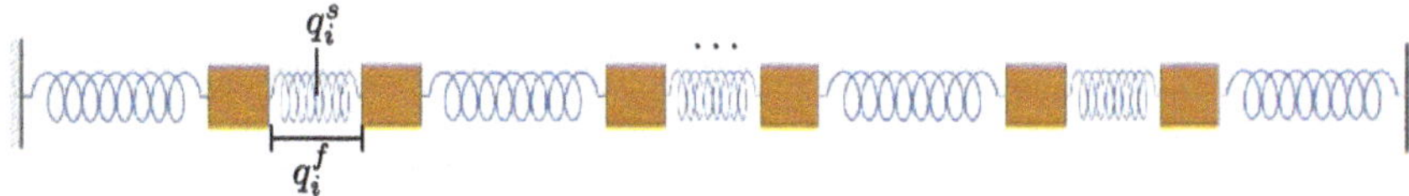

Fig. 2 Fermi-Pasta-Ulam problem: $2l$ point masses that are chained together by soft and stiff springs

fast variable q_i^f, $i = 1, \dots, l$. The Lagrangian is composed by the kinetic energy of slow and fast variables and the spring potentials

$$L = \frac{1}{2}\sum_{i=1}^{l}((\dot{q}_i^s)^2 + (\dot{q}_i^f)^2)$$
$$- \frac{1}{4}\left[(q_1^s - q_1^f)^4 + \sum_{i=1}^{l-1}(q_{i+1}^s - q_{i+1}^f - q_i^s - q_i^f)^4 + (q_l^s + q_l^f)^4\right]$$
$$- \frac{\omega^2}{2}\sum_{i=1}^{l}(q_i^f)^2$$

where the second term is the soft spring potential $V((q^s, q^f))$ depending on the complete configuration variable, while the third term is the stiff potential $W(q^f)$ that depends on the spring lengths only and includes the stiffness $\omega \in \mathbb{R}$ which is supposed to be large. For this system, no constraints are present. The Fermi-Pasta-Ulam problem is a multirate system, i.e. it shows different behaviour on different time scales (confirm [12]). The vibration of the stiff linear springs takes place on the time scale ω^{-1}, while ω^0 is the time scale of the soft nonlinear springs' motion. Furthermore, on the time scale ω, energy exchanges among the stiff springs. For the simulations, we consider 6 point masses (i.e. $l = 3$) with mass $m = 1$ and the stiffness of the stiff springs is $\omega = 50$. The system has an initial displacement $q_1^s(0) = 1$ and an initial extension $q_1^f(0) = \omega^{-1}$, initial velocities are $\dot{q}_1^s(0) = 1$ and $\dot{q}_1^f(0) = 1$. All remaining initial values are zero.

In this simulation, the quadrature (16) with $\alpha = 1$ is used for the slow potential and the midpoint rule (14) for the fast one. As a reference solution, a standard variational integrator ($p = 1$) with the time step $\Delta T = 0.01$ is used. This time step is small enough to resolve the fast oscillations of the stiff springs' extensions. In the left hand side plot in Fig. 3, the configuration and momentum of the first slow and the first fast variable (i.e. the first stiff spring's centre and the length of the first stiff spring) are shown. Using a bigger time step $\Delta T = 0.3$, the fast motion cannot be captured anymore as can be seen on the right hand side of Fig. 3.

Keeping a macro time step of $\Delta T = 0.3$, the multirate variational integrator is used for a different number of intermediate micro steps. In Fig. 4, micro (red solid) and macro (blue dashed) solutions for configuration and momentum of the first slow and the first fast variable are shown for $p = 10$ micro steps on the left and for $p = 30$

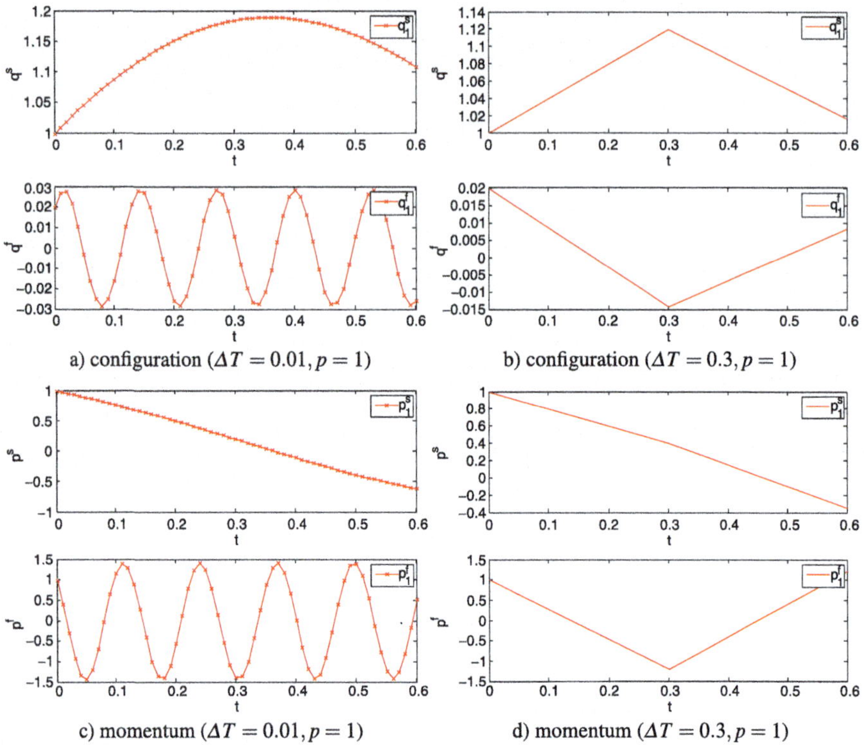

Fig. 3 FPU problem. Simulation results using a standard variational integrator ($p = 1$) with time step $\Delta T = 0.01$ (*left*) and $\Delta T = 0.3$ (*right*). Configuration (**a**, **b**) and momentum (**c**, **d**) of first slow (*top*) and first fast (*bottom*) variable

micro steps on the right. For an increasing number of micro steps, the approximation of the fast variables becomes better. For $p = 30$, the micro step size $\Delta t = 0.01$ is equal to the step size of the standard variational integrator in the reference solution. As a result, the discrete solution of the fast variable nicely coincides with the reference solution although the macro solution alone (red solid) does not resolve the fast dynamics.

In Fig. 5, the exchange of energy between the stiff springs (blue dashed, black dash-dotted, cyan dashed) is shown. The total oscillatory energy, i.e. the sum of the stiff springs' energy (red solid) remains close to a constant value (this is called an adiabatic invariant of the Hamiltonian system, see [12]) which is nicely visible in Fig. 5(a) for the reference solution. Using the macro time step $T = 0.3$ and $p = 1$ (Fig. 5(b)), the total energy oscillates much more and cannot be considered a constant value anymore. However, for $p = 10$ (Fig. 5(c)) and $p = 30$ (Fig. 5(d)) micro steps the oscillations become smaller, and for $p = 30$ the same qualitative long term energy behaviour as for the reference solution is obtained.

Computational costs for different examples of quadrature rules are depicted in Fig. 6. Computation times for the simulation of $t_N = 30$ seconds are shown in the

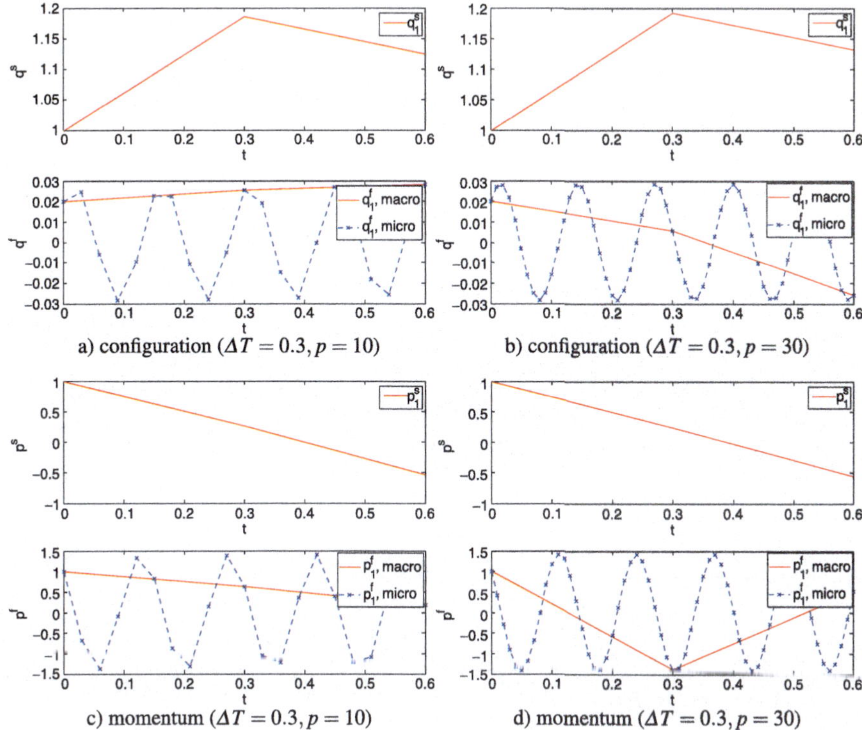

Fig. 4 FPU problem. Simulation results using a multirate variational integrator with macro time step $\Delta T = 0.3$ and $p = 10$ (*left*) and $p = 30$ (*right*) micro steps. Configuration (**a, b**) and momentum (**c, d**) of first slow (*top*) and first fast (*bottom*) variable

left hand side plot for the fully implicit scheme (Example 5: (14) and (15)) and on the right hand side for a quadrature leading to an explicit treatment of the fast potential and an implicit treatment of the slow potential (Example 4: (14) and (13)). All simulations are based on a constant micro time step of $\Delta t = 0.01$. For an increasing number p of micro steps per macro time step (thus for an increasing macro step $\Delta T = p\Delta t$), the computational costs decrease as expected. Thus, the resolution for the fast dynamics stays constant and fine enough, while the overall computational costs get lower since the number of slow potential evaluations decreases.

5.2 *Triple Spherical Pendulum*

For the triple spherical pendulum in Fig. 7, the slow variable $q^s = q_1 \in \mathbb{R}^3$ is the placement of the large mass ($m_1^{\text{slow}} = 100$), while $q^f = (q_2, q_3) \in \mathbb{R}^6$ contains the placements of the two smaller masses ($m_2^{\text{fast}} = m_3^{\text{fast}} = 2$). The slow potential energy reads $V(q) = q^T \cdot M \cdot \bar{g}$ with the constant mass matrix $M \in \mathbb{R}^{9\times 9}$ and the gravity

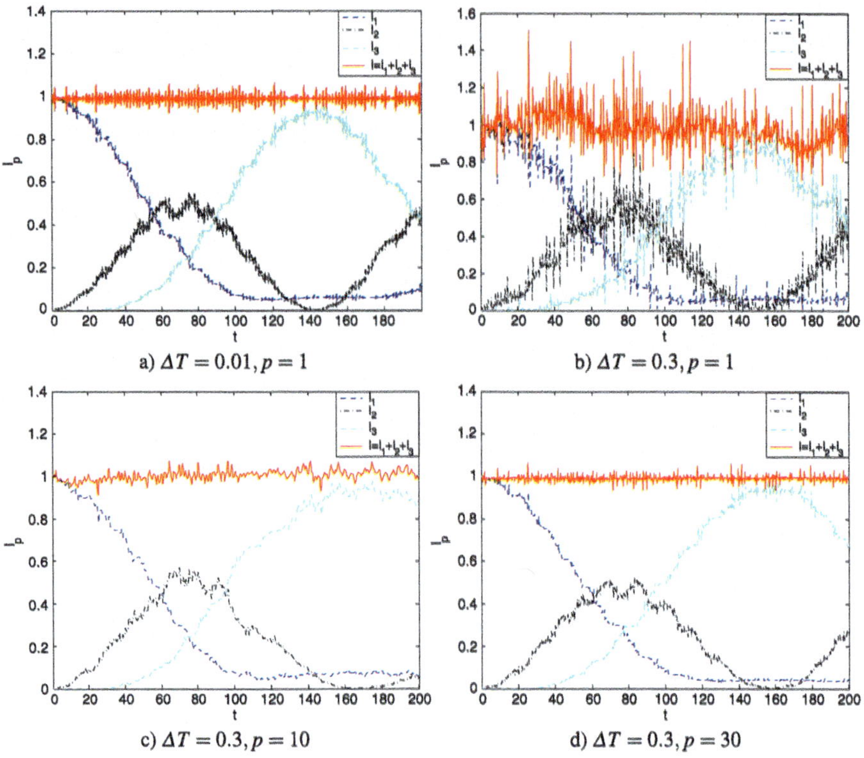

Fig. 5 FPU problem. Energy of the three stiff springs (*blue dashed*, *black dash-dotted*, *cyan dashed*) and the total oscillatory energy (*red solid*)

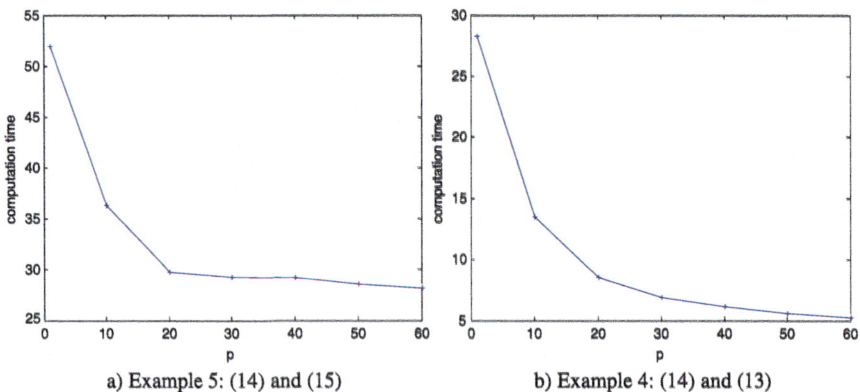

Fig. 6 FPU problem. Computation time for the simulation of $t_N = 30$ seconds based on a constant micro time step $\Delta t = 0.01$ and an increasing number p of micro nodes per macro step ΔT

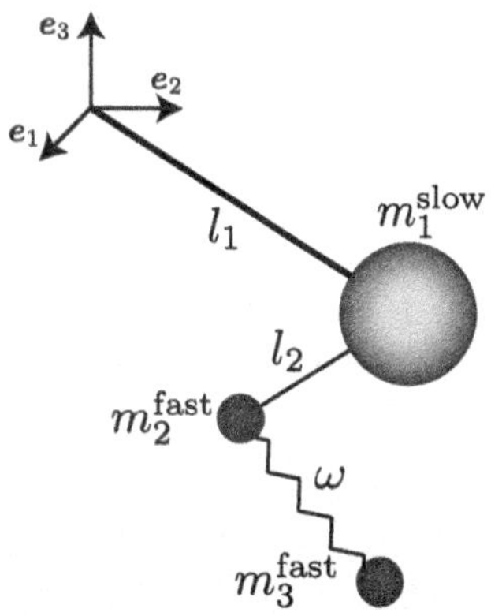

Fig. 7 Triple pendulum consisting of one large slow and two small fast masses

vector $\bar{g} \in \mathbb{R}^9$ acting in the negative e_3-direction with acceleration 9.81. Massless rigid links of lengths $l_1 = 20$ and $l_2 = 3$ connect the large mass to the origin and the first small mass to the large one, respectively. They give rise to a purely slow constraint $g^s(q^s) = \frac{1}{2}(q_1^2 - l_1^2)$ and a constraint $g^{sf}(q) = \frac{1}{2}((q_2 - q_1)^2 - l_2^2)$ coupling the slow and the first fast mass. Both constraints are combined into the vector valued constraint function $g = (g^s, g^{sf})$. The second small mass is connected to the first one by a linear spring with the stiffness $\omega = 5000$, thus the fast potential takes the form $W(q^f) = \frac{1}{2}\omega((q_3 - q_2)^2 - l_3^2)$ where $l_3 = 3$ is the length of the unstretched spring. Initially, the triple pendulum is aligned with the e_1-axis and the spring is prestretched by 2. The slow mass has an initial velocity of $\dot{q}^s(0) = (0, 2, -3)$ and the fast masses' initial velocity is $\dot{q}^f(0) = (0, 3l_2, -l_2, l_2 + l_3, 5(l_2 + l_3), -(l_2 + l_3))$.

In the simulation of the triple pendulum's dynamics, the midpoint evaluation (14) is used in the fast potential. Since the gravity potential is a linear function, it yields a constant force vector, which is independent of the choice of quadrature. The two left hand side plots in Fig. 8 show the evolution of the configuration and conjugate momentum of the second fast mass, being computed via a standard variational integrator ($p = 1$) with $\Delta T = 0.001$ as a reference solution. The right hand side plots show the results from the variational multirate scheme with $\Delta T = 0.08$ and $p = 5$, while the corresponding results for $p = 10$ and $p = 20$ are depicted in Fig. 9. The lines connect the values at the macro nodes and the intermediate micro node values are indicated by little crosses. One can see clearly, that the macro grid with $\Delta T = 0.08$ is too coarse to resolve the fast motion. For an increasing number of micro nodes, the fast oscillations of the second small mass become more and more visible. Finally, a numerical indicator for the variational character of the proposed method is given in Fig. 10. The triple pendulum's Lagrangian is invariant with respect to rotation about the gravitational axis, thus the corresponding angular momentum component L_3 is conserved exactly along the trajectory. The algorithm does conserve L_3 to numerical accuracy, independent of the macro or micro time step size.

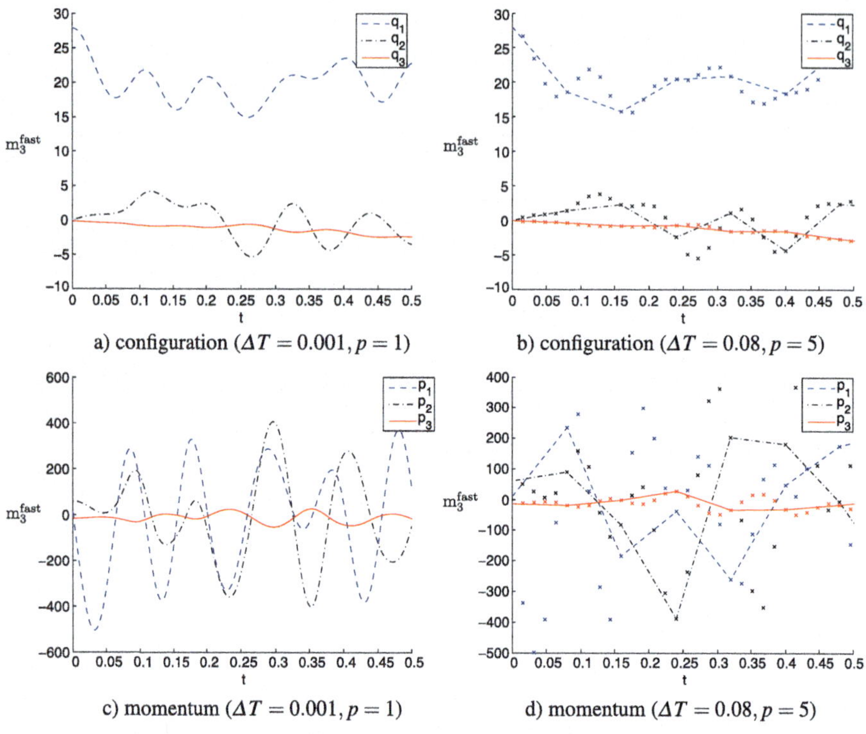

Fig. 8 Triple pendulum. Simulation results using a standard variational integrator ($p = 1$) with time step $\Delta T = 0.001$ (*left*) and a multirate variational integrator with macro time step $\Delta T = 0.08$ and $p = 5$ (*right*) micro steps. Configuration (**a**, **b**) and momentum (**c**, **d**) of second fast mass m_3^{fast}

6 Conclusion

A unified framework for the derivation of different multirate integrators for constrained dynamical systems is presented. All schemes are derived in closed form via a discrete variational principle on a time grid consisting of macro and micro time nodes. Being based on a discrete version of Hamilton's principle, the resulting variational multirate integrators are symplectic and momentum preserving integration schemes and also exhibit good energy behaviour. The choice of quadrature in the slow and fast potentials of the system can be adapted to the simulation goal like e.g. a low number of function evaluations of a costly potential or obtaining a partly of fully explicit scheme. In particular, if the number of micro nodes is large enough, fast oscillations can be resolved without solving for the slow variables on the micro grid. This leads to savings in the computational costs. This unified variational framework allows the analysis of a large class of multirate schemes, which has to be done in future work with particular focus on stability problems caused by resonance phenomena.

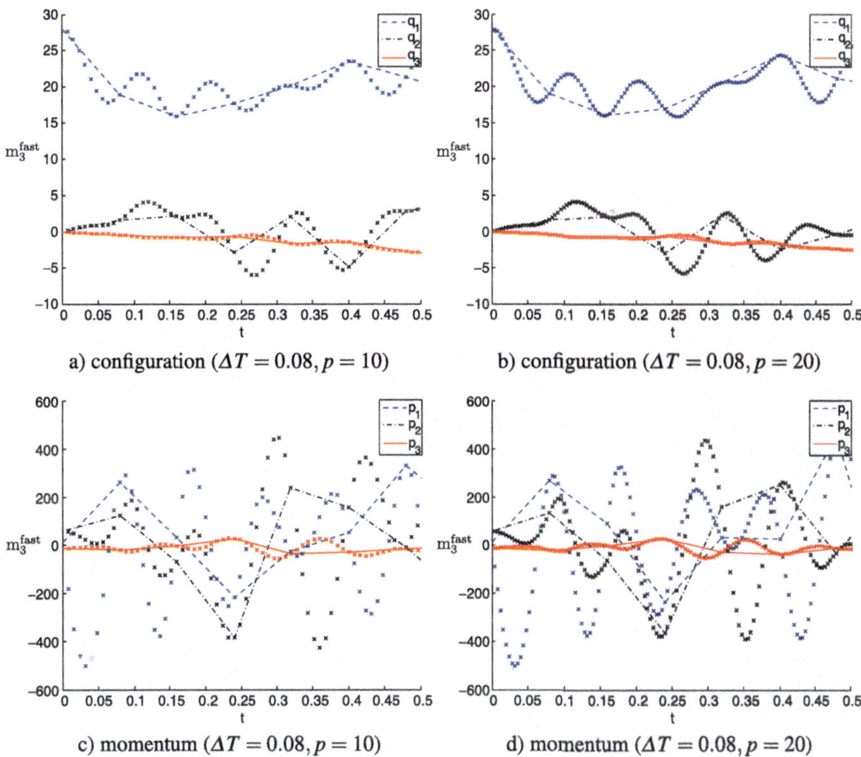

Fig. 9 Triple pendulum. Simulation results using a multirate variational integrator with macro time step $\Delta T = 0.08$ and $p = 10$ (*left*) and with $p = 20$ (*right*) micro steps. Configuration (**a**, **b**) and momentum (**c**, **d**) of second fast mass m_3^{fast}

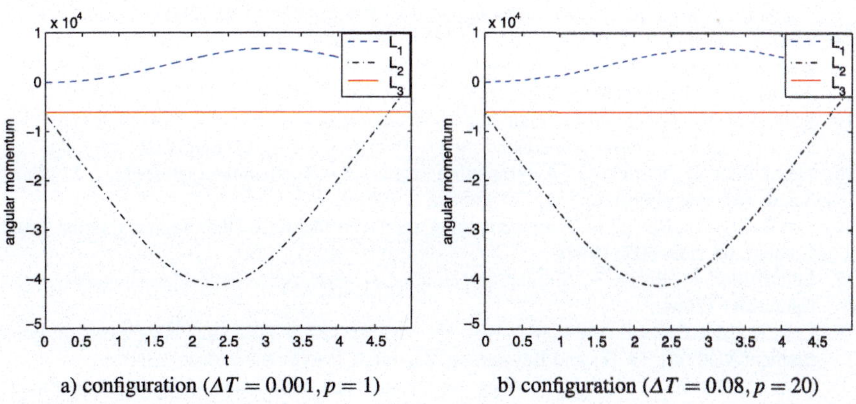

Fig. 10 Triple pendulum. Evolution of angular momentum using a standard variational integrator ($p = 1$) with time step $\Delta T = 0.001$ (*left*) and a multirate variational integrator with macro time step $\Delta T = 0.08$ and $p = 20$ (*right*) micro steps

Acknowledgements This chapter was partly developed and published in the course of the Collaborative Research Centre 614 "Self-Optimizing Concepts and Structures in Mechanical Engineering" funded by the German Research Foundation (DFG) under grant number SFB 614.

References

1. Abraham, R., Marsden, J.E., Ratiu, T.: Manifolds, Tensor Analysis, and Applications. Springer, New York (1988)
2. Arnold, M.: Multi-rate time integration for large scale multibody system models. In: Proceedings of the IUTAM Symposium on Multiscale Problems in Multibody System Contacts, Stuttgart, Germany (2006)
3. Barth, E., Schlick, T.: Extrapolation versus impulse in multiple-timestepping schemes. II. Linear analysis and applications to Newtonian and Langevin dynamics. J. Chem. Phys. **109**, 1633–1642 (1998)
4. Betsch, P., Leyendecker, S.: The discrete null space method for the energy consistent integration of constrained mechanical systems. Part II: Multibody dynamics. Int. J. Numer. Methods Eng. **67**(4), 499–552 (2006)
5. Bou-Rabee, N., Owhadi, H.: Stochastic variational integrators. IMA J. Numer. Anal. **29**, 421–443 (2008)
6. Cohen, D., Jahnke, T., Lorenz, K., Lubich, C.: Numerical integrators for highly oscillatory Hamiltonian systems: a review. In: Analysis, Modeling and Simulation of Multiscale Problems, pp. 553–576 (2006)
7. Fetecau, R.C., Marsden, J.E., Ortiz, M., West, M.: Nonsmooth Lagrangian mechanics and variational collision integrators. SIAM J. Appl. Dyn. Syst. **2**(3), 381–416 (2003)
8. Fong, W., Darve, E., Lew, A.: Stability of asynchronous variational integrators. J. Comput. Phys. **227**, 8367–8394 (2008)
9. Ge, Z., Marsden, J.E.: Lie-Poisson Hamilton-Jacobi theory and Lie-Poisson integrators. Phys. Lett. A **133**(3), 134–139 (1988)
10. Gear, C.W., Wells, R.R.: Multirate linear multistep methods. BIT **24**, 484–502 (1984)
11. Günther, M., Kærnø, A., Rentrop, P.: Multirate partitioned Runge-Kutta methods. BIT Numer. Math. **41**(3), 504–514 (2001)
12. Hairer, E., Wanner, G., Lubich, C.: Geometric Numerical Integration: Structure-Preserving Algorithms for Ordinary Differential Equations. Springer, New York (2004)
13. Kane, C., Marsden, J.E., Ortiz, M., West, M.: Variational integrators and the Newmark algorithm for conservative and dissipative mechanical systems. Int. J. Numer. Methods Eng. **49**(10), 1295–1325 (2000)
14. Kobilarov, M., Marsden, J.E., Sukhatme, G.S.: Geometric discretization of nonholonomic systems with symmetries. Discrete Contin. Dyn. Syst., Ser. S **1**(1), 61–84 (2010)
15. Leimkuhler, B., Patrick, G.: A sympletic integrator for Riemannian manifolds. J. Nonlinear Sci. **6**, 367–384 (1996)
16. Leimkuhler, B., Reich, S.: Symplectic integration of constrained Hamiltonian systems. Math. Comput. **63**, 589–605 (1994)
17. Leimkuhler, B., Reich, S.: Simulating Hamiltonian Dynamics. Cambridge University Press, Cambridge (2004)
18. Lew, A., Marsden, J.E., Ortiz, M., West, M.: An overview of variational integrators. In: Finite Element Methods: 1970's and Beyond, pp. 85–146. CIMNE, Barcelona (2003)
19. Lew, A., Marsden, J.E., Ortiz, M., West, M.: An overview of variational integrators. In: Franca, L.P., Tezduyar, T.E., Masud, A. (eds.) Finite Element Methods: 1970's and Beyond, pp. 98–115. CIMNE, Barcelona (2004)
20. Lew, A., Marsden, J.E., Ortiz, M., West, M.: Variational time integrators. Int. J. Numer. Methods Eng. **60**(1), 153–212 (2004)

21. Leyendecker, S., Marsden, J.E., Ortiz, M.: Variational integrators for constrained dynamical systems. Z. Angew. Math. Mech. **88**, 677–708 (2008)
22. Leyendecker, S., Ober-Blöbaum, S., Marsden, J.E., Ortiz, M.: Discrete mechanics and optimal control for constrained systems. Optim. Control Appl. Methods **31**(6), 505–528 (2010)
23. Marsden, J.E., Ratiu, T.S.: Introduction to Mechanics and Symmetry. Springer, New York (1994)
24. Marsden, J.E., West, M.: Discrete mechanics and variational integrators. Acta Numer. **10**, 357–514 (2001)
25. McLachlan, R., Quispel, G.: Geometric integrators for ODEs. J. Phys. A **39**(19), 5251–5286 (2006)
26. Ober-Blöbaum, S., Junge, O., Marsden, J.E.: Discrete mechanics and optimal control: an analysis. ESAIM Control Optim. Calc. Var. **17**(2), 322–352 (2010)
27. Reich, S.: Momentum conserving symplectic integrations. Physica D **76**(4), 375–383 (1994)
28. Stern, A., Grinspun, E.: Implicit-explicit integration of highly oscillatory problems. Multiscale Model. Simul. **7**, 1779–1794 (2009)
29. Striebel, M., Bartel, A., Günther, M.: A multirate ROW-scheme for index-1 network equations. Appl. Numer. Math. **59**, 800–814 (2009)
30. Tao, M., Owhadi, H., Marsden, J.E.: Nonintrusive and structure preserving multiscale integration of stiff ODEs, SDEs, and Hamiltonian systems with hidden slow dynamics via flow averaging. Multiscale Model. Simul. **8**(4), 1269–1324 (2010)
31. Verhoeven, A., Tasić, B., Beelen, T.G.J., ter Maten, E.J.W, Mattheij, R.M.M.: BDF compound-fast multirate transient analysis with adaptive stepsize control. J. Numer. Anal. Ind. Appl. Math. **3**(3–4), 275–297 (2008)
32. Weinan, E., Engquist, B., Li, X., Ren, W., Vanden-Eijnden, E.: Heterogeneous multiscale methods: a review. Commun. Comput. Phys. **2**(3), 367–450 (2007)

Symbolic Sensitivity Analysis of Multibody Systems

Joydeep M. Banerjee and John McPhee

Abstract Sensitivity analysis is the process of apportioning changes in the response of the system into the perturbations of the system parameters. In this chapter, we will present an overview of various issues regarding sensitivity analysis of multibody systems using symbolic formulations. Symbolic formulation for multibody system simulation has been demonstrated to have a number of benefits (Samin and Fisette in Symbolic Modeling of Multibody Systems, 2004), and has proven itself to be a very important tool for efficient sensitivity analysis. We present a detailed literature review of the subject, highlighting the challenges and the diverse applications of sensitivity analysis. We identify direct differentiation as a key approach towards symbolic sensitivity for its simplicity and accuracy. We will discuss software implementation and issues regarding the efficiency of the process of sensitivity analysis. We will present an overview of generation of sensitivity equations and using numerical examples, we will outline different approaches towards the evaluation of sensitivity information.

1 Introduction

Sensitivity analysis refers to the study of changes in system behavior brought about by the changes in entities inherent to the system. Mathematically, it is a problem of finding the derivative of a function with respect to the system parameters. To illustrate the importance of sensitivity analysis, it is worthwhile to look into the range of engineering applications that uses sensitivity analysis.

Importance Analysis The knowledge about the effect of parameter perturbation on a predefined objective function or a performance measure is important for many industrial applications and manufacturing processes. For example, during the manufacturing process, stricter quality control needs to be imposed on the components

J.M. Banerjee (✉) · J. McPhee
Department of Systems Design Engineering, University of Waterloo, Waterloo, Ontario, Canada
e-mail: jbanerjee83@gmail.com

J. McPhee
e-mail: mcphee@uwaterloo.ca

J.-C. Samin, P. Fisette (eds.), *Multibody Dynamics*,
Computational Methods in Applied Sciences 28,
DOI 10.1007/978-94-007-5404-1_6,

that have greater influence on the overall performance of the system. Importance analysis can be used to determine which parameters have greater effect on a particular criterion than others. In other words, importance analysis captures the essence of sensitivity analysis, which is to relate uncertainties in the values of the model parameters to the uncertainties in a desired objective function.

Process Sensitivity Efficient management of any process requires knowledge of the key issues that affect the output of the process. For processes where an analytical model is available, symbolic sensitivity studies can provide invaluable insight about the critical parameters and optimization strategies.

Design and Optimization of Physical Systems From the simple mechanism that closes the door to the complex spacecraft that sends humans to the moon, proper design makes the difference between a complete failure and "a giant leap for mankind". For most practical systems of the modern age, iterative analysis and design computation is expensive in terms of human time and resources. For this reason, an optimization process is highly desired. Efficient optimization routines require gradient information, which are essentially sensitivity data.

Model Simplification Model simplification is an attempt to capture the important system behaviors by using a bare minimum of model entities and ignoring the rest of the features. It is important for analysis, control design and simulation of complex large scale systems. Essentially the model simplification problem reduces to the problem of knowing which feature is important for the purpose. Sensitivity analysis provides a systematic way to evaluate the effect of each entity on the desired behavior.

Robust Design To achieve robust design for dynamic systems, one has to minimize the deviation of the system from its intended behavior due to changes in the varying parameters. The problem essentially becomes a minimization problem where gradient information or sensitivity data greatly improves efficiency and accuracy.

Parameter Identification Parameter identification finds the parameters for a given system model to make it match a measured behavior as closely as possible. Mathematically it is a problem of minimization of the difference between the model prediction and the measured data. Sensitivity data enables the implementation of efficient techniques to deal with this problem.

Optimal Control Sensitivity analysis is also used while designing optimal controls, to identify control parameters and to study the behavior or the system with changes in control parameters.

In light of the above discussion, it is quite clear how important it is to have an efficient and accurate algorithm for sensitivity analysis. Unfortunately, for practical systems, the sensitivity study is a very complicated problem. To illustrate further on the topic, a mathematical description of the process is required at this point.

2 The Formulation of Sensitivity Analysis

For a multibody system, the most general form of the governing equation is a set of differential-algebraic equations or DAEs as shown in Eq. (1).

$$\begin{aligned} \mathbf{M}\ddot{\mathbf{q}} + \boldsymbol{\Phi}_{\mathbf{q}}^{\mathrm{T}}\boldsymbol{\lambda} &= \mathbf{Q} \\ \boldsymbol{\Phi} &= 0 \end{aligned} \tag{1}$$

In the above equation, $\mathbf{q}$ is the vector of state variables, $\boldsymbol{\Phi}$ is the vector of constraint equations, $\boldsymbol{\Phi}_{\mathbf{q}}$ is the Jacobian of the constraint vector and $\mathbf{Q}$ is the vector of forcing functions, which are functions of the state variables, its derivatives and the model parameters. The vector of Lagrange multipliers enforcing the constraints that evolve from the closed loops of multibody systems is denoted by $\boldsymbol{\lambda}$.

To perform sensitivity analysis on a system, we need to define some sort of a measure for the physical characteristics that we want to study. For this reason we define what is known as an objective function. For a general multibody system governed by Eq. (1), a performance measure can be formulated as shown in Eq. (2).

$$\psi = G\big((\mathbf{q}, \dot{\mathbf{q}})|_{t=T}, \mathbf{p}, T\big) + \int_{t_0}^{T} F(\mathbf{q}, \dot{\mathbf{q}}, \boldsymbol{\lambda}, \mathbf{p}, t)\,\mathrm{d}t \tag{2}$$

In the above equation G and F represent arbitrary functions with sufficient smoothness, $\mathbf{p}$ represents the set of parameters, and $\mathbf{q}$, $\dot{\mathbf{q}}$ and $\boldsymbol{\lambda}$ are the generalized displacements, velocities and the Lagrange multipliers respectively. The final time for simulation is assumed to be dependent on the model parameters, i.e. $T = y(\mathbf{p})$.

The term G is a function of the final time T, the model parameters $\mathbf{p}$ and the values of $\mathbf{q}$ and $\dot{\mathbf{q}}$ evaluated at $t = T$. The term F is an objective function in an integrated form where the integration is performed over a time span with the final time being a function of the model parameters $\mathbf{p}$.

The sensitivity of the objective function ψ can be obtained by differentiating the expression with respect to $\mathbf{p}$. Using the chain rule of differentiation on Eq. (2) we obtain an expression for the sensitivity $\mathbf{S}$. For convenient illustration, we use subscripts to denote partial differentiation operation.

$$f_b = \frac{\partial f}{\partial b} \quad \text{and} \quad \mathbf{S} = \frac{\mathrm{d}\psi}{\mathrm{d}\mathbf{p}} \tag{3}$$

$$\begin{aligned} \mathbf{S} = {} & G_{\mathbf{p}} + G_T y_{\mathbf{p}} + G_{\mathbf{q}}\big(\mathbf{q}_{\mathbf{p}}|_{t=T} + (\dot{\mathbf{q}}|_{t=T})y_{\mathbf{p}}\big) + G_{\dot{\mathbf{q}}}\big(\dot{\mathbf{q}}_{\mathbf{p}}|_{t=T} + (\ddot{\mathbf{q}}|_{t=T})y_{\mathbf{p}}\big) \\ & + \int_{t_0}^{T} (F_{\mathbf{q}}\mathbf{q}_{\mathbf{p}} + F_{\dot{\mathbf{q}}}\dot{\mathbf{q}}_{\mathbf{p}} + F_{\lambda}\boldsymbol{\lambda}_{\mathbf{p}} + F_{\mathbf{p}})\,\mathrm{d}t + (F|_{t=T})y_{\mathbf{p}} \end{aligned} \tag{4}$$

Thus, to evaluate $\psi_{\mathbf{p}}$, one needs to evaluate $y_{\mathbf{p}}$, $\mathbf{q}_{\mathbf{p}}$, $\dot{\mathbf{q}}_{\mathbf{p}}$ and $\boldsymbol{\lambda}_{\mathbf{p}}$, i.e. the derivatives of T, $\mathbf{q}$, $\dot{\mathbf{q}}$ and $\boldsymbol{\lambda}$ with respect to the parameters respectively.

There are different numerical, analytical and hybrid methods which can evaluate these derivatives. The divided difference method, automatic differentiation techniques, direct differentiation and adjoint variable method have been used by many researchers to perform sensitivity analysis. Each of these methods has its own sets of advantages and disadvantages. For proper application of sensitivity analysis, it is important to correctly identify the advantages and disadvantages of these methods and consequently the applications where they can be used efficiently.

It is also necessary to address the issues encountered by these methods to improve their suitability in different engineering applications. In the subsequent sections, details of some of the existing methods are discussed with examples and simulation results from kinematic and dynamic problems.

Many researchers who work in this area often face questions from individuals, not acquainted with the challenges of sensitivity analysis, regarding the justification of research efforts towards better methods of sensitivity analysis, which after all is just a process that evaluates derivatives of certain quantities. To address this query, it is important to emphasize the ever increasing complexity of the situation. With the rapid increase in the size and complexity of the models being analyzed, the process of providing efficient and accurate sensitivity information suitable for diverse engineering applications is never "*just*" a differentiation.

3 Literature Review

Sensitivity analysis can be classified into two basic categories. Based on the domain in which their results stay valid, sensitivity analysis can be described as either a global or a local study.

The term "global sensitivity analysis" was introduced by noted econometrician Edward Leamer [34]. In successful global sensitivity analysis, the conclusions remain valid for the entire range of values of the parameters [32]. Mathematically, global sensitivity analysis can be performed by variance based methods like high dimensional model representation [45] or sampling based methods like elementary effect method and Monte Carlo filtering.

However, for practical multibody systems, these methods can be extremely inefficient. Due to the requirement of large number of simulations, sampling and statistical sensitivity analysis is often not suitable and much better performance can be obtained by employing what is known as local sensitivity analysis.

On the other hand, local sensitivity analysis gives results that are valid only in the neighborhood of the point at which they are evaluated. Since multibody systems are non-linear in nature, local sensitivity information are generally not valid for the entire range of possible parameter values. As a result, conclusions based on local sensitivity analysis need a specification of operating point to become meaningful.

Direct differentiation, the adjoint variable method, automatic differentiation and other analytical formulations can be used to evaluate local sensitivity information.The fact that these methods yield locally valid information makes these approaches suitable for applications where exp These methods are suitable for optimization algorithms and have been used in a wide range of engineering applications. Haug and Serban [49] have stated

> Dynamic design sensitivity analysis of multi-body systems represents the link between optimization tools and simulation tools.

Finite Difference Formulation One of the easiest methods of local sensitivity analysis for multibody systems is to use a finite difference scheme to evaluate the gradient information of the objective function with respect to the parameters. This method is probably the simplest to implement and can easily be extended to evaluate derivatives of higher orders. However it is often not worthwhile for practical implementation because of its dependence on the perturbation size. Furthermore, researchers [2, 5, 8] have shown that extra computational cost is necessary to determine optimal perturbations for the parameters for different scenarios.

Direct Differentiation The objective of local sensitivity analysis for a multibody system is to evaluate the derivative of the state variables representing the system. For models where symbolic equations are available, this can be accomplished by a method known as the direct differentiation.

To perform sensitivity analysis on these models, one approach is to find a stepping stone between the system equations governing the state variables and the sensitivity equations governing the sensitivities of the state variables. The simplest method to derive the set sensitivity equations is to differentiate the governing equations symbolically. This is precisely the approach followed by the method of direct differentiation [16, 33, 49].

In direct differentiation, a set of auxiliary equations known as the sensitivity equations are generated from the original system equations by differentiating them with respect to the model parameters. By solving these sensitivity equations, the corresponding sensitivity information is obtained as functions of time.

To illustrate this method, we consider a system governed by a set of nonlinear equations as shown in Eq. (5), where $\mathbf{q}$ is a vector of state variables, $\mathbf{p}$ is the vector of model parameters and $\mathbf{f}$ is a vector of sufficiently smooth functions.

$$\mathbf{f}(\mathbf{q}, \mathbf{p}, t) = 0 \tag{5}$$

We also introduce a set of objective functions as given in Eq. (6).

$$\mathbf{g}(\mathbf{q}, \mathbf{p}, t) = 0 \tag{6}$$

In Eqs. (5) and (6), the number of state variables and the number of model parameters are n and m respectively i.e., $\mathbf{q} \in \mathrm{R}^n$ and $\mathbf{p} \in \mathrm{R}^m$. The vectors $\mathbf{f}$ and $\mathbf{g}$ have n and k functions respectively, $\mathbf{f} : \mathrm{R}^{m+n} \to \mathrm{R}^n$ and $\mathbf{g} : \mathrm{R}^{m+n} \to \mathrm{R}^k$.

For this scenario, the objective of sensitivity analysis is the evaluation of the derivative of $\mathbf{g}$ with respect to the model parameters $\mathbf{p}$. Using matrix notations, the expression for this derivative is shown in Eq. (7).

$$\mathbf{S} = \frac{\mathrm{d}\mathbf{g}}{\mathrm{d}\mathbf{p}} = \mathbf{g}_{\mathbf{q}}\mathbf{q}_{\mathbf{p}} + \mathbf{g}_{\mathbf{p}} \tag{7}$$

The terms $\mathbf{g}_{\mathbf{q}}$ and $\mathbf{g}_{\mathbf{p}}$ in Eq. (7) can be derived from the structure of the vector $\mathbf{g}$. However, the $n \times m$ matrix $\mathbf{q}_{\mathbf{p}}$ is unknown and can only be evaluated from the original model. To calculate the matrix $\mathbf{q}_{\mathbf{p}}$ we differentiate Eq. (5) with respect to the model parameter vector $\mathbf{p}$ and obtain the following equation.

$$\mathbf{f}_{\mathbf{q}}\mathbf{q}_{\mathbf{p}} + \mathbf{f}_{\mathbf{p}} = 0 \tag{8}$$

Equation (8) is matrix of $n \times m$ equations that can be solved to evaluate the $n \times m$ elements of the matrix $\mathbf{q_p}$. The matrix $\mathbf{q_p}$ can then be substituted in Eq. (7) to evaluate the required sensitivity information.

Direct differentiation is portable, easy to implement, stable, and produces results that are numerically exact. This method has been used to perform sensitivity analysis for systems governed by kinematic equations [38], ordinary differential equations [14, 43, 47] and also differential-algebraic equations [48]. Serban and Freeman [48] have performed simultaneous sensitivity analysis with respect to multiple parameters. It is widely used to perform parameter identification [47] and design optimization [20].

The drawback of direct differentiation is also apparent from the presented example. To evaluate a sensitivity matrix $\mathbf{S}$ of $k \times m$ elements, direct differentiation requires the solution of $n \times m$ equations, (8). Usually, for practical scenarios, the number k is much smaller than n or m. This means, using direct differentiation method, one needs to solve a much larger set of equations to get a smaller set of required information.

The size and the complexity of Eq. (8) depends on the nature of the governing equation $\mathbf{f}$, the size of the vector of state variables $\mathbf{q}$ and the number of model parameters $\mathbf{p}$. For large systems, especially while dealing with a large number of parameters, this makes direct differentiation unsuitable for implementation. To address this issue, an alternate method using adjoint variables was developed.

Adjoint Variable Method The process of finding the gradient of a function involving the state variables of a system with respect to a system parameter can be described as the process of optimizing an objective function, where the quantities are constrained by the set of governing system equations. Lagrange's method of optimization can be used to convert the constrained optimization problem into an unconstrained optimization problem using a set of *adjoint variables*. This is known as the adjoint variable method.

Mathematically, the adjoint variable method can be demonstrated by considering the system governed by Eq. (5). To evaluate the sensitivity of the objective function vector given in Eq. (7), the first step in adjoint variable method is to introduce the adjoint variables by multiplying Eq. (8) with the transpose of a $n \times k$ matrix $\boldsymbol{\lambda}$.

$$\boldsymbol{\lambda}^{\mathrm{T}}\mathbf{f_q}\mathbf{q_p} + \boldsymbol{\lambda}^{\mathrm{T}}\mathbf{f_p} = 0 \tag{9}$$

The adjoint equations are formed as shown in Eq. (10). The terms $\mathbf{f_q}$ and $\mathbf{g_q}$ are derived from the vectors $\mathbf{f}$ and $\mathbf{g}$ using symbolic differentiation.

$$\boldsymbol{\lambda}^{\mathrm{T}}\mathbf{f_q} = \mathbf{g_q} \tag{10}$$

By multiplying Eq. (10) with $\mathbf{q_p}$ we obtain

$$\boldsymbol{\lambda}^{\mathrm{T}}\mathbf{f_q}\mathbf{q_p} = \mathbf{g_q}\mathbf{q_p} \tag{11}$$

Combining Eqs. (11) and (9) we obtain

$$\mathbf{g_q}\mathbf{q_p} = -\boldsymbol{\lambda}^{\mathrm{T}}\mathbf{f_p} \tag{12}$$

The required sensitivity matrix can be then written as

$$\mathbf{S} = \frac{\mathbf{dg}}{\mathbf{dp}} = -\boldsymbol{\lambda}^{\mathrm{T}}\mathbf{f_p} + \mathbf{g_p} \tag{13}$$

With the values of $\boldsymbol{\lambda}$ evaluated by solving Eq. (10), the sensitivity information $\mathbf{S}$ can be easily evaluated by substitution.

Using adjoint variable method, the $k \times m$ elements of $\mathbf{S}$ are evaluated through a set of $n \times k$ adjoint variables. The number of adjoint variables required is independent of the number of model parameters. This makes this approach suitable for scenarios where a large number of model parameters are under study.

Physically, the adjoint variable method uses the Lagrange multipliers as a stepping stone to avoid the problems of direct differentiation. The adjoint variables force the governing equations to hold. Thus the goal switches from finding the derivatives of the governing equations to finding the adjoint variables which would make the system satisfy the governing equations.

The implementation of adjoint variable method is more complicated than that of direct differentiation. In case of dynamic systems, i.e. systems that are governed by ODEs or DAEs the generated adjoint systems become sets of ODEs and DAEs themselves.

Often, for practical applications, the objective functions are defined in an integrated form. For dynamic systems, this leads to a situation, where the adjoint system becomes a terminal value problem, instead of an initial value problem.

Theoretically, this can be solved by simulating the governing equations in the forward time direction and use the simulation data to solve the adjoint system in the backward direction. This poses a challenge for the practical implementation of this method.

Most efficient numerical integrators use adaptive step size selection method. If adaptive step size solvers are used, the time points, were the actual quadrature is evaluated, never match up for the forward and backward simulations. To address this problem, the interpolation polynomials associated with every time points of the forward problem need to be stored along with the response of the system.

From this discussion, the drawbacks of adjoint variable method can be summarized. Its complex implementation and requirement of data storage makes it unsuitable for large systems, and for longer simulations. Also, for complicated objective functions, adjoint variable method generates more complicated equations that need to be solved.

The adjoint variable method has been extensively used for optimal control and optimal design by Haug [28] and Bestle et al. [5, 6]. For control applications, the model size is usually small, which makes adjoint variable method appropriate for these applications.

Sandu et al. [41], Hindmarsh et al. [29] and Petzold et al. [11, 12, 39] have worked extensively on this subject and have developed numerical packages to perform sensitivity analysis using the adjoint variable method on systems governed by different types of DAEs and ODEs.

Ding et al. [21] have extended this method to perform second-order sensitivity studies on DAEs. Serban [46] has presented a parallel computational model based on the adjoint variable method.

Recursive Formulation Recursive formulations were originally used to generate efficient sets of governing equations for multibody systems [31].

To avoid the complicated and resource-hungry implementation of the adjoint variable method, Anderson and other researchers [1, 2, 7, 30, 36] have extended the formulation to generate sensitivity equations using recursive methods.

However, the implementation of the recursive approach is still complicated and shows its benefits mainly for systems with open kinematic chains. It is also suggested that there is a critical number of bodies that must be present in a serial chain to make the recursive algorithm effective. For most multibody systems, especially those encountered in vehicular systems, this critical number is higher than the maximum number of bodies connected in series.

Automatic Differentiation To simulate models of physical systems, numerical methods are often used to calculate the values of the state variables that describe the configuration of the system as functions of time. For most practical systems, the numerical methods involve some sort of an integrator to solve the different types of differential equations encountered in the models.

Automatic differentiation is a numerical tool which, when applied to any numerical method, can calculate the derivatives of the output of the numerical method with respect to any relevant quantity.

It is based on the fact that any numerical process, however complicated, can be broken down in smaller elemental operations. Therefore, using the chain rule of differentiation and a table of standard differentiation results, it is possible to construct the derivative of the original numerical process.

In case of sensitivity analysis, when automatic differentiation is applied to the integrator that simulates the system, it calculates the values of the state variables and their derivatives at different points of time.

This method of differentiation is completely different from the finite difference approach where the entire numerical process is performed repeatedly with perturbed values of the parameters.

Over the years, many numerical integration schemes have been developed to simulate systems governed by kinematic, differential [19] and differential-algebraic equations [15, 17]. Automatic differentiation has been successfully used to produce the state variables and their derivatives [4, 8, 10, 22, 27].

It is to be noted that in automatic differentiation the evaluation of the derivative is a numerical process. It's an augmentation of the numerical simulation routine that generates the sensitivity information. Therefore it does not benefit from the advantages of the symbolic approach towards sensitivity analysis and cannot formally be classified as a symbolic method.

Bond Graph-Based Sensitivity Analysis Bond graphs represent an unified approach toward modeling multi-domain physical systems. Since the bond graph method is capable of generating the governing equations in an automated fashion, sensitivity analysis based on a bond-graph formulation has received considerable attention. Cabanellas et al. [9] have formulated sensitivity analysis for bond graph models using the concept of pseudo-bond graphs. Gawthrop [25] has developed the "Sensitivity Bond Graph" to perform sensitivity analysis. Ronco and Gawthrop [26] and Perry et al. [37] have used similar methods to perform parameter estimation, optimization and uncertainty analysis for multi-domain mechatronic systems. Unfortunately, the bond graph method is not very effective when it comes to three dimensional multi-body systems [42]. Its application to sensitivity analysis of multibody systems is limited due to the complexities involved.

Sensitivity Analysis Based on Linear Graph Theory Savage [43] has presented a method to automatically generate sensitivity models of first and higher orders from a linear graph representing the system. Carr and Savage [14] have extended this method to include cases of nonlinear constitutive equations. These methods were restricted to steady-state problems and scalars were used to represent the variables.

From the discussion of the existing literature it is clear that as the size of the system increases, the established methods start to become more and more inefficient. The finite difference method becomes unstable for different perturbation sizes, direct differentiation becomes computationally expensive for large number of parameters, the adjoint variable method becomes inefficient for large systems, and automatic differentiation misses out on the benefits of symbolic simplification procedures. Furthermore, most of these methods require system specific considerations.

It is obvious that automated, accurate and efficient sensitivity analysis for large scale multibody and multidomain systems is still an open topic for research.

4 Software Implementation

There are commercial software packages that can perform sensitivity analysis on systems governed by ordinary differential equations (ODEs) [29] and differential-algebraic equations (DAEs) [11, 29].

Although these packages have been used to perform design and analysis of multibody systems, they are not geared towards multibody systems in general. Consequently, implementation of sensitivity analysis for multibody systems using these packages is somewhat complicated.

So far the process of symbolic sensitivity analysis of multibody systems has been a sequential procedure. Specialized softwares are used to generate the symbolic equations for the system, which are then solved using the available software packages to evaluate the system response and sensitivity information. The information generated can then be used for further analysis and post-processing like optimization or importance analysis. To improve the effectiveness of the process, a unified method of efficient sensitivity analysis is desired for multibody systems.

The automated generation of symbolic governing equations is a desirable feature for efficient sensitivity analysis of multibody systems. During the design phase of development, system models undergo numerous alterations e.g., addition or removal of components, changes in structure or assembly, changes in component properties etc. To perform accurate sensitivity analysis, the corresponding sensitivity equations need to be updated constantly. An automated generation of sensitivity equations can greatly improve the effectiveness of this process.

In this section, brief overviews of some of the existing software packages that are used for modeling multibody and/or multi-domain systems will be presented.

ROBOTRAN ROBOTRAN is a commercial software package developed by the Multibody Research Group at the Université Catholique de Louvain. It can automatically generate symbolic governing equations for any multibody system given a description of the system as the input [24].

The program was written in the C language and all the symbolic operators are hard-coded into the processing engine. It allows the final mathematical models to be exported as Fortran, C or Matlab routines.

A joint coordinate formulation is used extensively in ROBOTRAN. The coordinate partitioning method is used in case of algebraic constraints. Topological description of the system provides the algebraic constraints as the loop closure equations.

It has been successfully used for multibody modeling and simulation (articulated tramway bogie system, human body motion, cam follower systems and flexible multibody systems), optimization of closed-loop mechanisms and parameter identification. ROBOTRAN has proved itself to be effective in both research and industrial scenarios.

Maple Maple is a general-purpose computer algebra system developed and marketed by Waterloo Maple Inc. It incorporates a dynamically typed programming language which resembles Pascal. There are provisions of interfacing with C, Fortran, Java and Matlab. The heart of Maple is a kernel written in C. This provides the Maple language. Most mathematical functionalities are provided by libraries. The usual user interface is written in Java.

The main aspect of Maple is the ability to manipulate symbolic equations and expressions. It contains a very large library of symbolic operations. Simplification and modification of symbolic expression are routinely done by Maple.

Apart from the symbolic capabilities, Maple can also be used for numerical simulations. Advanced numerical routines allow users to solve complicated large systems of ODEs and DAEs using a variety of different algorithms.

Because of its excellent symbolic and numeric capabilities, Maple can be used to simulate multibody systems effectively. Mathematically it becomes the process of solving a set of ODEs or DAEs depending on the nature of the system. Also there are software packages specifically written for multibody system modeling that are based on Maple and Maple language.

DynaFlexPro DynaFlexPro was developed by McPhee and Schmitke at the University of Waterloo, Canada. It is capable of generating symbolic governing equations for multi-domain engineering systems. The formulation is based on linear graph theory and the principle of orthogonality [44]. The input to the program is a topological description of the system using predefined blocks.

DynaFlexPro is implemented using Maple. It extensively uses Maple's symbolic processing libraries to generate and simplify symbolic equations. It can also be used to generate very efficient simulation codes for different languages. The "CodeGeneration" package can produce optimized simulation codes for Maple, C, Matlab and Fortran.

MapleSim MapleSim is a multi-domain modeling and simulation tool developed by Maplesoft Inc. It is capable of simulating electrical, electronic, mechanical, hydraulic and magnetic systems. The input to the software is the description of the system using the components from a central library that can be drag-dropped on to a worksheet.

MapleSim is built on Maple, which enables it to perform very effective symbolic simplifications of the generated equations. Also, using Maple's DAE and ODE solvers, it can simulate and present the output of the models in an interactive three-dimensional environment. It can also perform post-processing on the generated data using predefined templates.

The multibody package of MapleSim is based on the DynaFlexPro engine and uses a linear graph based formulation. Other components (e.g. electrical, magnetic, thermal, hydraulic etc.) are based on Modelica codes.

Apart from the built-in library of standard components, MapleSim allows users to create custom components for user specific implementation. These custom components are based on Maple language and can be readily included in models created using MapleSim's standard components. It is also possible to create custom components based on Modelica codes.

5 Formulation of Direct Differentiation Equations

The main advantage of direct differentiation is its simplicity. It is easy to implement direct differentiation for systems governed by algebraic equations, ordinary differentiation equations or differential-algebraic equations. Direct differentiation is especially suitable for applications where explicit evaluation of the sensitivities of the state variables are required. In this section, application of direct differentiation to different types of equations will be considered.

Algebraic Equations For a system governed by kinematic equations, direct differentiation can be demonstrated by the following equations. In terms of the state variables $\mathbf{q}$, the set of model parameters $\mathbf{p}$ and time t, the set of equations governing the position, velocity and acceleration variables is given as

$$\begin{aligned} &\text{Position} && \boldsymbol{\Phi}(\mathbf{q}, \mathbf{p}, t) = 0 \\ &\text{Velocity} && \boldsymbol{\Phi}_{\mathbf{q}}\dot{\mathbf{q}} = -\boldsymbol{\Phi}_t = \boldsymbol{\nu} \\ &\text{Acceleration} && \boldsymbol{\Phi}_{\mathbf{q}}\ddot{\mathbf{q}} = -(\boldsymbol{\Phi}_{\mathbf{q}}\dot{\mathbf{q}})_{\mathbf{q}}\dot{\mathbf{q}} - 2\boldsymbol{\Phi}_{\mathbf{q}t}\dot{\mathbf{q}} - \boldsymbol{\Phi}_{tt} = \boldsymbol{\gamma} \end{aligned} \tag{14}$$

By differentiating (14) with respect to a single parameter b, we obtain three sets of linear equations governing the sensitivity of the position, velocity and acceleration variables.

$$\begin{aligned} &\boldsymbol{\Phi}_{\mathbf{q}}\mathbf{q}_b = -\boldsymbol{\Phi}_b \\ &\boldsymbol{\Phi}_{\mathbf{q}}\dot{\mathbf{q}}_b = -(\boldsymbol{\Phi}_{\mathbf{q}}\dot{\mathbf{q}})_{\mathbf{q}}\mathbf{q}_b - (\boldsymbol{\Phi}_{\mathbf{q}})_b\dot{\mathbf{q}} + \boldsymbol{\nu}_{\mathbf{q}}\mathbf{q}_b + \boldsymbol{\nu}_b \\ &\boldsymbol{\Phi}_{\mathbf{q}}\ddot{\mathbf{q}}_b = -(\boldsymbol{\Phi}_{\mathbf{q}}\ddot{\mathbf{q}})_{\mathbf{q}}\mathbf{q}_b - (\boldsymbol{\Phi}_{\mathbf{q}})_b\ddot{\mathbf{q}} + \boldsymbol{\gamma}_{\mathbf{q}}\mathbf{q}_b + \boldsymbol{\gamma}_{\dot{\mathbf{q}}}\dot{\mathbf{q}}_b + \boldsymbol{\gamma}_b \end{aligned} \tag{15}$$

In the above equations, the subscripts refer to partial differentiation operation.

By solving Eqs. (14) and (15) the state variables and all their derivatives can be directly evaluated.

Ordinary Differential Equations The general form for a set of ODEs can be written in terms of a mass matrix $\mathbf{M}$, force vector $\mathbf{Q}$, a vector of state variables $\mathbf{q}$, $\dot{\mathbf{q}}$ and $\ddot{\mathbf{q}}$ and a set of model parameter $\mathbf{p}$.

$$\mathbf{M}(\mathbf{q}, \mathbf{p})\ddot{\mathbf{q}}(\mathbf{p}, t) = \mathbf{Q}(\mathbf{q}, \dot{\mathbf{q}}, \mathbf{p}, t) \tag{16}$$

Using the chain rule of differentiation, we differentiate Eq. (16) with respect to a single model parameter $b \in \mathbf{p}$ and obtain

$$\mathbf{M}\ddot{\mathbf{q}}_b + \mathbf{M}_b\ddot{\mathbf{q}} + (\mathbf{M}\bar{\ddot{\mathbf{q}}})_{\mathbf{q}}\mathbf{q}_b = \mathbf{Q}_b + \mathbf{Q}_{\mathbf{q}}\mathbf{q}_b + \mathbf{Q}_{\dot{\mathbf{q}}}\dot{\mathbf{q}}_b \tag{17}$$

In Eq. (17), the subscripts denote the partial differentiation operation with respect to the quantities in the subscript. Also the symbol $\bar{\ddot{\mathbf{q}}}$ means that the quantity is being kept constant with respect to the parameter b, during the differentiation process. By rearranging terms we obtain

$$\mathbf{M}\ddot{\mathbf{q}}_b + (\mathbf{M}_b + \mathbf{P})\ddot{\mathbf{q}} = \mathbf{Q}_b + \mathbf{Q}_{\mathbf{q}}\mathbf{q}_b + \mathbf{Q}_{\dot{\mathbf{q}}}\dot{\mathbf{q}}_b \tag{18}$$

where we have used the following expressions.

$$\begin{gathered} (\mathbf{M}\bar{\ddot{\mathbf{q}}})_{\mathbf{q}}\mathbf{q}_b = \mathbf{P}\ddot{\mathbf{q}} \quad \text{where } P_{ij} = \sum_{k=1}^{n} \frac{\partial m_{ij}}{\partial \mathbf{q}_k}(\mathbf{q}_k)_b \\ m_{ij}\text{: elements of matrix } \mathbf{M} \end{gathered} \tag{19}$$

While performing sensitivity analysis with respect to multiple parameters, separate sensitivity equations can be generated for each of the parameters under study.

For m parameters, those equations can be written by combining Eqs. (16) and (18) into a more convenient matrix form as given below.

$$\begin{bmatrix} \mathbf{M}_{b_1}+\mathbf{P} & \mathbf{M} & \cdots & 0 \\ \vdots & \vdots & \ddots & \vdots \\ \mathbf{M}_{b_m}+\mathbf{P} & 0 & \cdots & \mathbf{M} \\ \mathbf{M} & 0 & \cdots & 0 \end{bmatrix} \begin{Bmatrix} \ddot{\mathbf{q}} \\ \ddot{\mathbf{q}}_{b_1} \\ \vdots \\ \ddot{\mathbf{q}}_{b_m} \end{Bmatrix} = \begin{Bmatrix} \mathbf{Q}_{b_1}+\mathbf{Q}_{\mathbf{q}}\mathbf{q}_{b_1}+\mathbf{Q}_{\dot{\mathbf{q}}}\dot{\mathbf{q}}_{b_1} \\ \vdots \\ \mathbf{Q}_{b_m}+\mathbf{Q}_{\mathbf{q}}\mathbf{q}_{b_m}+\mathbf{Q}_{\dot{\mathbf{q}}}\dot{\mathbf{q}}_{b_m} \\ \mathbf{Q} \end{Bmatrix} \tag{20}$$

If the total number of state variables is n, Eq. (20) will have a total of $n \times (m+1)$ differential equations.

Differential-Algebraic Equations Equation (1) shows the general form of a set of differential-algebraic equations frequently encountered while modeling multibody systems. By applying direct differentiation to Eq. (1) and using a similar grouping and rearrangement as shown in Eq. (19), we obtain the following equations.

$$\begin{gathered} \mathbf{M}\ddot{\mathbf{q}}_b+(\mathbf{M}_b+\mathbf{P})\ddot{\mathbf{q}}+\mathbf{\Phi}_{\mathbf{q}}^{\mathrm{T}}\boldsymbol{\lambda}_b+\left((\mathbf{\Phi}_b)_{\mathbf{q}}^{\mathrm{T}}+(\mathbf{\Phi}_{\mathbf{q}}\bar{\mathbf{q}}_b)_{\mathbf{q}}^{\mathrm{T}}\right)\boldsymbol{\lambda}=\mathbf{Q}_b+\mathbf{Q}_{\mathbf{q}}\mathbf{q}_b+\mathbf{Q}_{\dot{\mathbf{q}}}\dot{\mathbf{q}}_b \\ \mathbf{\Phi}_{\mathbf{q}}\mathbf{q}_b+\mathbf{\Phi}_b=0 \end{gathered} \tag{21}$$

Equation (21) is a set of differential and algebraic equations involving the state vector $\mathbf{q}$, the Lagrange multipliers $\boldsymbol{\lambda}$ and their corresponding sensitivities $\mathbf{q}_b$ and $\boldsymbol{\lambda}_b$.

Serban and Freeman have argued that, although Eq. (21) is a valid set of relationships, it is not a proper system of DAEs [48].

However, it is their opinion that, if Eqs. (21) and Eq. (1) are combined, the complete set becomes a proper set of differential-algebraic equations that can be numerically solved to generate the system response and the sensitivity information. The combined equations can be written in a compact form as shown below.

$$\begin{gathered} \tilde{\mathbf{M}}\ddot{\mathbf{r}}+\mathbf{\Pi}_{\mathbf{r}}^{\mathrm{T}}\boldsymbol{\mu}=\tilde{\mathbf{Q}} \\ \mathbf{\Pi}=0 \end{gathered} \tag{22}$$

where

$$\tilde{\mathbf{M}}=\begin{bmatrix} \mathbf{M}_b+\mathbf{P} & \mathbf{M} \\ \mathbf{M} & 0 \end{bmatrix}, \qquad \mathbf{\Pi}_{\mathbf{r}}^{\mathrm{T}}=\begin{bmatrix} \mathbf{\Phi}_{\mathbf{q}} & 0 \\ (\mathbf{\Phi}_b)_{\mathbf{q}}+(\mathbf{\Phi}_{\mathbf{q}}\bar{\mathbf{q}}_b)_{\mathbf{q}} & \mathbf{\Phi}_{\mathbf{q}} \end{bmatrix}^{\mathrm{T}} \tag{23}$$

And

$$\tilde{\mathbf{Q}}=\begin{Bmatrix} \mathbf{Q}_b+\mathbf{Q}_{\mathbf{q}}\mathbf{q}_b+\mathbf{Q}_{\dot{\mathbf{q}}}\dot{\mathbf{q}}_b \\ \mathbf{Q} \end{Bmatrix} \tag{24}$$

Also

$$\begin{gathered} \mathbf{\Pi}=\begin{bmatrix} \mathbf{\Phi} \\ \mathbf{\Phi}_{\mathbf{q}}\mathbf{q}_b+\mathbf{\Phi}_b \end{bmatrix}, \qquad \ddot{\mathbf{r}}=\begin{Bmatrix} \ddot{\mathbf{q}} \\ \ddot{\mathbf{q}}_b \end{Bmatrix}, \qquad \boldsymbol{\mu}=\begin{Bmatrix} \boldsymbol{\lambda}_b \\ \boldsymbol{\lambda} \end{Bmatrix} \quad \text{and} \\ P_{ij}=\sum_{k=1}^{n}\frac{\partial m_{ij}}{\partial \mathbf{q}_k}(\mathbf{q}_k)_b \end{gathered} \tag{25}$$

Serban and Freeman have also generalized this derivation for multiple parameters, and have presented the modified expressions for the matrices $\tilde{\mathbf{M}}$ and $\boldsymbol{\Pi}_{\mathbf{r}}$ and vectors $\tilde{\mathbf{Q}}$, $\boldsymbol{\mu}$ and $\mathbf{r}$ [48]. Unfortunately, by close observation, it can be seen that their formulation is not accurate.

According to their derivation, for sensitivity analysis with respect to 'm' number of parameters, the expressions are

$$\tilde{\mathbf{M}} = \begin{bmatrix} \mathbf{M}_{b_1} + \mathbf{P} & \mathbf{M} & \cdots & 0 \\ \vdots & \vdots & \ddots & \vdots \\ \mathbf{M}_{b_m} + \mathbf{P} & 0 & \cdots & \mathbf{M} \\ \mathbf{M} & 0 & \cdots & 0 \end{bmatrix}, \quad \tilde{\mathbf{Q}} = \begin{Bmatrix} \mathbf{Q}_{b_1} + \mathbf{Q}_{\mathbf{q}}\mathbf{q}_{b_1} + \mathbf{Q}_{\dot{\mathbf{q}}}\dot{\mathbf{q}}_{b_1} \\ \vdots \\ \mathbf{Q}_{b_m} + \mathbf{Q}_{\mathbf{q}}\mathbf{q}_{b_m} + \mathbf{Q}_{\dot{\mathbf{q}}}\dot{\mathbf{q}}_{b_m} \\ \mathbf{Q} \end{Bmatrix} \tag{26}$$

$$\boldsymbol{\Pi} = \begin{bmatrix} \boldsymbol{\Phi} \\ \boldsymbol{\Phi}_{\mathbf{q}}\mathbf{q}_{b_1} + \boldsymbol{\Phi}_{b_1} \\ \vdots \\ \boldsymbol{\Phi}_{\mathbf{q}}\mathbf{q}_{b_m} + \boldsymbol{\Phi}_{b_m} \end{bmatrix} \tag{27}$$

The modified vector of state variables and the Lagrange multipliers are given by

$$\begin{aligned} \mathbf{r} &= \left\{ \mathbf{q}^{\mathrm{T}} \quad \mathbf{q}_{b_1}^{\mathrm{T}} \quad \cdots \quad \mathbf{q}_{b_m}^{\mathrm{T}} \right\}^{\mathrm{T}} \\ \boldsymbol{\mu} &= \left\{ \boldsymbol{\lambda}_{b_1}^{\mathrm{T}} \quad \cdots \quad \boldsymbol{\lambda}_{b_m}^{\mathrm{T}} \quad \boldsymbol{\lambda}^{\mathrm{T}} \right\}^{\mathrm{T}} \end{aligned} \tag{28}$$

The Jacobian matrix $\boldsymbol{\Pi}_{\mathbf{r}}$ is given by

$$\boldsymbol{\Pi}_{\mathbf{r}} = \begin{bmatrix} \boldsymbol{\Phi}_{\mathbf{q}} & 0 & \cdots & 0 \\ \boldsymbol{\Phi}_1 & \boldsymbol{\Phi}_{\mathbf{q}} & \cdots & 0 \\ \vdots & \vdots & \ddots & \vdots \\ \boldsymbol{\Phi}_m & 0 & \cdots & \boldsymbol{\Phi}_{\mathbf{q}} \end{bmatrix} \tag{29}$$

where

$$\boldsymbol{\Phi}_j = (\boldsymbol{\Phi}_{b_j})_{\mathbf{q}} + (\boldsymbol{\Phi}_{\mathbf{q}}\bar{\mathbf{q}}_{b_j})_{\mathbf{q}}, \quad j = 1 \ldots m \tag{30}$$

If Eq. (29) is substituted in Eq. (22), the resulting sensitivity equations would reveal that the set of equations governing the sensitivities with respect to parameter b_1 is dependent on the sensitivities with respect to b_2 and so on. Also, one can make the observation that, in (26)–(27) the nature of the individual equations change with the number of parameters under study. This is not a plausible situation.

To correct the expressions for the matrices, symbolic sensitivity equations as derived in Eq. (21) are generated for the model parameters $b_1 \ldots b_m$. Subsequently, they are arranged in a matrix format which results in the following equations.

$$\begin{aligned} \tilde{\mathbf{M}}\ddot{\mathbf{r}} + \tilde{\boldsymbol{\Pi}}_{\mathbf{r}}^{\mathrm{T}}\boldsymbol{\mu} &= \tilde{\mathbf{Q}} \\ \boldsymbol{\Pi} &= 0 \end{aligned} \tag{31}$$

The matrix $\tilde{\mathbf{M}}$ and the vectors $\tilde{\mathbf{Q}}$, $\mathbf{\Pi}$, $\mathbf{r}$ and $\boldsymbol{\mu}$ are given by Eqs. (26)–(28) respectively. The matrix $\tilde{\mathbf{\Pi}}_{\mathbf{r}}$ is given by the following equation.

$$\tilde{\mathbf{\Pi}}_{\mathbf{r}} = \begin{bmatrix} \mathbf{\Phi}_{\mathbf{q}} & 0 & \cdots & 0 & 0 \\ 0 & \mathbf{\Phi}_{\mathbf{q}} & 0 & \cdots & 0 \\ \vdots & & \ddots & & \vdots \\ 0 & \cdots & 0 & \mathbf{\Phi}_{\mathbf{q}} & 0 \\ \mathbf{\Phi}_1 & \mathbf{\Phi}_2 & \cdots & \mathbf{\Phi}_m & \mathbf{\Phi}_{\mathbf{q}} \end{bmatrix} \tag{32}$$

where

$$\mathbf{\Phi}_j = (\mathbf{\Phi}_{b_j})_{\mathbf{q}} + (\mathbf{\Phi}_{\mathbf{q}}\bar{\mathbf{q}}_{b_j})_{\mathbf{q}}, \quad j = 1 \ldots m \tag{33}$$

Equations (31)–(33), when solved using numerical methods, will yield the correct sensitivity information for any number of parameters.

6 Sequential Sensitivity Analysis

One approach to solving the sensitivity equations is to store the dynamic simulation results and use them as a input to solve the sensitivity equations at a later stage. Early researchers have criticized this approach of solving sensitivity equations.

In case of ODEs, the generation of sensitivity equations e.g., (18) is straight forward. However, the approach of solving them using stored simulation results is ill-advised; since the stiffness of the original set of ODEs and the individual sets of sensitivity ODEs are different, the automatic step size selector routine in variable step size integrators would generally produce different time points for integration for the two systems. As a result, the accuracy of the solution of the sensitivity ODEs will be inaccurate.

In this section, we will demonstrate these concepts using a simple example.

6.1 Ordinary Differential Equations

For ODEs we take up the example of a single body pendulum represented by joint coordinates. In terms of the joint coordinate θ, acceleration due to gravity g, and the length of the link L, the governing system equation is

$$\begin{aligned} \ddot{\theta} + \frac{g}{L}\sin(\theta) &= 0 \\ \theta(0) &= \frac{\pi}{6} \end{aligned} \tag{34}$$

By differentiating this equation with respect to the parameter L we obtain the sensitivity equation.

$$\begin{aligned} \ddot{\theta}_L - \frac{g}{L^2}\sin(\theta) + \frac{g}{L}\cos(\theta)\theta_L &= 0 \\ \theta_L(0) &= 0 \end{aligned} \tag{35}$$

To solve the sensitivity equations in a sequential fashion the following steps can be taken. First the system is simulated by solving Eq. (34) and the values of θ are stored in a matrix. A computer procedure is written to extract the value of θ at any time point using linear interpolation of the stored values of θ. The quantity θ in Eq. (35) is replaced by this interpolation function and an ODE solver is used to computer $\theta_L(t)$.

Source of Error One source of numerical error in this setup is the accuracy of the interpolation. At any particular time point, the difference between the actual value and the interpolated value of $\theta(t)$ depends on how coarse or fine the grid is for the stored values. If the interpolation error is too much, or the grid is too coarse, the sensitivity results would definitely be affected.

In Sect. 7, the effect of grid resolution on the solution accuracy will be demonstrated by a numerical example.

6.2 Differential-Algebraic Equations

Although the sequential solution of the sensitivity equations for DAEs is quite complicated and direct attempts towards the solution of sensitivity DAEs have been demonstrated to be problematic [48], there are some efficient methods that have been developed to take advantages of the unique properties of the Jacobian matrices of the DAE systems.

In the "*Simultaneous Corrector Method*" presented by Maly and Petzold [35], trapezoidal rule is used to discretize the equations which are then solved by Newton-Raphson method. In this method the Jacobian matrix is approximated with its diagonal elements and unlike the usual approach of updating the Jacobian matrix at each time step, this method keeps the Jacobian matrix constant for a number of time steps without significant loss in solution accuracy.

The "*Staggered Direct Method*" presented by Caracotsios and Stewart [13] simplifies the solution of the sensitivity equations by reusing the Jacobian matrix used to solve the system equations. However this method requires that the time steps used for the two integrations are identical and the Jacobian matrix is calculated at each one of them.

To negotiate the problems of these methods, methods like "*Staggered Corrector Method*" [23] and "*Staggered Hybrid Method*" [18] have been applied with variable effectiveness.

The problem with differential-algebraic systems is the index of the DAEs. Most of the above mentioned methods were developed for index-1 systems. In multibody systems in general, index-3 systems are quite common. As such the applicability of these methods for sensitivity analysis of multibody system is somewhat restrictive.

However, it is conceivable that by using different formulations or by employing index reduction methods, it might be possible to use these methods for certain types of multibody systems effectively.

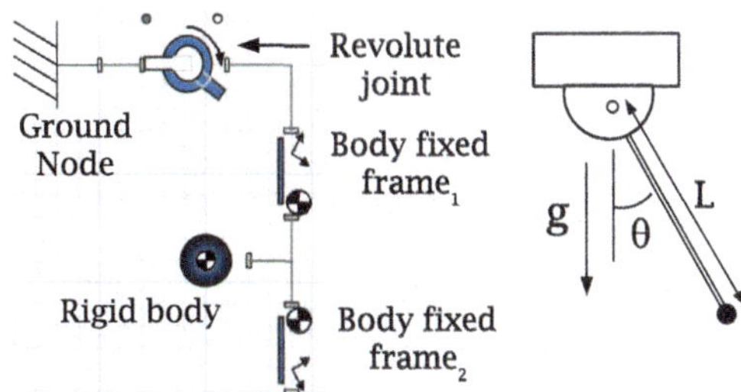

Fig. 1 Model description in MapleSim

7 Numerical Examples

In this section we present numerical examples to illustrate some of the methods and related issues, discussed in this chapter. Since, sensitivity analysis of ODEs and DAEs are structurally different, they are presented separately.

To illustrate the process of sensitivity analysis on a system governed by ODEs, we choose an example of a simple pendulum. For the example of a system governed by DAEs, a slider crank mechanism is chosen.

7.1 Sensitivity Analysis of a System Governed by ODEs

First we use MapleSim to model the system and extract symbolic equations from the model. Figure 1 shows the topological description of the system that acts as an input to the MapleSim program.

From MapleSim, we extract the governing equation in terms of the model parameters, (i.e. mass m, length L) and the state variable associated with the revolute joint $\theta(t)$. For the system shown in Fig. 1, MapleSim generates Eq. (34) as the governing equation.

The next step is to use computer algebra packages to simplify the generated equations and apply direct differentiation to generate the sensitivity equations. Using Maple as the tool for algebraic manipulations we obtain the sensitivity equations with respect to the model parameter L, Eq. (35). The final step is to solve Eqs. (34) and (35) in conjunction with proper initial conditions to evaluate the required sensitivity information.

Effect of Grid Resolution As mentioned in the previous section, while performing sequential sensitivity analysis using stored values of simulation data, storage grid resolution is expected to have an influence on solution accuracy. To demonstrate the effect of the resolution of the storage grid on the accuracy of the sensitivity analysis, the following numerical experiments were conducted.

Equations (34) and (35) were solved simultaneously. Next, only Eq. (34) was simulated and the results were stored in a matrix with different grid resolutions. The coarse grid was created by sampling the solution with a frequency of 1 Hz and the

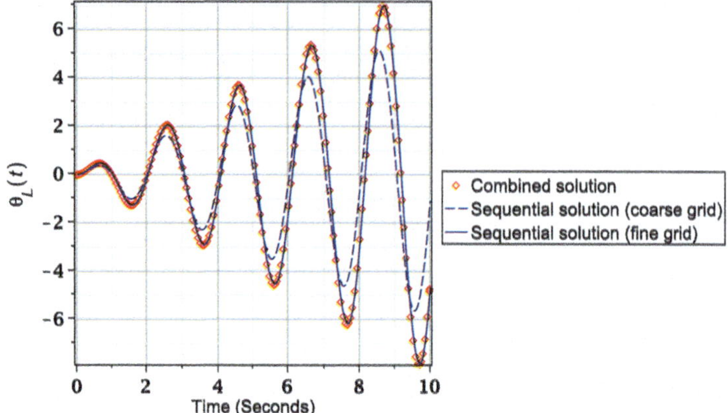

Fig. 2 Comparison of solution methods and effect of grid resolution

finer grid was created by sampling at 10 Hz. Using these stored values, Eq. (35) was solved and the results were compared with those obtained from the combined solution.

We have used Maple and its built in numerical ODE solver "dsolve [numeric]" implementing the *Runge-Kutta-Feldberg 4th and 5th order* algorithm.

In the first stage, while solving Eq. (34), "dsolve" was programmed to generate the solutions at specific values of the independent variable t. The data was stored in a matrix, which was used to generate an interpolation function for the quantity θ. In the second stage, while solving Eq. (35), the symbol θ was replaced by the interpolation function and "dsolve" was programmed to treat it as a known function.

The parameter values used for the simulation were [$g = 9.81\ \text{m/s}^2$, $L = 1$ m]. The integrator properties were set as follows [$abserr = 10^{-7}$, $relerr = 10^{-6}$, $maxfun = 3 \times 10^4$].

Figure 2 shows that for the coarse grid, the solution accuracy is quite poor. However, for a finer grid the two solution methods are found to be in good agreement with each other.

7.2 Sensitivity Analysis of a System Governed by DAEs

A slider crank mechanism is shown in Fig. 3. The angles θ_1 and θ_2 are selected as the generalized coordinates and the governing equations are generated in terms of these two state variables. Due to the closed kinematic chain, the system has one constraint equation which reduces the total degrees of freedom to one.

The centers of mass of the crank and the connecting rods are assumed to be at a distance b_1 and b_2 from the points A and B respectively. The masses, lengths and the centroidal moments of inertia of the elements are shown in the figure. The system is driven by an applied torque τ and is subjected to a linear rotational damper with

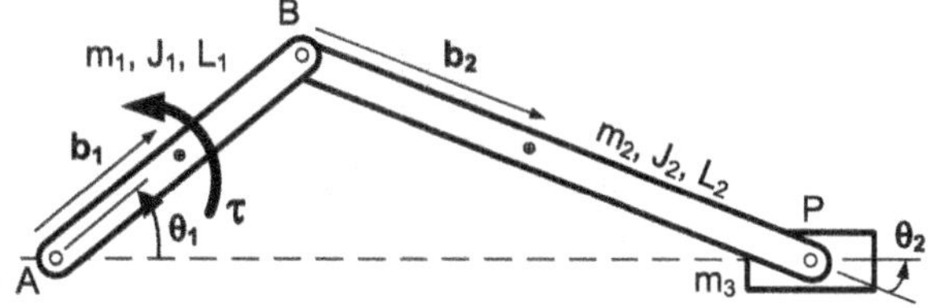

Fig. 3 Slider crank mechanism governed by DAEs

damping coefficient c at the hinge point A. The governing equations are generated using MapleSim and Maple is used for the analysis.

The set of system equations consists of two differential equations and one algebraic constraint equation. The dynamic equations can be expressed in a matrix format as described by Eq. (1). The expressions for the matrices are shown in Eq. (36).

$$\begin{bmatrix} m_{11} & m_{12} \\ m_{21} & m_{22} \end{bmatrix} \begin{bmatrix} \ddot{\theta}_1 \\ \ddot{\theta}_2 \end{bmatrix} + \begin{bmatrix} L_1 \cos\theta_1 \\ -L_2 \cos\theta_2 \end{bmatrix} \lambda = \begin{bmatrix} F_1 \\ F_2 \end{bmatrix} \tag{36}$$

where

$$\begin{aligned} m_{11} &= J_1 + (m_2 + m_3) {L_1}^2 + m_1 {b_1}^2 \\ m_{22} &= m_2 {b_2}^2 + m_3 {L_2}^2 + J_2 \\ m_{12} &= m_{21} = -L_1 (m_2 b_2 + m_3 L_2) \cos(\theta_1 + \theta_2) \\ F_1 &= \tau - c\dot{\theta}_1 - L_1 \dot{\theta}_2^2 \sin(\theta_1 + \theta_2)(m_3 L_2 + m_2 b_2) \\ &\quad - (m_1 b_1 + m_3 L_1 + m_2 L_1) \cos\theta_1 \\ F_2 &= \big(\cos\theta_2 - L_1 \dot{\theta}_1^2 \sin(\theta_1 + \theta_2)\big)(m_3 L_2 + m_2 b_2) \end{aligned} \tag{37}$$

The algebraic constraint equation, or the vector $\mathbf{\Phi}$ described in Eq. (1), is given by

$$L_1 \sin\theta_1 - L_2 \sin\theta_2 = 0 \tag{38}$$

In Eq. (36), λ is the Lagrange multiplier, which is a measure of the joint reaction force at the point P. The following parameter values were used for the simulation [$b_1 = 0.5$ m, $L_1 = L_2/2 = 1$ m, $b_2 = 0.66666$ m, $m_1, m_2, m_3 = 1$ kg, $c =$ 0.5 N s/rad, $\tau = 1.5$ N m, $g = 1$ m/s^2, $J_1 = 0.08333$ kg m^2, $J_2 = 0.44444$ kg m^2].

To demonstrate the method of sensitivity analysis using direct differentiation on DAEs, we try to evaluate $\theta_{1_\mathbf{p}}$, where $\mathbf{p}$ is the vector of model parameters. Equations (26)–(30) are derived for the system for different selections of vector $\mathbf{p}$. The generated set of DAEs are numerically solved in Maple v. 15 using built-in numerical routines. The "Modified Extended Backward Difference Method" is used for the integration and the following integrator parameters are used to achieve convergence [$abserr = 10^{-12}$, $relerr = 10^{-11}$].

Figure 4 plots the sensitivity of the crank angle θ_1 with respect to the model parameter b_2 on the vertical axis and the number of revolution of the crank on the horizontal axis. The plot shows that the result of the sensitivity analysis is different for different choices of the set of parameters. Also, it was observed that the results vary if the order of the parameters is changed. This clearly shows that Eqs. (26)–(30) do not yield correct results. As mentioned before, it is due to the incorrect formulation of the Jacobian matrix given in Eq. (29).

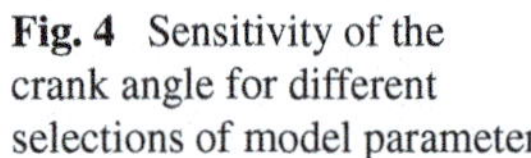

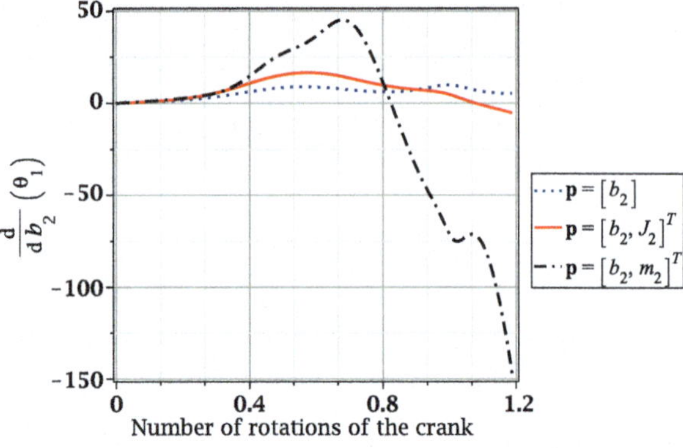

Fig. 4 Sensitivity of the crank angle for different selections of model parameter

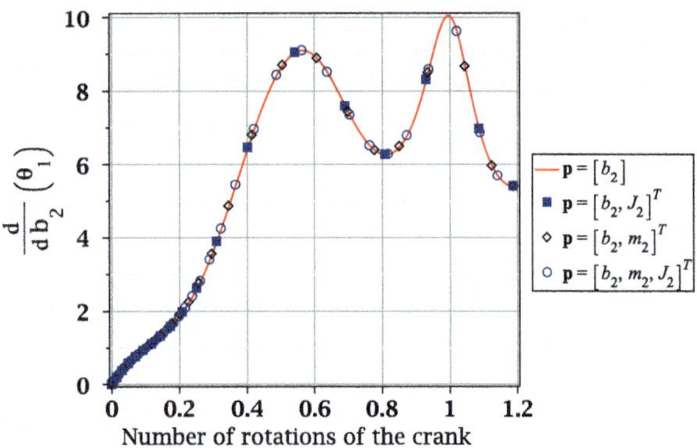

Fig. 5 Sensitivity of the crank angle for different selections of model parameters

To correct this problem we evaluate the sensitivities by numerically solving Eq. (31) and using the expression given in Eq. (32) for the Jacobian matrix.

Figure 5 demonstrates that Eq. (32) does produce accurate evaluation of sensitivity information. To validate the results generated by direct differentiation, we use finite difference formulation to evaluate the sensitivities and compare them to those evaluated using direct differentiation.

Figure 6 shows the difference between the direct differentiation results and finite difference results for the sensitivity of angle θ_1 with respect to the parameter b_2. The plot clearly demonstrates the validity of the method presented in this chapter.

8 Conclusion

Sensitivity analysis for multibody systems is an involved, highly specialized and fascinating area of research. At any point of time, the objective is to provide efficient, accurate and preferably automated evaluation of sensitivity information. Due to the ever increasing complexity of the systems being processed, the proverbial "bar" is always set at a higher level.

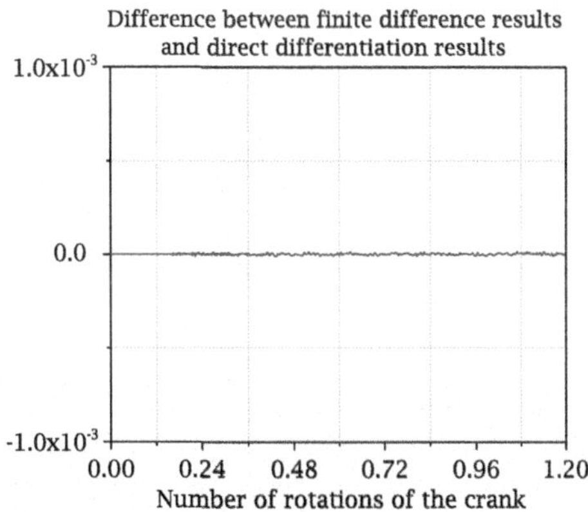

Fig. 6 Validation of results obtained by direct differentiation

To plan towards better sensitivity analysis, it is important to identify the steps involved. Generally, the process can be divided into three main stages.

The first stage is the generation of the underlying model of the system. Better and efficient models lead to more efficient post-processing. So, one way to improve sensitivity analysis is to improve the process that generates the governing equations. This is precisely where the modeling software come into the picture. Software packages like ROBOTRAN, MapleSim and DynaFlexPro have made huge improvements to the process of modeling multibody systems. Not only do they cut down on the difficulties of modeling and hence the time involved, they are also capable of simplifying the generated equations and contribute significantly towards improved sensitivity analysis.

The second stage is where sensitivity analysis is formulated. Over the years, various methods and formulations have been developed for this purpose. Different formulations have their own sets of advantages and disadvantages and often require system-specific considerations. That is why there is still a lot of scope for improvements in this subject.

The third stage is solving the formulated equations and using the simulation results for further analysis and processing. Better numerical solution algorithms lead to efficient sensitivity analysis. This is where the commercial solvers and integrators come in.

In practice, the second and third stages are often combined together. There are many software packages that formulate and solve the sensitivity equations. However, at the current state, there is a sizable discrepancy between the required and available amount of automation and efficiency in this process and it continues to be a research challenge.

We believe that some of the existing problems of the formulation of sensitivity analysis can be addressed by extending the modeling techniques to the process of generation of sensitivity equations. Some of the commercial modeling softwares use linear graph theory to efficiently model and simplify multibody system equations. The entire multibody component package of MapleSim is based on a graph-theoretic formulation. In our opinion, if we use graph-theory to generate sensitivity equations, there will be significant improvement in the process involved.

Graph-theoretic sensitivity analysis will be based on existing software tools and therefore it can be automated. Furthermore, the approach can benefit from the existing simplification techniques that are being used by these software packages to simplify system equations.

Automated graph-theoretic sensitivity analysis is an ongoing topic of research [3]. Future work is focused on the performance evaluation of this approach for various large scale multibody systems.

It seems unlikely that just by improving one particular stage of the process, it would be possible to reach the required level of efficiency and accuracy from sensitivity analysis. The prudent approach appears to be to improve the overall efficiency by coming up with better methods for modeling, formulation and numerical processing in a combined and well-coordinated fashion.

References

1. Anderson, K., Hsu, Y.: Order $(n+m)$ direct differentiation determination of design sensitivity for constrained multibody dynamic systems. Struct. Multidiscip. Optim. **26**, 171–182 (2004). doi:10.1007/s00158-003-0336-1
2. Anderson, K.S., Hsu, Y.: Analytical fully-recursive sensitivity analysis for multibody dynamic chain systems. Multibody Syst. Dyn. **8**, 1–27 (2002). doi:10.1023/A:1015867515213
3. Banerjee, J.M., Dao, T.S., McPhee, J.: Mathematical modeling and symbolic sensitivity analysis of Ni-MH batteries. In: Advanced Battery Technology, SP-2310. SAE International, Warrendale (2011)
4. Barrio, R.: Sensitivity analysis of ODES/DAES using the Taylor series method. SIAM J. Sci. Comput. **27**, 1929–1947 (2005)
5. Bestle, D., Eberhard, P.: Analyzing and optimizing multi-body systems. Mech. Struct. Mach. **62**, 181–190 (1992)
6. Bestle, D., Seybold, J.: Sensitivity analysis of constrained multibody systems. Arch. Appl. Mech. **62**, 181–190 (1992)
7. Bhalerao, K., Poursina, M., Anderson, K.: An efficient direct differentiation approach for sensitivity analysis of flexible multibody systems. Multibody Syst. Dyn. **23**, 121–140 (2010)
8. Bischof, C.H.: On the automatic differentiation of computer programs and an application to multibody systems. In: ICIAM/GAMM 95: Issue 1: Numerical Analysis, Scientific Computing, Computer Science, pp. 267–272 (1995)
9. Cabanellas, J.M., Félez, J., Vera, C.: A formulation of the sensitivity analysis for dynamic systems optimization based on pseudo bond graphs. In: Cellier, F.E., Granda, J.J. (eds.) Proceedings of the International Conference on Bond Graph Modeling and Simulation. Simulation Series, vol. 27, pp. 135–144. Society of Computer Simulation, Las Vegas (1995)
10. Campbell, S.L., Hollenbeck, R.: Automatic differentiation and implicit differential equations. In: Computational Differentiation: Techniques, Applications and Tools, pp. 215–227 (1996)
11. Cao, Y., Li, S., Petzold, L.: Adjoint sensitivity analysis for differential-algebraic equations: algorithms and software. J. Comput. Appl. Math. **149**(1), 171–191 (2002)
12. Cao, Y., Li, S., Petzold, L., Serban, R.: Adjoint sensitivity analysis for differential-algebraic equations: the adjoint DAE system and its numerical solution. SIAM J. Sci. Comput. **24**(3), 1076–1089 (2003)
13. Caracotsios, M., Stewart, W.E.: Sensitivity analysis of initial value problems with mixed odes and algebraic equations. Comput. Chem. Eng. **9**(4), 359–365 (1985)
14. Carr, S., Savage, G.J.: Symbolic sensitivity analysis of nonlinear physical systems using graph theoretic modeling. In: Mathematical Computation with Maple V. Ideas and Applications.

Proceedings of the Maple Summer Workshop and Symposium, pp. 118–127. University of Michigan, Ann Arbor (1995)

15. Cash, J.R.: Modified extended backward differentiation formulae for the numerical solution of stiff initial value problems in ODEs and DAEs. J. Comput. Appl. Math. **125**, 117–130 (2000)
16. Chang, C.O., Nikravesh, P.E.: Optimal design of mechanical systems with constraint violation stabilization method. J. Mech. Transm. Autom. Des. **107**, 493–498 (1985)
17. Chang, Y., Corliss, G.: ATOMFT: solving ODEs and DAEs using Taylor series. Comput. Math. Appl. **28**(1012), 209–233 (1994)
18. Chaniotis, D., Pai, M., Hiskens, I.: Sensitivity analysis of differential-algebraic systems using the GMRES method-application to power systems. In: The 2001 IEEE International Symposium on Circuits and Systems, ISCAS 2001, vol. 3, pp. 117–120 (2001)
19. Corliss, G., Chang, Y.F.: Solving ordinary differential equations using Taylor series. ACM Trans. Math. Softw. **8**, 114–144 (1982)
20. Dias, J., Pereira, M.: Sensitivity analysis of rigid-flexible multibody systems. Multibody Syst. Dyn. **1**, 303–322 (1997)
21. Ding, J.Y., Pan, Z.K., Chen, L.Q.: Second-order sensitivity analysis of multibody systems described by differential-algebraic equations: adjoint variable approach. Int. J. Comput. Math. **85**, 899–913 (2008)
22. Eberhard, P., Bischof, C.: Automatic differentiation of numerical integration algorithms. Math. Comput. **68**(226), 717–731 (1999)
23. Feehery, W.F., Tolsma, J.E., Barton, P.I.: Efficient sensitivity analysis of large-scale differential-algebraic systems. Appl. Numer. Math. **25**(1), 41–54 (1997)
24. Fisette, P., Postiau, T., Sass, L., Samin, J.C.: Fully symbolic generation of complex multibody models. Mech. Struct. Mach. **30**(1), 31–82 (2002)
25. Gawthrop, P.J.: Sensitivity bond graphs. J. Franklin Inst. **337**(7), 907–922 (2000)
26. Gawthrop, P.J., Ronco, E.: Estimation and control of mechatronic systems using sensitivity bond graphs. Control Eng. Pract. **8**(11), 1237–1248 (2000)
27. Griewank, A.: On automatic differentiation. In: Proceedings of Mathematical Programming: Recent Developments and Applications (1989)
28. Haug, E.J.: Design sensitivity analysis of dynamic systems. In: Computer Aided Optimal Design: Structural and Mechanical Systems. NATO ASI Series, vol. F27, pp. 705–755. Springer, London (1987)
29. Hindmarsh, A.C., Brown, P.N., Grant, K.E., Lee, S.L., Serban, R., Shumaker, D.E., Woodward, C.S.: SUNDIALS: suite of nonlinear and differential/algebraic equation solvers. ACM Trans. Math. Softw. **31**, 363–396 (2005)
30. Hsu, Y., Anderson, K.: Recursive sensitivity analysis for constrained multi-rigid-body dynamic systems design optimization. Struct. Multidiscip. Optim. **24**, 312–324 (2002)
31. Hwang, R.S., Haug, E.J.: Topological analysis of multibody systems for recursive dynamics formulations. Mech. Struct. Mach. **17**(2), 239–258 (1989)
32. Ingalls, B.: Sensitivity analysis: from model parameters to system behaviour. Essays Biochem. **45**, 177–193 (2008)
33. Krishnaswami, P., Wehage, R.A., Haug, E.J.: Design sensitivity analysis of constrained dynamic systems by direct differentiation. Tech. Rep. 83-5, Center for Computer-Aided Design, The University of Iowa, Iowa City (1983)
34. Leamer, E.E.: Let's take the con out of econometrics, and sensitivity analysis would help. In: Granger, C. (ed.) Modeling Economics. Clarendon, Oxford (1990)
35. Maly, T., Petzold, L.R.: Numerical methods and software for sensitivity analysis of differential-algebraic systems. Appl. Numer. Math. **20**(12), 57–79 (1996)
36. Mukherjee, R., Bhalerao, K., Anderson, K.: A divide-and-conquer direct differentiation approach for multibody system sensitivity analysis. Struct. Multidiscip. Optim. **35**, 413–429 (2008)
37. Perry, M., Atherton, M.A., Bates, R.A., Wynn, H.P.: Bond graph based sensitivity and uncertainty analysis modelling for micro-scale multiphysics robust engineering design. J. Franklin Inst. **345**(3), 282–292 (2008)

38. Pesch, V.J., Hinkle, C.L., Tortorelli, D.A.: Optimization of planar mechanism kinematics with symbolic computation. In: Proceedings of the IUTAM Symposium on Optimization of Mechanical Systems, pp. 221–230 (1996)
39. Petzold, L., Li, S., Cao, Y., Serban, R.: Sensitivity analysis of differential-algebraic equations and partial differential equations. Comput. Chem. Eng. **30**(10–12), 1553–1559 (2006)
40. Samin, J.C., Fisette, P.: Symbolic Modeling of Multibody Systems. Solid Mechanics and Its Applications, vol. 112. Springer, Berlin (2004)
41. Sandu, A., Daescu, D.N., Carmichael, G.R.: Direct and adjoint sensitivity analysis of chemical kinetic systems with KPP. Part 1: Theory and software tools. Atmos. Environ. **37**(36), 5083–5096 (2003)
42. Sass, L., McPhee, J., Schmitke, C., Fisette, P., Grenier, D.: A comparison of different methods for modelling electromechanical multibody systems. Multibody Syst. Dyn. **12**, 209–250 (2004)
43. Savage, G.J.: Automatic formulation of higher order sensitivity models. Civ. Eng. Syst. **10**(4), 335–350 (1993)
44. Schmitke, C., McPhee, J.: Using linear graph theory and the principle of orthogonality to model multibody, multi-domain systems. Adv. Eng. Inform. **22**, 147–160 (2008)
45. Schwieger, V.: Sensitivity analysis as a general tool for model optimisation—examples for trajectory estimation. J. Appl. Geodesy **1**(1), 27–34 (2007)
46. Serban, R.: A parallel computational model for sensitivity analysis in optimization for robustness. Optim. Methods Softw. **24**, 105–121 (2009)
47. Serban, R., Freeman, J.: Identification and identifiability of unknown parameters in multibody dynamic systems. Multibody Syst. Dyn. **5**, 335–350 (2001)
48. Serban, R., Freeman, J.S.: Direct differentiation methods for the design sensitivity of multibody dynamic systems. In: Proceedings of the 1996 ASME Design Engineering Technical Conferences and Computers in Engineering Conference, pp. 18–22 (1996)
49. Serban, R., Haug, E.J.: Kinematic and kinetic derivatives in multibody system analysis. Mech. Struct. Mach. **26**(2), 145–173 (1998)

Efficient Coarse-Grained Molecular Simulations in the Multibody Dynamics Scheme

Mohammad Poursina and Kurt S. Anderson

Abstract The numerical simulation of highly complex biomolecular systems such as DNAs, RNAs, and proteins become intractable as the size and fidelity of these systems increase. Herein, efficient techniques to accelerate multibody-based coarse-grained simulations of such systems are presented. First, an adaptive coarse-graining framework is explained which is capable of determining when and where the system model needs to change to achieve an optimal combination of speed and accuracy. The metrics to guide these on-the-fly instantaneous model adjustments and the issues associated with post-transition system's states are addressed in this book chapter. Due to its highly modular and parallel nature, the Generalized Divide-and-Conquer Algorithm (GDCA) forms the bases for a suite of dynamics simulation tools used in this work. For completeness, the fundamental aspects of the GDCA are presented herein. Finally, a novel method for the efficient and accurate approximation of far-field force and moment terms are developed. This aspect is key to the success of any large molecular simulation since more than 90 % of the computational load in such simulations is associated with pairwise force calculations. The presented approximations are efficient, accurate, and highly compatible with multibody-based coarse-grained models.

1 Introduction

Development and application of the efficient techniques to model, simulate, and analyze highly complex biomolecular systems such as DNAs (Deoxyribonucleic acid), RNAs (Ribonucleic acid), enzymes, and proteins have been gaining attention by scientists and engineers in an effort to predict and understand different structural, mechanical, and thermodynamic properties of such systems [17, 22, 44, 56, 71].

M. Poursina (✉) · K.S. Anderson
Computational Dynamics Laboratory (CDL), Department of Mechanical, Aerospace and Nuclear Engineering, Rensselaer Polytechnic Institute (RPI), Troy, NY 12180, USA
e-mail: poursm2@rpi.edu

K.S. Anderson
e-mail: anderk5@rpi.edu

J.-C. Samin, P. Fisette (eds.), *Multibody Dynamics*,
Computational Methods in Applied Sciences 28,
DOI 10.1007/978-94-007-5404-1_7,

These molecular simulations provide important information about the relationships between the structure and function of biopolymers, and provide insight into various biological processes. Fully atomistic representation of such systems [33, 43] results in detailed information on the underlying physics of these biopolymers and captures all scales of the problem. However, these simulations are greatly limited because they cannot be accomplished in a timely manner for models possessing the desired size and fidelity. This is due the existence of high frequency motions which impose a tight constraint on the temporal integration step size (0.5–2 fs) of explicit integrators [19, 75], while biologically important processes occur on time scales as slow as milliseconds to seconds [55]. Furthermore, these models with large number of atoms ($n_a \approx 10^6$ [23]) suffer from the cumbersome pairwise force field calculations with the computational complexity of $O(n_a{}^2)$ at each time step. As such, different methods have been developed to improve the temporal integration step size and reduce the computational cost per integration time step.

In biomolecular systems, high frequency and low amplitude motions of the atoms are responsible for local motions, while low frequency and high amplitude motions of the system's subdomains are dominant in representing the global conformation. Hence, the simulation performance may be improved significantly via the intelligent use of specialized coarse-grained models in which high frequency modes of motion of the system are removed from consideration. These models still capture the overall conformational motion of the biopolymers while allowing the application of larger integration time steps (e.g., 20–50 fs [19, 31, 75]).

The coarse-grained model may be realized by treating a group of atoms as a spherical bead (pseudo-atom) [16, 31, 66]. For instance, each nucleotide in a protein chain may be modeled using one to six beads [74]. Other applications of such spherical beads in dissipative particle dynamics and solvent lipid interactions are reported in [48, 73].

Alternatively, a group of atoms may be represented by a rigid or flexible body connected to its parent and child bodies via kinematic joints. Using internal coordinates (i.e., generalized coordinates which describe the relative motions of the child bodies with respect to the parent bodies at the connecting joints), the geometric constraints such as fixing bond lengths can be enforced exactly. In the finest coarse-grained articulated body model, dihedral angles (torsion dynamics) are used to describe the dynamics of the system [1, 36, 40, 72], while the bond stretch and bond angles are frozen. Unlike the spherical beads, the mass distribution and geometry of each articulated pseudo-atom is expressed in terms of the associated inertia tensor, and the distance from the corresponding mass center to the joints of the pseudo-atom. In such models, which may be particularly applicable to simulate the dynamic behavior of polymer chains, both translational and rotational motion [11, 18] of each cluster [36] are considered in forming the equations of motion. As such, the effect of Coriolis and centrifugal inertial forcing terms are considered in the equations of motion. As the length of the polymer chain increases, the role of these terms becomes more important in capturing the dynamics of the system due to scaling effects. Since these articulated models address the rotational motion of the rigid and flexible subdomains of the system in the equations of motion, they can capture the geometry of biopolymers more accurately than most of the bead-based coarse-grained models which ignore the rotational dynamics of the spherical pseudo-atoms.

It is demonstrated in [57, 63] that the dynamic behavior of biopolymers is highly nonlinear, and significantly affected by the change in the kinematics and dynamics of the boundary conditions of the system. As such, static (time-invariant) coarse-grained models may not appropriately capture the dynamics of the system for the entire course of the simulation. This requires the development of adaptive machinery to perform such simulations, particularly when the non-equilibrium behavior of these systems is of interest.

In the adaptive multiscale strategy presented here, some degrees of freedom (internal coordinates) have their definitions/meanings adjusted "on-the-fly" at different instants and different locations of the system based on the values of knowledge-, math-, and/or physics-based metrics. Herein, the appropriate metrics to guide these model transitions are investigated. Each model adjustment towards the lower or higher fidelity system model may be viewed as the instantaneous application or release of system's internal constraints. As such, the generalized momentum of the system must be conserved to arrive at the physically meaningful post-transition system's states. It is also demonstrated that within the transitions to the finer-scale models, some issues arise which are associated with the proper amount and placement of the energy within the system.

Given the central role that multibody dynamics plays in the presented framework, a suite of Generalized Divide-and-Conquer Algorithm-based approaches is employed to this end. These methods offer a good combination of computational efficiency and modular structure. Furthermore, the computational complexity of the algorithm is $O(n)$ and $O(\log n)$ in serial and parallel implementations, respectively, where n denotes the number of degrees of freedom of the system.

A key aspect of this book chapter is associated with pairwise force calculations in molecular simulations. More than 90 % of the computational cost per temporal integration step in modeling biopolymers is associated with calculating these forcing terms. Focusing on long-range (far-field) interactions in the system, various methods such as Barnes-Hut [5, 10], Edwald summation [21, 24], and the Fast Multipole Method [32] have been developed to reduce the cost of these forcing terms evaluations. A review on these methods is presented in [57]. Herein, a novel approximation for the far-field force and moment calculations that is well suited for use in an articulated body modeling of biopolymers is presented. This technique may be viewed as a generalization and extension of the method presented in [38] to approximate the gravitational force used in modeling spacecraft dynamics. The resultant force and moment due to the pairwise interactions are approximated between: the atoms (particles) embedded in a pseudo-atom (body) in the coarse-scale region and an atom (particle) which resides in the fine-scale domain; and between the particles embedded in two different pseudo-atoms (bodies) in the coarse-scale regions of the system. The value of the moment ignored in bead-based models can produce significant errors which is more accurately captured using multibody-based coarse-grained simulations. These low order multipole approximations and Taylor expansions are expressed in terms of the geometric and physical properties of the pseudo-atoms (subdomains). Since these properties are constant for rigid domains of the system, these approximations significantly accelerate the evaluation of the far-field forcing terms.

2 The Need for Adaptive Simulation of Molecular Systems

The dynamics of biopolymers is highly nonlinear and chaotic. Although these systems have highly chaotic components, their coarser scale conformational behavior may tend towards a specific structure. As such, it may be possible to remove high frequency modes of motion which contribute little to the final structure, while maintaining more important lower frequency components. However, given the complex nature of the system behavior, it may not be possible to identify which modes of motion can be removed a priori. For instance, the simulations conducted on articulated RNAs with various sequences in [57, 63] show that the dynamic behavior of each joint angle is highly time variant, and is significantly affected by the changes in the dynamics of the rest of the system. Those results demonstrate that the current coarse-grained model which is potentially correct for a specific time interval may not provide accurate and reliable information about the conformational motion of the system for the entire simulation. This results in the inadequacy of the static coarse graining in which the system model does not change within the course of the simulation. As such, adaptive multiscale methods which perform the coarse graining in time and space must be developed to better model the dynamics of biopolymers.

The adaptive machinery is capable of identifying the critical locations of the system to remove and/or add fidelity from/to the system model as necessary. In this process, some degrees of freedom of the system are adaptively constrained or released at different instants and different locations of the system. As such, this framework automatically adjusts the coarseness of the model, in an effort to more optimally increase the simulation speed, while maintaining accuracy. In the following, the necessary machinery to implement the adaptive multiscale simulation of biopolymers is presented. Development of the metrics to guide the model adjustments, efficiently modeling the forward dynamics, as well as appropriately handling the dynamics of the transitions are important aspects in this scheme. More detail on the adaptive multiscale framework to model biopolymers is found in [13, 57, 63].

3 Metrics to Steer Transitions

In situations where modification of the dynamics model is appropriate, these adjustments should be guided by suitable internal metrics. These metrics may be physics-based (derived directly from physical laws), knowledge-based (derived empirically) and/or math-based (derived from strictly mathematical relations). Herein, two different metrics which respectively guide the model transitions to the coarser and finer models are investigated.

3.1 Fine-to-Coarse Transitions

The transition to the coarser-scale model can be achieved via removing the less significant modes of motion of the system. This can be realized by monitoring the behavior of the individual internal coordinates of the system model to assess which of them may be removed, while not adversely changing the conformational behavior of the system. In molecular simulations, high frequency modes of motion do not contribute significantly to the global conformation of the system, while providing large instantaneous relative velocities and accelerations. As such, velocity- and acceleration-based metrics are not well-suited for identifying the more significant degrees of freedom of the system in the overall conformational motion. Furthermore, these complex systems are highly nonlinear and chaotic; therefore, the instantaneous values of the states of the system are not expected to (and have been shown not to [63]) work well for guiding model transitions.

Monitoring the moving-window statistical properties of the internal coordinates of the system is proposed in [63] as a math-based metric to assess and guide the coarse graining process instead of the instantaneous velocity- and acceleration-based measures described in [67, 70]. The standard deviation of the generalized coordinates defined at the joints (internal coordinates) collected within the moving window as given by

$$S_w = \sqrt{\frac{\sum_{k=1}^{n_w} (q_k - \bar{q}_w)^2}{n_w}} \tag{1}$$

is suggested as the metric of choice to determine if an existing joint should be kept or removed. In this relation, $\bar{q}_w$ is the moving-window average of the sequence of data $\{q_k\}_{k=1}^{n_w}$ within the window of the size of n_w.

It should be mentioned that the overall conformation of the system may be more sensitive to some specific internal coordinates, while this sensitivity varies with time [13, 57]. As such, if the scaled (weighted) moving-window standard deviation of any internal coordinate of the system is less than a predefined threshold, then the associated degree of freedom is eligible to be frozen.

3.2 Coarse-to-Fine Transitions

Since the static coarse-grained models may not appropriately predict the overall conformational motion of the biopolymer, it may be necessary to add fidelity to the system model within the course of the simulation. The system's constraints and internal loads as shown in Fig. 1 arise from the kinematic constraints imposed on adjacent body-to-body motions by the connecting joints, the interactions between the bodies, and the imposed boundary conditions. Furthermore, the constraint load indicates the degree to which the body or joint in question is attempting to be deformed at its location. Therefore, monitoring the spatial constraint loads (forces and torques)

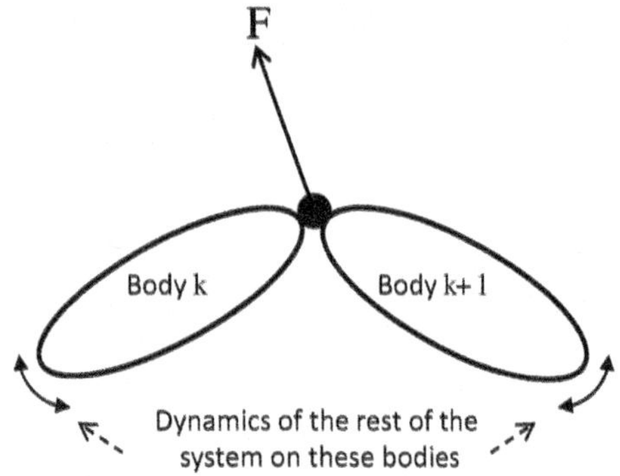

Fig. 1 The constraint load F gives an estimate of the error introduced because of locking the joint

acting on all kinematic joints of the system, and intermediate locations within the rigid and flexible bodies is the proposed metric to assess the validity of the selected coarse-grained model, and to guide the model refinement. In other words, the joint is released (or added) if the spatial constraint force at the associated location exceeds the nominal load which figuratively causes a mechanical failure [63].

4 Generalized Divide-and-Conquer Based Adaptive Framework

The simulation of the biopolymer in an adaptive framework should appropriately and efficiently addresses the forward dynamics, model adjustments, and dynamics of the transitions. In all of these steps, multibody dynamics plays an important role since herein the pseudo-atoms may be viewed as rigid and/or flexible domains of the system connected together via kinematic joints. As the complexity of these multibody systems (manifesting itself in the form of modes of motion) increases, a prohibitive computational burden can be imposed on the simulation due to the kinematic coupling which exists in most articulated multibody formulations. Different so-called $O(n)$ algorithms (as opposed to traditional $O(n^3)$ algorithms) in which the simulation turnaround time scales linearly with the increase in the system's degrees of freedom (i.e., n) have been developed in [2, 6, 8, 15, 25, 26, 35, 45, 54, 68, 69, 76, 77] as an effort to reduce this undesirable scaling in computational effort with problem size. These $O(n)$ methods are less costly for large n, but due to their underlying serial recursive under-pinnings, they generally do not lend themselves well to massive parallelization. As such, different algorithms have been designed to model and simulate multibody systems which better exploit the parallel computing capability [7, 9, 20, 29, 30, 34, 39, 42].

The Divide-and-Conquer Algorithm (DCA) is a recursive method of modeling multibody systems first demonstrated by Featherstone [27, 28]. Its recursive structure is not serial, but that of a binary tree. Different extensions and applications of this method in modeling and sensitivity analysis of the multi-rigid/flexible-body systems, studying the impulsive behavior and contact problems in such systems, as well as the parallel implementation of this algorithm are reported in [12–14, 46, 47, 49–53]. In this scheme, a complete set of both absolute and generalized coordinates are used to form the equations of motion. As such, the analyst can select a set of appropriate coordinates to integrate and work with. Applying both spatial and generalized coordinates, and directly imposing the constraints describing the kinematics of the joints to form the equations of motion of the system, the DCA provides a robust framework to address the dynamics of the kinematically closed-loop systems in singular configurations [50]. A Generalized Divide-and-Conquer Algorithm (GDCA) developed in [57, 60] is an extension to the DCA which can be easily used to model multibody systems in which a part of the forcing information is provided in a generalized force format. This may occur in modeling the systems in which a set of known/unknown generalized forces must be considered in the equations of motion due to the application of the control law or the imposition of the algebraic constraints [58, 59]. For instance, this method is used in [57, 58] to perform the constant temperature simulation of biopolymers in which the feedback forces from the thermostat are provided in the generalized format.

Herein, the GDCA-based methods are suggested to be used in the context of the large-scale adaptive molecular problems because:

1. They are relatively efficient for the large-scale sequential computer implementation. The computational complexity of these algorithms for unconstrained systems is $O(n)$ in the serial implementation.
2. These methods are highly parallelizable which provide a time optimal order $\log n$ computational performance achieved with a processor optimal order n processors.
3. The implementation and use of these formulations within an adaptive framework are relatively straightforward due to the algorithm's highly modular structure.

4.1 Forward Dynamics

Similar to the DCA, the dynamics of each body in the GDCA scheme is expressed in terms of the handle equations of motion [50]. A handle is a point of the body through which it has an interaction with its surroundings. Herein, the handle equations in the GDCA scheme are presented to illustrate how they accommodate generalized forces.

Consider an arbitrary body k shown in Fig. 2 connected to bodies $k-1$ and $k+1$ via kinematic joints J^k and J^{k+1}, respectively. Each degree of freedom of this system is defined as the relative motion of the child body with respect to its parent body. Let the column matrix $\hat{f}$ contain the known/unknown generalized forces associated with some specific degrees of freedom which must be included in the equations of motion. The two-handle equations of motion for body k in the GDCA scheme are presented by the following relations [57, 60]

$$\mathscr{A}_1^k = \phi_{11}^k\big(\mathscr{F}_{1c}^k + \mathscr{P}_s^{J^k}\hat{f}_k\big) + \phi_{12}^k\big(\mathscr{F}_{2c}^k - \mathscr{P}_s^{J^{k+1}}\hat{f}_{k+1}\big) + \phi_{13}^k, \tag{2}$$

$$\mathscr{A}_2^k = \phi_{21}^k\big(\mathscr{F}_{1c}^k + \mathscr{P}_s^{J^k}\hat{f}_k\big) + \phi_{22}^k\big(\mathscr{F}_{2c}^k - \mathscr{P}_s^{J^{k+1}}\hat{f}_{k+1}\big) + \phi_{23}^k. \tag{3}$$

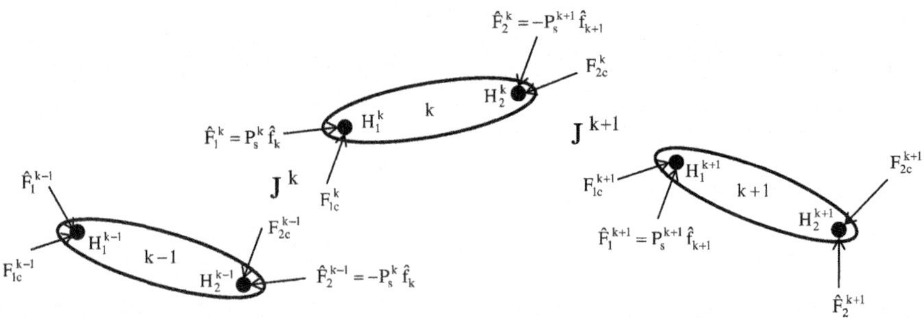

Fig. 2 Interactions between consecutive bodies in a multibody system

In the above equations, $\mathscr{A}_i^k$ $(i = 1, 2)$ represent the spatial accelerations of the handles of the body. The terms $\mathscr{F}_{ic}^k$ $(i = 1, 2)$ are the spatial constraint forces due to the kinematic constraint associated with the connecting joint, whiles $\hat{f}_k$ is the column matrix of the generalized forces associated with those degrees of freedom of the joint which are represented by the normalized subspace of the joint free-motion map $\mathscr{P}_s^{J^k}$ [57]. It is proven in [57, 60] that $\mathscr{P}_s^{J^k}\hat{f}_k$ is the dynamically equivalent spatial force [57] due to the corresponding generalized force $\hat{f}_k$. As such, the handle equations of motion in a Generalized-DCA can accommodate the known/unknown generalized forces, as well as the unknown spatial forces. Different applications of these equations are reported in [57].

Similar to the DCA, the Generalized-DCA is implemented using a series of recursive assembly and disassembly processes [50] to respectively form and solve the equations of motion of the system. The main goal of the assembly pass is to recursively generate larger encompassing subsystems by assembling the adjacent articulated bodies/subsystems of a multibody system as shown in Fig. 3. It is seen that the information flow in the underlying recursive operations is not serial, but in the structure of a binary tree. In the disassembly process, these equations of motion are then solved for the spatial accelerations and constraint forces of the handles of all nodes of the binary tree of Fig. 3. The detailed information on how the new terms due to the application of the generalized forces are treated in these processes are explained in [57, 59, 60].

4.2 Model Adjustment

In the adaptive scheme, it may deemed necessary to change the definition (i.e. locking or releasing) of the joints based on the value of the applied metrics. This model transition is realized by adjusting the joint free-motion map $\mathscr{P}^{J^k}$ [50] of the associated joint, and the corresponding orthogonal complement map $\mathscr{D}^{J^k}$ at the leaf level of the binary tree. For instance, if a revolute joint is to be locked because it is determined to be making an insignificant contribution to the overall conformation

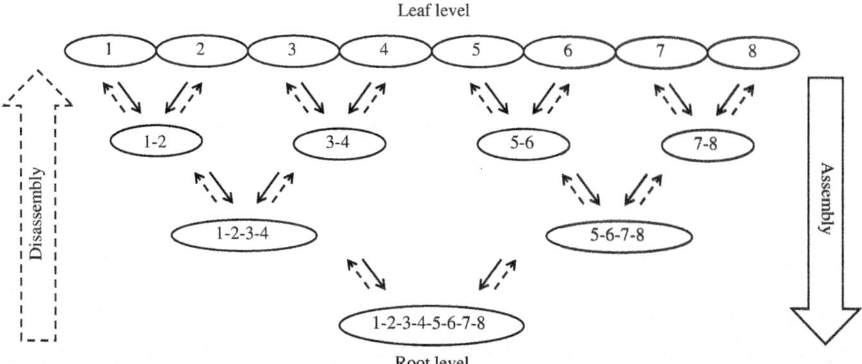

Fig. 3 Assembly and disassembly passes to recursively form and solve the equations of motion of the nodes of the binary tree

of the system, the associated spatial joint free-motion map $\mathscr{P}^{J^k} = [1\ 0\ 0\ 0\ 0\ 0]^T$ is replace by the null matrix after the transition. Similarly, the current orthogonal complement of the joint free-motion map

$$D^{J^k} = \begin{bmatrix} 0 & 0 & 0 & 0 & 0 \\ 1 & 0 & 0 & 0 & 0 \\ 0 & 1 & 0 & 0 & 0 \\ 0 & 0 & 1 & 0 & 0 \\ 0 & 0 & 0 & 1 & 0 \\ 0 & 0 & 0 & 0 & 1 \end{bmatrix} \tag{4}$$

is replaced by the identity matrix if expressed in the joint coordinate system.

Using the appropriate matrices characterizing the kinematics of the joints of the desired model after the transition, the assembly and disassembly processes are then performed as described previously to form and solve the forward dynamics equations of motion of the revised system model. More detail on these model adjustments is presented in [57, 63].

4.3 Dynamics of the System Within the Transition

All of the adjustments between different system models including coarse-to-fine and fine-to-coarse transitions are incurred without the influence of any external load. Furthermore, the configuration of the system does not change within each model adjustment. Therefore, any violation in the conservation of the generalized momentum of the system in these transitions leads to nonphysical results. In other words, the integration of the momentum of each differential element projected onto the space of admissible motions permitted by the more restrictive model (whether pre- or post-transition) over the entire system must be conserved across the model transition [37]. For instance, consider the process of instantaneous coarse graining of a fine-scale model. The coarse-scale model at $t = t^c$ is realized by imposing constraints on the fine-scale model at $t = t^f$ where $|t^c - t^f| = \varepsilon$ where ε is vanishingly small. In this transition, the number of generalized speeds of the system reduces from $\{u_i^f\}_{i=1}^n$ to $\{u_i^c\}_{i=1}^{n-m}$, where n and $n-m$ represent the number of degrees of freedom of the fine and coarse models, respectively. As such, the conservation of the generalized momentum of the system is expressed as

$$\mathscr{L}^{c/c} = \mathscr{L}^{f/c}, \tag{5}$$

where $\mathscr{L}$ is the generalized momentum of the system. The terms $\mathscr{L}^{c/c}$ and $\mathscr{L}^{f/c}$ represent the momenta of the coarse and fine models, respectively, projected onto the space of admissible motions (partial velocity vectors) of the coarse model at the time of transition.

Hence, the adaptive framework should be equipped with the machinery to efficiently form the generalized momenta balance equations and solve for the generalized speeds corresponding to the new set of degrees of freedom after each model adjustment. The formation of these impulse-momentum equations within each transition can be efficiently performed with a divide-and-conquer based scheme [52, 57]. As with the forward dynamics DCA assembly process, the handle equations which address the impulse-momentum equations of the bodies/assemblies of the system are recursively formed and combined together to find the associated equations of the resulting assemblies. These handle equations are then recursively solved for the jumps in the spatial velocities of the handles of the assemblies, as well as the constraint impulses applied to these locations when the generalized momentum balance is enforced.

5 Issues in Transitioning to Higher Fidelity Models

The transition towards the coarser-scale model which is effectively solving Eq. (5) for $\{u_i^c\}_{i=1}^{n-m}$ always results in a unique solution. Unlike real mechanical systems, the conservation of the generalized momentum across the model transition to the finer-scale model of a biomolecular system may not result in a unique solution. Because, in the transitions to the coarser model, naturally existing higher modes of motion are ignored since the internal metric had previously indicated these modes as less relevant. As such, in the transition to the higher fidelity models, the kinetic energy of these ignored modes must be estimated and considered appropriately. The generalized momentum conserving distribution (Eq. (5)) of the kinetic energy among the modes of the fine model is not unique even if the value of the lost kinetic energy within this transition is known [62]. More specifically, an underdetermined set of equations must be solved for $\{u_i^f\}_{i=1}^n$ when more than a single degree of freedom of the system is released to achieve a higher fidelity model [57].

To solve the problem of the transition to the finer fidelity model, an optimization-based technique may be used to arrive at the "best" solutions from an infinite pool of possible physically meaningful solutions. This optimization problem which is implemented on a knowledge-, math-, or physics-based cost function is heavily constrained since it must additionally satisfy the impulse-momentum equations i.e. Eq. (5). For instance, one may minimize the L^2 norm of the difference between the generalized speeds of the fine-scale model and those of the coarse-scale model. In this situation, the generalized speeds are prevented from deviating greatly from the ones before unlocking [3, 62]. Alternatively, it may be desired to perform the optimization on the energy transferred to the solvent [62].

The application of the traditional methods such as Lagrange multipliers to form and solve this constrained optimization problem is computationally expensive [57]. The computational complexity of this problem reduces via changing the constrained optimization problem to an unconstrained one which effectively reduces the optimization parameters. However, in this scheme, using the coordinate partitioning to find the relations between the dependent and independent design parameters (generalized speeds) may become very costly [57, 78] if not performed wisely. The mathematical framework of forming the impulse-momentum equations (constraints), and efficiently finding the relations between the dependent generalized speeds and the independent ones in the DCA scheme for rigid and flexible body systems are presented in [4, 41, 57, 64]. It is demonstrated in these works that the application of this algorithm can significantly reduce the computational expenses associated with the manipulations performed to derive the dynamics of the transition, as well as those performed as parts of the optimization problem.

6 Preliminaries for Efficient Pairwise Force Calculations

1. Consider body B (not necessarily a rigid body) containing N particles, and the individual particle $\bar{P}$ shown in Fig. 4. In general, the pairwise force interaction between an arbitrary particle P_i embedded in B and $\bar{P}$ may be expressed in the following format:

$$\mathbf{F}_{\bar{P}P_i} = \frac{\beta\bar{\lambda}\lambda_i}{(|\mathbf{r}'_i|)^s}\mathbf{e}_{\mathbf{r}'_i} = \beta\bar{\lambda}\lambda_i\mathbf{r}'_i\left(\mathbf{r}'^2_i\right)^{-\frac{s+1}{2}}, \tag{6}$$

where, β is the constant associated with the force field of interest, $s > 1$ is an integer, $\mathbf{e}_{\mathbf{r}'_i}$ is the unit vector from $\bar{P}$ to P_i. Additionally, λ_i is the quantity corresponding to the force field which is associated with particle P_i, and similarly, $\bar{\lambda}$ is the same quantity associated with particle $\bar{P}$. This general formulation may be used to address the gravitational, Coulombic or London forces. For instance, if one is interested in the pairwise interactions due to the Coulomb's law [65], λ_i represents the charge of the particle, s becomes 2, and β is replaced by the Coulomb force constant.

2. For body (pseudo-atom) B, the *pseudo-center* denoted by C_λ is defined as the center of the body corresponding to the quantity of interest λ provided that the pseudo-mass (lumped quantity) of the body defined as

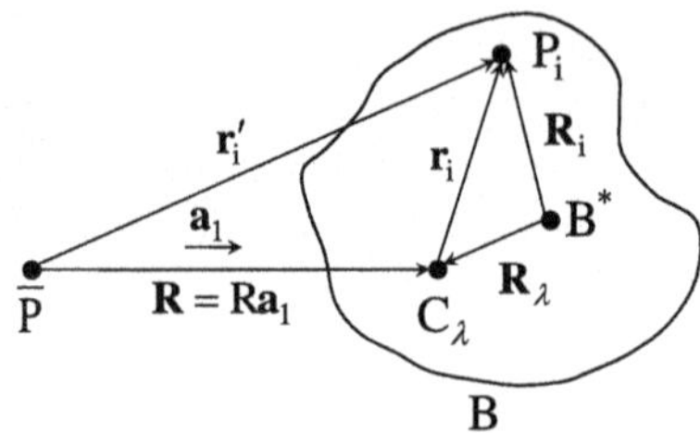

Fig. 4 Pairwise interactions between the particle $\bar{P}$ and the particles embedded in $\bar{B}$ (particle-body interactions)

$$\Lambda \triangleq \sum_{i=1}^{N} \lambda_i \tag{7}$$

is not zero. The position of this point with respect to the mass center of the body (i.e., B^*) is calculated using the relation

$$\mathbf{R}_\lambda = \frac{\sum_{i=1}^{N} \mathbf{R}_i \lambda_i}{\Lambda}, \tag{8}$$

where $\mathbf{R}_i$ is the position vector of the particle (atom) P_i measured from the center of mass of the body. Discussions on the subdomains of the system for which the pseudo-center is not defined ($\Lambda = 0$) are provided in Sect. 9.

3. For body B, the *pseudo-inertia* tensor associated with the quantity λ with respect to the pseudo-center of the body is defined as

$$\mathscr{I}_\lambda^{B/C_\lambda} \triangleq \sum_{i=1}^{N} (\mathscr{U}\mathbf{r}_i^2 - \mathbf{r}_i\mathbf{r}_i)\lambda_i. \tag{9}$$

In this definition, $\mathscr{U}$ denotes the identity tensor, and $\mathbf{r}_i$ is the position vector of the particle P_i relative to the pseudo-center of the body. This tensor represents the dyadic of the moment of inertia (second moment) of the body if one studies the gravitational force [37]. It should be mentioned that for *rigid* subdomains of the system, this dyadic is constant if expressed in the axes fixed in the domain of interest.

7 Force Approximation

In the following, the approximation of the net force applied to a body due to the interactions between a single particle and the particles in the body is presented. Additionally, the resultant force interactions between the particles in two different domains (bodies) of the system is also approximated.

7.1 Particle-Body Force Interactions

Consider the interaction between particle $\bar{P}$ and an arbitrary particle P_i embedded in body B shown in Fig. 4 is expressed by Eq. (6). Also assume that the origin of the body-fixed frame is located at the pseudo-center of the body. According to the geometry shown in Fig. 4, the position vector from $\mathbf{r}'_i$ in Eq. (6) can be replaced by $\mathbf{R}+\mathbf{r}_i$ to arrive at

$$\mathbf{F}_{\bar{P}P_i} = \beta\bar{\lambda}\lambda_i\left(\mathbf{R}^2+\mathbf{r}_i^2+2\mathbf{R}\cdot\mathbf{r}_i\right)^{-\frac{s+1}{2}}(\mathbf{R}+\mathbf{r}_i). \tag{10}$$

The above equation can be rewritten as

$$\mathbf{F}_{\bar{P}P_i} = \frac{\beta\bar{\lambda}\lambda_i}{R^s}\left(1+\left(\frac{r_i}{R}\right)^2+2\mathbf{a}_1\cdot\frac{\mathbf{r}_i}{R}\right)^{-\frac{s+1}{2}}\left(\mathbf{a}_1+\frac{\mathbf{r}_i}{R}\right), \tag{11}$$

where, r_i, and R are the lengths of the vectors $\mathbf{r}_i$ and $\mathbf{R}$, respectively. In this relation, $\mathbf{a}_1$ is the unit vector from $\bar{P}$ to the pseudo-center of body B as shown in Fig. 4.

Using the binomial series expansion, the term $(1+(\frac{r_i}{R})^2+2\mathbf{a}_1\cdot\frac{\mathbf{r}_i}{R})^{-\frac{s+1}{2}}$ in Eq. (11) is expanded as

$$\begin{aligned}
&\left(1+\left(\frac{r_i}{R}\right)^2+2\mathbf{a}_1\cdot\frac{\mathbf{r}_i}{R}\right)^{-\frac{s+1}{2}} \\
&=1-\frac{s+1}{2}\left(\left(\frac{r_i}{R}\right)^2+2\mathbf{a}_1\cdot\frac{\mathbf{r}_i}{R}\right)+\frac{(s+1)(s+3)}{8}\left(\left(\frac{r_i}{R}\right)^2+2\mathbf{a}_1\cdot\frac{\mathbf{r}_i}{R}\right)^2 \\
&\quad-\frac{(s+1)(s+3)(s+5)}{48}\left(\left(\frac{r_i}{R}\right)^2+2\mathbf{a}_1\cdot\frac{\mathbf{r}_i}{R}\right)^3+\cdots
\end{aligned} \tag{12}$$

provided that $|(\frac{r_i}{R})^2+2\mathbf{a}_1\cdot\frac{\mathbf{r}_i}{R}|<1$.

The total force experienced by body B due to the pairwise interactions between its own particles and $\bar{P}$ is expressed as

$$\mathbf{F}_{\bar{P}B} = \sum_{i=1}^{N}\mathbf{F}_{\bar{P}P_i}. \tag{13}$$

Using the expression provided in Eq. (12), this net force is rewritten as

$$\begin{aligned}
\mathbf{F}_{\bar{P}B} = \frac{\beta\bar{\lambda}}{R^s}\Bigg[&\sum_{i=1}^{N}\lambda_i\mathbf{a}_1-\sum_{i=1}^{N}(s+1)\lambda_i\left(\mathbf{a}_1\cdot\frac{\mathbf{r}_i}{R}\right)\mathbf{a}_1+\sum_{i=1}^{N}\lambda_i\frac{\mathbf{r}_i}{R} \\
&-\sum_{i=1}^{N}\frac{(s+1)}{2}\lambda_i\left(\frac{r_i}{R}\right)^2\mathbf{a}_1+\sum_{i=1}^{N}\frac{(s+1)(s+3)}{2}\lambda_i\left(\mathbf{a}_1\cdot\frac{\mathbf{r}_i}{R}\right)^2\mathbf{a}_1 \\
&-\sum_{i=1}^{N}(s+1)\lambda_i\left(\mathbf{a}_1\cdot\frac{\mathbf{r}_i}{R}\right)\frac{\mathbf{r}_i}{R}+O\left(\frac{r}{R}\right)^3\Bigg].
\end{aligned} \tag{14}$$

In the above relation, r is the length of the position vector of a generic point on B with respect to the pseudo-center of the body.

Elaborating on different terms in the above expression based on their orders with respect to $\frac{r_i}{R}$ [57, 61], Eq. (14) provides the net force applied to the body as

$$\mathbf{F}_{\bar{P}B} = \frac{\beta\bar{\lambda}}{R^s}\left\{\Lambda\mathbf{a}_1 + \left[\frac{s(s+1)}{4R^2}tr(\mathscr{I}_\lambda^{B/C_\lambda}) - \frac{(s+1)(s+3)}{2R^2}\mathbf{a}_1\cdot\mathscr{I}_\lambda^{B/C_\lambda}\cdot\mathbf{a}_1\right]\mathbf{a}_1 + \frac{s+1}{R^2}\mathbf{a}_1\cdot\mathscr{I}_\lambda^{B/C_\lambda} + O\left(\frac{r}{R}\right)^3\right\}, \tag{15}$$

where $tr(\mathscr{I}_\lambda^{B/C_\lambda})$ is the trace of the pseudo-inertia tensor. This equation can be expressed as

$$\mathbf{F}_{\bar{P}B} = \frac{\beta\bar{\lambda}\Lambda}{R^s}\left[\mathbf{a}_1 + \sum_{i=2}^{\infty}\mathbf{f}_i\left(\frac{r}{R}\right)^i\right], \tag{16}$$

where $\mathbf{f}_i \triangleq \mathbf{f}_i(\frac{r}{R})^i$ is the collection of terms associated with the ith degree of $\frac{r}{R}$.

Ignoring the third and higher order terms in this relation, the net force is approximated as

$$\mathbf{F}_{\bar{P}B} \approx \tilde{\mathbf{F}}_{\bar{P}B} = \frac{\beta\bar{\lambda}\Lambda}{R^s}(\mathbf{a}_1 + \mathbf{f}_2), \tag{17}$$

provided that

$$\max_{i\in B} r_i \ll R. \tag{18}$$

Introducing a dextral, orthogonal set of unit vectors, $\mathbf{a}_1$, $\mathbf{a}_2$, and $\mathbf{a}_3$ with the origin passing through the pseudo-center of B, and defining the elements of the pseudo-inertia tensor in **a**-basis as

$$I_{ij} = \mathbf{a}_i\cdot\mathscr{I}_\lambda^{B/C_\lambda}\cdot\mathbf{a}_j \quad (i,j = 1,2,3), \tag{19}$$

the term $\mathbf{f}_2$ may be written as

$$\mathbf{f}_2 = \frac{s+1}{\Lambda R^2}\left\{\left[\frac{s}{4}tr(\mathscr{I}_\lambda^{B/C_\lambda}) - \frac{s+1}{2}I_{11}\right]\mathbf{a}_1 + I_{21}\mathbf{a}_2 + I_{31}\mathbf{a}_3\right\}. \tag{20}$$

7.2 Body-Body Force Interactions

Due to the pairwise interactions between particles $\{P_i\}_{i=1}^N$ belonging to body B (Fig. 5), and $\{\bar{P}_j\}_{j=1}^{\bar{N}}$ embedded in body $\bar{B}$, the resultant force $\mathbf{F}_{\bar{B}B}$ is applied to B by $\bar{B}$. This force can be approximated ($\tilde{\mathbf{F}}_{\bar{B}B}$) by summing over all approximate forces applied to B by all particles $\bar{P}_j$ on $\bar{B}$. As such, using the second order approximation of Eq. (15), this net force is approximated as

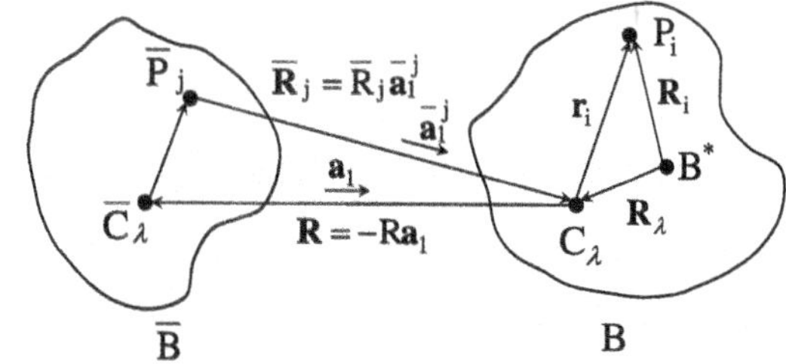

Fig. 5 Pairwise interactions between the particles embedded in B and $\bar{B}$ (body-body interaction)

$$\tilde{\mathbf{F}}_{\bar{B}B} = \sum_{j=1}^{\bar{N}} \frac{\beta\Lambda\lambda_j}{\bar{R}_j^s}\bar{\mathbf{a}}_1^j + \frac{\beta\lambda_j}{\bar{R}_j^s}\sum_{j=1}^{\bar{N}}\left\{\left[\frac{s(s+1)}{4\bar{R}_j^2}tr\left(\mathscr{I}_\lambda^{B/C_\lambda}\right)\right.\right.$$
$$\left.\left. - \frac{(s+1)(s+3)}{2\bar{R}_j^2}\bar{\mathbf{a}}_1^j \cdot \mathscr{I}_\lambda^{B/C_\lambda} \cdot \bar{\mathbf{a}}_1^j\right]\bar{\mathbf{a}}_1^j + \frac{s+1}{\bar{R}_j^2}\bar{\mathbf{a}}_1^j \cdot \mathscr{I}_\lambda^{B/C_\lambda}\right\}, \quad (21)$$

where $\bar{R}_j$ denotes the distance from $\bar{P}_j$ to C_λ, and $\bar{\mathbf{a}}_1^j$ is the corresponding unit vector.

Since the second summation involves the terms of second or higher degrees in r_i, $\bar{R}_j$ and $\bar{\mathbf{a}}_1^j$ may be replaced by R and $\mathbf{a}_1$, respectively. No terms of interest for the purpose at hand are lost through this replacement. This substitution is effectively the application of the Taylor series expansion in the approximation.

Let us define $\mathbf{a}_2$ and $\mathbf{a}_3$ such that $\mathbf{a}_1$, $\mathbf{a}_2$, and $\mathbf{a}_3$ establish a dextral, orthogonal set of unit vectors. Defining the moments and products of inertia of B and $\bar{B}$ for axes parallel to $\mathbf{a}_1$, $\mathbf{a}_2$, $\mathbf{a}_3$, and passing through the pseudo-center of the individual bodies as

$$I_{ij} = \mathbf{a}_i \cdot \mathscr{I}_\lambda^{B/C_\lambda} \cdot \mathbf{a}_j \quad (i, j = 1, 2, 3), \quad (22)$$
$$\bar{I}_{ij} = \mathbf{a}_i \cdot \mathscr{I}_\lambda^{\bar{B}/\bar{C}_\lambda} \cdot \mathbf{a}_j \quad (i, j = 1, 2, 3), \quad (23)$$

and elaborating on the first summation of Eq. (21), this equation is simplified as

$$\tilde{\mathbf{F}}_{\bar{B}B} = \frac{\beta\Lambda\bar{\Lambda}}{R^s}(\mathbf{a}_1 + \mathbf{g}_2 + \bar{\mathbf{g}}_2), \quad (24)$$

where

$$\mathbf{g}_2 = \frac{s+1}{\Lambda R^2}\left\{\left[\frac{s}{4}tr\left(\mathscr{I}_\lambda^{B/C_\lambda}\right) - \frac{(s+1)}{2}I_{11}\right]\mathbf{a}_1 + I_{21}\mathbf{a}_2 + I_{31}\mathbf{a}_3\right\}, \quad (25)$$
$$\bar{\mathbf{g}}_2 = \frac{s+1}{\bar{\Lambda}R^2}\left\{\left[\frac{s}{4}tr\left(\mathscr{I}_\lambda^{\bar{B}/\bar{C}_\lambda}\right) - \frac{(s+1)}{2}\bar{I}_{11}\right]\mathbf{a}_1 + \bar{I}_{21}\mathbf{a}_2 + \bar{I}_{31}\mathbf{a}_3\right\}. \quad (26)$$

8 Torque Approximation

Since the resultant forces calculated in Sects. 7.1 and 7.2 do not necessarily act through the center of mass of the body, they create moments about the mass center.

In the following, these resultant torques due to the long-range particle-body and body-body interactions are calculated.

8.1 Particle-Body Torque Interactions

Based on the geometry shown in Fig. 4, the following cross product is used to calculate the torque which body B experiences about its mass center B^* due to the interaction between $\bar{P}$ and P_i

$$\mathbf{M}^{B^*}_{\bar{P}P_i} = \mathbf{R}_i \times \mathbf{F}_{\bar{P}P_i}. \tag{27}$$

Replacing $\mathbf{R}_i$ by $\mathbf{R}_\lambda - \mathbf{R} + \mathbf{r}'_i$, this torque can be rewritten as

$$\mathbf{M}^{B^*}_{\bar{P}P_i} = (\mathbf{R}_\lambda - \mathbf{R}) \times \mathbf{F}_{\bar{P}P_i} + \underbrace{\mathbf{r}'_i \times \mathbf{F}_{\bar{P}P_i}}_{\mathbf{0}}. \tag{28}$$

The last term in the above relation disappears since both $\mathbf{r}'_i$ and $\mathbf{F}_{\bar{P}P_i}$ are collinear vectors. As such, body B experiences the following moment about B^* due to the interactions between its own particles and $\bar{P}$

$$\mathbf{M}^{B^*}_{\bar{P}B} = \sum_{i=1}^{N} (\mathbf{R}_\lambda - \mathbf{R}) \times \mathbf{F}_{\bar{P}P_i} = (\mathbf{R}_\lambda - \mathbf{R}) \times \mathbf{F}_{\bar{P}B}. \tag{29}$$

Using the second order approximation of the net force from Eq. (15), this moment is approximated as

$$\tilde{\mathbf{M}}^{B^*}_{\bar{P}B} = \frac{\beta\Lambda\bar{\lambda}}{R^s}\mathbf{R}_\lambda \times \left\{\mathbf{a}_1 + \frac{s+1}{\bar{\lambda}R^2}\left\{\left[\frac{s}{4}tr\left(\mathscr{I}^{B/C_\lambda}_\lambda\right) - \frac{(s+3)}{2}\mathbf{a}_1 \cdot \mathscr{I}^{B/C_\lambda}_\lambda \cdot \mathbf{a}_1\right]\mathbf{a}_1 \right.\right.$$
$$\left.\left. + \mathbf{a}_1 \cdot \mathscr{I}^{B/C_\lambda}_\lambda\right\}\right\} - \frac{\beta\bar{\lambda}(s+1)}{R^{(s+1)}}\mathbf{a}_1 \times \left(\mathbf{a}_1 \cdot \mathscr{I}^{B/C_\lambda}_\lambda\right). \tag{30}$$

Defining the elements of the pseudo-inertia tensor in **a**-basis as defined in Eq. (19), and using $\mathbf{f}_2$ from Eq. (20), this expression is simplified as

$$\tilde{\mathbf{M}}^{B^*}_{\bar{P}B} = \frac{\beta\Lambda\bar{\lambda}}{R^s}\mathbf{R}_\lambda \times (\mathbf{a}_1 + \mathbf{f}_2) - \frac{\beta\bar{\lambda}(s+1)}{R^{(s+1)}}(I_{21}\mathbf{a}_3 - I_{31}\mathbf{a}_2). \tag{31}$$

8.2 Body-Body Torque Interactions

Using Eq. (30), the resultant moment applied to B about B^* from $\bar{B}$ due to the interactions between the particles in these bodies can be approximated by summing over the approximate torques which body B experiences about B^* by all particles $\bar{P}_j$ on $\bar{B}$ as

$$\tilde{\mathbf{M}}^{B^*}_{\bar{B}B} = \sum_{j=1}^{\bar{N}} \frac{\beta\bar{\lambda}_j \Lambda}{\bar{R}_j^s} \mathbf{R}_\lambda \times \bar{\mathbf{a}}_1^j + \sum_{j=1}^{\bar{N}} \frac{\beta\bar{\lambda}_j}{\bar{R}_j^s} \mathbf{R}_\lambda \times \frac{s+1}{\bar{R}_j^2} \left\{ \left[\frac{s}{4} tr(\mathscr{I}_\lambda^{B/C_\lambda}) - \frac{(s+3)}{2} \bar{\mathbf{a}}_1^j \cdot \mathscr{I}_\lambda^{B/C_\lambda} \cdot \bar{\mathbf{a}}_1^j \right] \bar{\mathbf{a}}_1^j + \bar{\mathbf{a}}_1^j \cdot \mathscr{I}_\lambda^{B/C_\lambda} \right\} - \sum_{j=1}^{\bar{N}} \frac{\beta\bar{\lambda}_j (s+1)}{\bar{R}_j^{(s+1)}} \bar{\mathbf{a}}_1^j \times \left(\bar{\mathbf{a}}_1^j \cdot \mathscr{I}_\lambda^{B/C_\lambda} \right). \tag{32}$$

Similar to the argument made in Sect. 7.2, $\bar{R}_j$ and $\bar{\mathbf{a}}_1^j$ may be replaced by R and $\mathbf{a}_1$, respectively. Using the same strategy provided in [61], the approximate moment about B^* from $\bar{B}$ due to the pairwise interactions between the particles embedded in B and $\bar{B}$ is expressed as

$$\tilde{\mathbf{M}}^{B^*}_{\bar{B}B} = \frac{\beta \Lambda \bar{\Lambda}}{R^s} \mathbf{R}_\lambda \times (\mathbf{a}_1 + \mathbf{g}_1 + \bar{\mathbf{g}}_2) - \frac{\beta \bar{\Lambda}(s+1)}{R^{(s+1)}} (I_{21}\mathbf{a}_3 - I_{31}\mathbf{a}_2), \tag{33}$$

where I_{ij}, $\mathbf{g}_1$, and $\bar{\mathbf{g}}_2$ have already been defined in Eqs. (22), (25) and (26), respectively.

9 Discussions on the Developed Approximations

1. The presented approximations contain the terms up to the quadrupole moment (quadrupole-quadrupole interactions). Furthermore, since the origin of the body-fixed frame is located at its pseudo-center, the first moment $\sum_{i=1}^{N} \lambda_i \mathbf{r}_i$ does not appear in these approximations [57, 61].
2. According to Eq. (8), the pseudo-center is not defined when the pseudo-mass Λ of the body (subdomain) is zero. Similarly, if Λ is very close to zero, the pseudo-center may be located far away from the body. As such, the center of mass of the body may be considered as the origin of the body-fixed frame to derive the resultant forces and torques. In these situations, the pseudo-inertia tensor used in all the derived approximations is defined about B^*. Furthermore, due to locating the origin of the body-fixed frame at the mass center rather than the pseudo-center, the first moment appears in the approximations. However, it is proven in [57, 61] that the first moment is constant regardless of the choice of the origin. Consequently, when the pseudo-mass of the pseudo-atom is zero, it is necessary to express the first moment measured from any reference point defined by the analyst in the approximate force and torque. Furthermore, for rigid subdomains of the system, this term becomes time-invariant.
3. For *rigid* pseudo-atoms, the pseudo-inertia tensor which appears in all the approximations is a time-invariant quantity if expressed in body-fixed frame. As such, if this tensor is calculated for a rigid subdomain of the system at some time either before or during the simulation, no additional cost is incurred to form or use this dyadic during the course of the simulation. It is only needed to monitor the location and orientation of each pseudo-atom. Moreover, the trace of this

tensor which appears in these approximations is also a constant scalar for rigid subdomains.

4. Due to the symmetry observed in the body-body force approximation in Eqs. (24)–(26), the approximate net force does not violate Newton's third law of motion.
5. In the formulation presented for the particle-body and body-body torque approximations in Eqs. (31) and (33), if the pseudo-center and the center of mass of the body coincide, i.e., $\mathbf{R}_\lambda = \mathbf{0}$, the last term only contributes to the applied moment. In this case, the approximate moment formulation provides a zero value if $\mathbf{a}_1$ is aligned with one of the principal axes of the pseudo-inertia tensor of body B. Since the geometry of the system is known at each time step, one can avoid using these approximations by checking whether the orientation of the body is close to this specific configuration. In such situations, the analyst may use the exact calculations or a higher order approximation to find the moment. Moreover, in the body-body torque interaction, one may use the summation of the particle-body torque approximations over the entire particles of the body to calculate the torque applied to a body from another body.
6. Poursina and Anderson [61] demonstrate the efficiency of the method by comparing the operation counts for particle-body and body-body force interactions using the exact force calculations and the presented approximations. Additionally, for a system containing n_p particles, and N_r rigid subdomains ($N_r \ll n_p$), the computational complexity of the presented approximations is $O(N_r)^2$ as opposed to $O(n^2)$ complexity when all atoms are considered in pairwise calculations. Moreover, it is expected that the computational complexity will improve to $O(N_r \log N_r)$ if advanced algorithms are used to implement these approximations [57]. As such, an efficient implementation of these approximations which is well-suited in combination with the state-of-the-art multibody algorithms is an ongoing research by the authors [57].

10 Numerical Results

Consider particles with unit positive charges distributed equidistantly on two straight lines B and $\bar{B}$ with the length L as shown in Fig. 6. This could be a simple model of a rigidified residue of a DNA or an RNA. Since the Coulombic potential field is active between these charged particles, the values of the parameters in Eq. (6) are selected as $\beta = k_e = 8.9885518 \times 10^9\ \mathrm{N\,m^2/C^2}$, $\lambda_i = \bar{\lambda} = \lambda = 1.6021764 \times 10^{-19}$ C, and $s = 2$. Due to the symmetry in the mass and charge distribution, the mass center and the pseudo-center of each body coincide. The dextral orthogonal unit vectors $\mathbf{b}_1\mathbf{b}_2\mathbf{b}_3$ and $\bar{\mathbf{b}}_1\bar{\mathbf{b}}_2\bar{\mathbf{b}}_3$ are attached to the pseudo-centers of bodies B and $\bar{B}$, respectively. The angles θ_1 and θ_2 as shown in Fig. 6 are used to describe the orientation of these body-attached reference frames with respect to the

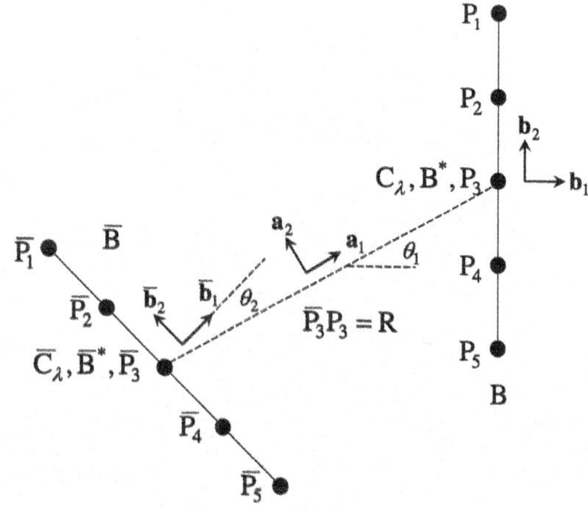

Fig. 6 Pairwise Coulombic interactions between the particles on two rigid bodies B and $\bar{B}$

a-frame. Further, the pseudo-inertia tensor of each body with respect to the associated pseudo-center is expressed in the corresponding body-basis as follows

$$\mathscr{I}_{\lambda}^{B/C_\lambda} = \mathscr{I}_{\lambda}^{\bar{B}/\bar{C}_\lambda} = \begin{bmatrix} \frac{5\lambda L^2}{8} & 0 & 0 \\ 0 & 0 & 0 \\ 0 & 0 & \frac{5\lambda L^2}{8} \end{bmatrix}. \tag{34}$$

To run the simulations, it is assumed that $L = 5$ Å which is approximately on the same order of magnitude of length of an RNA residue. Herein, various configurations of this planar system are formed by fixing $\theta_2 = 0$, and changing R from L to $4L$ and θ_1 from 0 to $\pi/2$. The resultant electrostatic force applied from $\bar{B}$ to B at each configuration is computed using three different methods. First, the exact net force due to the pairwise interactions is calculated. Then, the resultant force applied by each particle of $\bar{B}$ to the entire body B is found using the particle-body approximation derived in Eq. (17), and summed over all particles embedded in $\bar{B}$. Finally, the resultant force applied to B is calculated using the body-body approximation presented in Eq. (24). The torque experienced by B about B^* due to the interactions between the particles embedded in B and $\bar{B}$ is also calculated using these three methods.

Using the following definitions

$$E_{ij}^{\mathbf{F}} \triangleq \frac{\|\mathbf{F}_{approx.}(R_i, \theta_{1_j}) - \mathbf{F}_{exact}(R_i, \theta_{1_j})\|_2}{\|\mathbf{F}_{exact}(R_i, \theta_{1_j})\|_2}, \tag{35}$$

$$E_{ij}^{\mathbf{T}} \triangleq \frac{\|\mathbf{T}_{approx.}(R_i, \theta_{1_j}) - \mathbf{T}_{exact}(R_i, \theta_{1_j})\|_2}{\|\mathbf{T}_{exact}(R, \theta_1)\|_\infty}, \tag{36}$$

the values of the percentage error of the approximate resultant force and moment at each configuration (R_i, θ_{1_j}) are, respectively, calculated and depicted in Fig. 7. Since for this problem, the resultant force never becomes zero, the percentage error of the approximate force at each configuration is normalized by L^2 norm of the associated force. However, the percentage error of the approximate torque is normalized

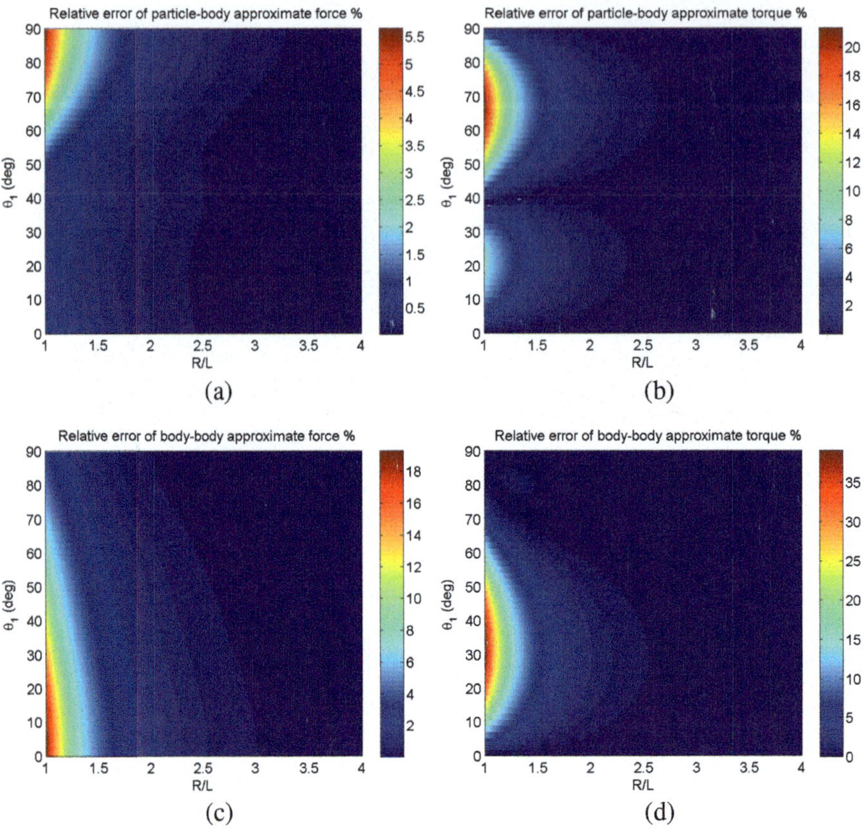

Fig. 7 (**a**) The percentage error of the approximate net force applied to B by summing over particle-body approximations. (**b**) The percentage error of the approximate resultant torque about B^* by summing over particle-body moment approximations. (**c**) The percentage error of the approximate net force applied to B using body-body approximation. (**d**) The percentage error of the approximate resultant moment about B^* using body-body torque approximation

by the maximum of the absolute value of the exact torque among all configurations since the net moment is zero in some configurations.

The results show that both particle-body and body-body formulas generate the acceptable approximations for the far-field interactions. Although in the entire configuration space sampled in this example, there exist very tiny regions for which the approximations provide large errors, these errors decay very quickly as the bodies become more distant. For instance, for a very conservative case with $\frac{R}{L} > 3$, Fig. 7(c) shows that the body-body force approximation provides the relative error less than 0.1 %. It is also observed that summing over the particle-body approximation to find the force and moment between two bodies, in general, provides less error than the application of the body-body approximation. Although the particle-body approximation is more accurate, the body-body formulation is faster, and provides

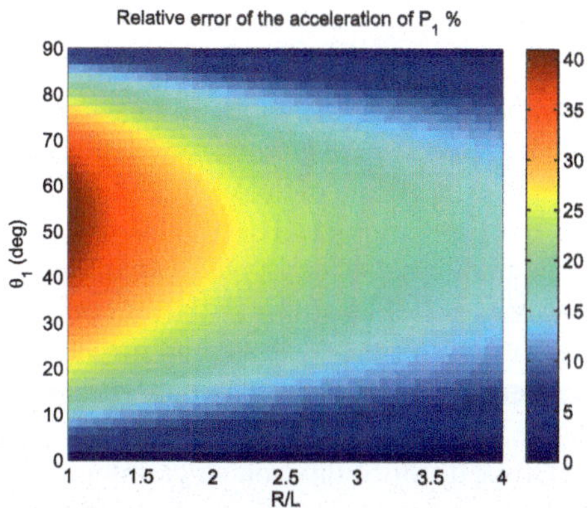

Fig. 8 The percentage relative error in the translational acceleration of P_1 when the angular motion of the body is ignored

acceptable approximation for the exact resultant force and torque for most configurations.

To analyze the importance of the angular motion in the determination of the configuration of the system, assume that body B is a segment of a very long chain (of a biopolymer), and the motion is transferred to the outboard handle of this body through point P_1. Without loss of generality and for simplicity, it is assumed that the angular velocity of B (i.e. ${}^N\boldsymbol{\omega}^B$) is zero. Therefore, the rotational motion of this body is reflected in its angular acceleration measured in the Newtonian frame of reference i.e. ${}^N\boldsymbol{\alpha}^B$. The exact translational acceleration of P_1 in the Newtonian frame (i.e. ${}^N\mathbf{a}^{P_1}_{exact}$) when the angular motion of the body is not ignored is expressed as

$$ {}^N\mathbf{a}^{P_1}_{exact} = {}^N\mathbf{a}^{B^*} + \underbrace{{}^N\boldsymbol{\omega}^B}_{\mathbf{0}} \times \big(\underbrace{{}^N\boldsymbol{\omega}^B}_{\mathbf{0}} \times \mathbf{r}^{B^*P_1}\big) + {}^N\boldsymbol{\alpha}^B \times \mathbf{r}^{B^*P_1}. \tag{37} $$

The term ${}^N\mathbf{a}^{B^*}$ in the above relation is the translational acceleration of B^* in the Newtonian frame, and $\mathbf{r}^{B^*P_1}$ denotes the position of P_1 with respect to B^*. This relation demonstrates that both P_1 and B^* have the same translational accelerations if the entire rotational motion of the body is ignored (i.e., ${}^N\boldsymbol{\omega}^B = {}^N\boldsymbol{\alpha}^B = 0$). As such, the percentage relative error of the translational acceleration of P_1 when the angular motion of this segment is neglected is defined as

$$ Err = \frac{\|{}^N\boldsymbol{\alpha}^B \times \mathbf{r}^{B^*P_1}\|_2}{\|{}^N\mathbf{a}^{P_1}_{exact}\|_2} \times 100. \tag{38} $$

This error is calculated and depicted in Fig. 8, assuming that the mass of each particle is 27.026 Daltons which is $\frac{1}{5}$ of the mass of the nucleotide Adenine. It is observed that this error is significant within the majority of the configuration space. As a result, the angular motion needs to be considered in modeling biopolymers such as DNAs and RNAs in which the geometry plays an important role in determining the conformational motion of the system.

11 Conclusions

Herein, different strategies to improve the computational efficiency of the multi-body-based coarse-grained simulations of biopolymers have been presented. The adaptive modeling of these systems in which the model is adjusted within the course of the simulation has been developed. This scheme which is much more accurate than traditional static (time-invariant) coarse-grained models is capable of identifying the critical locations of the system to add or remove fidelity to or from the system model on-the-fly. Potential metrics to direct these model transitions have been presented. Furthermore, the important issues associated with the implementation of these instantaneous model adjustments within the course of the simulation have been addressed. Since this coarse graining strategy is realized in a multibody dynamics scheme, a Generalized Divide-and-Conquer (GDCA) which is highly modular and lends itself well to adaptivity has been presented. The method for rigid body systems is exact, non-iterative and efficient, providing a time optimal order $\log n$ computational performance achieved with a processor optimal order n processors. Another aspect of this book chapter has been associated with the development of an efficient algorithm to approximate far-field interactions. The presented method approximates particle-body and body-body force and moment terms. The developed formulations are highly compatible with the state-of-the-art efficient multibody algorithms. The methods have provided relatively accurate results for the test case with Coulombic interactions. It has also been illustrated that the torque which is ignored in bead models plays an important role in more appropriately capturing the conformational motion of biopolymers.

Acknowledgements Support for this work received under National Science Foundation through award No. 0757936 is gratefully acknowledged. The author would like to thank Professor Alain Laederach from the University of North Carolina, Mr. Michael Sherman from Simbios: NIH Center for Biomedical Computation at Stanford University, and Dr. Kishor Bhalerao from the University of Melbourne for several useful discussions. They also would like to thank Jeremy Laflin who independently validated some of the simulation results.

References

1. Abagyan, R., Mazur, A.: New methodology for computer-aided modeling of biomolecular structure and dynamics. 2. Local deformations and cycles. J. Comput. Phys. **6**(4), 833–845 (1989)
2. Anderson, K.S.: Recursive derivation of explicit equations of motion for efficient dynamic/control simulation of large multibody systems. Ph.D. thesis, Stanford University (1990)
3. Anderson, K.S., Poursina, M.: Energy concern in biomolecular simulations with transition from a coarse to a fine model. In: Proceedings of the Seventh International Conference on Multibody Systems, Nonlinear Dynamics and Control, ASME Design Engineering Technical Conference 2009 (IDETC09), IDETC2009/MSND-87297, San Diego, CA (2009)
4. Anderson, K.S., Poursina, M.: Optimization problem in biomolecular simulations with DCA-based modeling of transition from a coarse to a fine fidelity. In: Proceedings of the Seventh International Conference on Multibody Systems, Nonlinear Dynamics and Control, ASME Design Engineering Technical Conference 2009 (IDETC09), IDETC2009/MSND-87319, San Diego, CA (2009)

5. Appel, A.W.: An efficient program for many-body simulation. SIAM J. Sci. Stat. Comput. **6**(1), 85–103 (1985)
6. Armstrong, W.W.: Recursive solution to the equations of motion of an n-link manipulator. In: Fifth World Congress on the Theory of Machines and Mechanisms, vol. 2, pp. 1342–1346 (1979)
7. Avello, A., Jiménez, J.M., Bayo, E., García de Jalón, J.: A simple and highly parallelizable method for real-time dynamic simulation based on velocity transformations. Comput. Methods Appl. Mech. Eng. **107**(3), 313–339 (1993)
8. Bae, D.S., Haug, E.J.: A recursive formation for constrained mechanical system dynamics: Part I. Open loop systems. Mech. Struct. Mach. **15**(3), 359–382 (1987)
9. Bae, D.S., Kuhl, J.G., Haug, E.J.: A recursive formulation for constrained mechanical system dynamics: Part III. Parallel processor implementation. Mech. Based Des. Struct. Mach. **16**(2), 249–269 (1988)
10. Barns, J., Hut, P.: A hierarchical $o(n \log n)$ force-calculation algorithm. Lett. Nature **324**(4), 446–449 (1986)
11. Becker, N.B., Everaers, R.: From rigid base pairs to semiflexible polymers: coarse-graining DNA. Phys. Rev. E **76**(2), 021923 (2007)
12. Bhalerao, K., Anderson, K.: Modeling intermittent contact for flexible multibody-rigid-body dynamics. Nonlinear Dyn. **60**(1–2), 63–79 (2010)
13. Bhalerao, K.D.: On methods for efficient and accurate design and simulation of multibody systems. Ph.D. thesis, Rensselaer Polytechnic Institute, Troy (2010)
14. Bhalerao, K.D., Poursina, M., Anderson, K.S.: An efficient direct differentiation approach for sensitivity analysis of flexible multibody systems. Multibody Syst. Dyn. **23**(2), 121–140 (2010)
15. Brandl, H., Johanni, R., Otter, M.: A very efficient algorithm for the simulation of robots and similar multibody systems without inversion of the mass matrix. In: IFAC/IFIP/IMACS Symposium, Vienna, Austria, pp. 95–100 (1986)
16. Chakrabarty, A., Cagin, T.: Coarse grain modeling of polyimide copolymers. Polymer **51**(12), 2786–2794 (2010)
17. Chen, S.J.: RNA folding: conformational statistics, folding kinetics, and ion electrostatics. Annu. Rev. Biophys. **37**(1), 197–214 (2008)
18. Chirikjian, G.S., Wang, Y.: Conformational statistics of stiff macromolecules as solutions to partial differential equations on the rotation and motion groups. Phys. Rev. E **62**(1), 880–892 (2000)
19. Chun, H.M., Padilla, C.E., Chin, D.N., Watenabe, M., Karlov, V.I., Alper, H.E., Soosaar, K., Blair, K.B., Becker, O.M., Caves, L.S.D., Nagle, R., Haney, D.N., Farmer, B.L.: MBO(N)D: a multibody method for long-time molecular dynamics simulations. J. Comput. Chem. **21**(3), 159–184 (2000)
20. Chung, S., Haug, E.J.: Real-time simulation of multibody dynamics on shared memory multiprocessors. J. Dyn. Syst. Meas. Control **115**(4), 627–637 (1993)
21. de Leeuw, S., Perram, J., Smith, E.: Simulation of electrostatic systems in periodic boundary conditions. I. Lattice sums and dielectric constants. Proc. R. Soc. Lond. Ser. A **373**(1752), 27–56 (1980)
22. Dill, K.A., Ozkan, S.B., Shell, M.S., Weikl, T.R.: The protein folding problem. Annu. Rev. Biophys. **37**(1), 289–316 (2008)
23. Ding, H.Q., Karasawa, N., Goddard, W.A. III: Atomic level simulation on a million particles: the cell multipole method for Coulomb and London nonbond interactions. J. Chem. Phys. **97**(6), 4309–4315 (1992)
24. Ewald, P.: Evaluation of optical and electrostatic lattice potentials. Ann. Phys. **64**, 253–287 (1921)
25. Featherstone, R.: The calculation of robotic dynamics using articulated body inertias. Int. J. Robot. Res. **2**(1), 13–30 (1983)
26. Featherstone, R.: Robot Dynamics Algorithms. Kluwer Academic, Boston (1987)

27. Featherstone, R.: A divide-and-conquer articulated body algorithm for parallel $O(\log(n))$ calculation of rigid body dynamics. Part 1: Basic algorithm. Int. J. Robot. Res. **18**(9), 867–875 (1999)
28. Featherstone, R.: A divide-and-conquer articulated body algorithm for parallel $O(\log(n))$ calculation of rigid body dynamics. Part 2: Trees, loops, and accuracy. Int. J. Robot. Res. **18**(9), 876–892 (1999)
29. Fijany, A., Bejczy, A.K.: Techniques for parallel computation of mechanical manipulator dynamics. Part II: Forward dynamics. In: Leondes, C. (ed.) Advances in Robotic Systems and Control, vol. 40, pp. 357–410. Academic Press, San Diego (1991)
30. Fijany, A., Sharf, I., D'Eleuterio, G.M.T.: Parallel $O(\log n)$ algorithms for computation of manipulator forward dynamics. IEEE Trans. Robot. Autom. **11**(3), 389–400 (1995)
31. Freddolino, P.L., Arkhipov, A., Shih, A.Y., Yin, Y., Chen, Z., Schulten, K.: Application of residue-based and shape-based coarse graining to biomolecular simulations. In: Voth, G.A. (ed.) Coarse-Graining of Condensed Phase and Biomolecular Systems, pp. 299–315. CRC Press, Boca Raton (2008)
32. Greengard, L., Rokhlin, V.: A fast algorithm for particle simulations. J. Comput. Phys. **135**(2), 280–292 (1997)
33. Haile, J.: Molecular Dynamics Simulation: Elementary Methods. Wiley Interscience, New York (1992)
34. Hwang, R.S., Bae, D., Kuhl, J.G., Haug, E.J.: Parallel processing for real-time dynamics systems simulations. J. Mech. Des. **112**(4), 520–528 (1990)
35. Jain, A.: Unified formulation of dynamics for serial rigid multibody systems. J. Guid. Control Dyn. **14**(3), 531–542 (1991)
36. Jain, A., Vaidehi, N., Rodriguez, G.: A fast recursive algorithm for molecular dynamics simulation. J. Comput. Phys. **106**(2), 258–268 (1993)
37. Kane, T.R., Levinson, D.A.: Dynamics: Theory and Application. McGraw-Hill, New York (1985)
38. Kane, T.R., Likins, P.W., Levinson, D.A.: Spacecraft Dynamics. McGraw-Hill, New York (1983)
39. Kasahara, H., Fujii, H., Iwata, M.: Parallel processing of robot motion simulation. In: Proceedings IFAC 10th World Conference (1987)
40. Kazerounian, K., Latif, K., Alvarado, C.: Protofold: a successive kinetostatic compliance method for protein conformation prediction. J. Mech. Des. **127**(4), 712–718 (2005)
41. Khan, I., Poursina, M., Anderson, K.S.: DCA-based optimization in transitioning to finer models in articulated multi-flexible-body modeling of biopolymers. In: Proceedings of the ECCOMAS Thematic Conference—Multibody Systems Dynamics, Brussels, Belgium (2011)
42. Lathrop, L.H.: Parallelism in manipulator dynamics. Int. J. Robot. Res. **4**(2), 80–102 (1985)
43. Leach, A.R.: Molecular Modelling Principles and Applications, 2nd edn. Prentice Hall, New York (2001)
44. Lebrun, A., Lavery, R.: Modeling the mechanics of a DNA oligomer. J. Biomol. Struct. Dyn. **16**(3), 593–604 (1998)
45. Luh, J.S.Y., Walker, M.W., Paul, R.P.C.: On-line computational scheme for mechanical manipulators. J. Dyn. Syst. Meas. Control **102**(2), 69–76 (1980)
46. Malczyk, P., Fraczek, J.: Lagrange multipliers based divide and conquer algorithm for dynamics of general multibody systems. In: Proceedings of the ECCOMAS Thematic Conference—Multibody Systems Dynamics, Warsaw, Poland (2009)
47. Malczyk, P., Fraczek, J.C.: Parallel index-3 formulation for real-time multibody dynamics simulations. In: Proceedings of the 1st Joint International Conference on Multibody System Dynamics, Lappeenranta, Finland (2010)
48. Marrink, S.J., de Vries, A.H., Mark, A.E.: Coarse grained model for semiquantitative lipid simulations. J. Phys. Chem. B **108**, 750–760 (2004)
49. Mukherjee, R., Anderson, K.S.: A logarithmic complexity divide-and-conquer algorithm for multi-flexible articulated body systems. J. Comput. Nonlinear Dyn. **2**(1), 10–21 (2007)

50. Mukherjee, R., Anderson, K.S.: An orthogonal complement based divide-and-conquer algorithm for constrained multibody systems. Nonlinear Dyn. **48**(1–2), 199–215 (2007)
51. Mukherjee, R.M.: Multibody dynamics algorithm development and multiscale modelling of materials and biomolecular systems. Ph.D. thesis, Rensselaer Polytechnic Institute, Troy (2007)
52. Mukherjee, R.M., Anderson, K.S.: Efficient methodology for multibody simulations with discontinuous changes in system definition. Multibody Syst. Dyn. **18**(2), 145–168 (2007)
53. Mukherjee, R.M., Bhalerao, K.D., Anderson, K.S.: A divide-and-conquer direct differentiation approach for multibody system sensitivity analysis. Struct. Multidiscip. Optim. **35**(5), 413–429 (2007)
54. Neilan, P.E.: Efficient computer simulation of motions of multibody systems. Ph.D. thesis, Stanford University (1986)
55. Niedermeier, C., Tavan, P.: A structure adapted multipole method for electrostatic interactions in protein dynamics. J. Chem. Phys. **101**(1), 734–748 (1994)
56. Parisien, M., Major, F.: The MC-fold and MC-Sym pipeline infers RNA structure from sequence data. Nature **452**(7183), 51–55 (2008)
57. Poursina, M.: Robust framework for the adaptive multiscale modeling of biopolymers. Ph.D. thesis, Rensselaer Polytechnic Institute, Troy (2011)
58. Poursina, M., Anderson, K.S.: Constant temperature simulation of articulated polymers using divide-and-conquer algorithm. In: Proceedings of the ECCOMAS Thematic Conference—Multibody Systems Dynamics, Brussels, Belgium (2011)
59. Poursina, M., Anderson, K.S.: Multibody dynamics in generalized divide and conquer algorithm (GDCA) scheme. In: Proceedings of the Eighths International Conference on Multibody Systems, Nonlinear Dynamics and Control, ASME Design Engineering Technical Conference 2011 (IDETC11), DETC2011-48383, Washington, DC (2011)
60. Poursina, M., Anderson, K.S.: An extended divide-and-conquer algorithm for a generalized class of multibody constraints. Multibody Syst. Dyn. (2012). doi:10.1007/s11044-012-9324-9
61. Poursina, M., Anderson, K.S.: Long-range force and moment in multiresolution simulation of molecular systems. J. Comput. Phys. **231**(21), 7237–7254 (2012)
62. Poursina, M., Bhalerao, K.D., Anderson, K.S.: Energy concern in biomolecular simulations with discontinuous changes in system definition. In: Proceedings of the ECCOMAS Thematic Conference—Multibody Systems Dynamics, Warsaw, Poland (2009)
63. Poursina, M., Bhalerao, K.D., Flores, S., Anderson, K.S., Laederach, A.: Strategies for articulated multibody-based adaptive coarse grain simulation of RNA. Methods Enzymol. **487**, 73–98 (2011)
64. Poursina, M., Khan, I., Anderson, K.S.: Model transitions and optimization problem in multiflexible-body modeling of biopolymers. In: Proceedings of the Eighths International Conference on Multibody Systems, Nonlinear Dynamics and Control, ASME Design Engineering Technical Conference 2011 (IDETC11), DETC2011-48383, Washington, DC (2011)
65. Poursina, M., Laflin, J., Anderson, K.S.: Fast electrostatic force and moment calculations in multibody-based simulations of coarse-grained biopolymers. In: Proceedings of the Eighths International Conference on Multibody Systems, Nonlinear Dynamics and Control, ASME Design Engineering Technical Conference 2011 (IDETC11), DETC2011-48383, Washington, DC (2011)
66. Praprotnik, M., Site, L., Kremer, K.: Adaptive resolution molecular-dynamics simulation: changing the degrees of freedom on the fly. J. Chem. Phys. **123**(22), 224106–224114 (2005)
67. Redon, S., Lin, M.C.: An efficient, error-bounded approximation algorithm for simulating quasi-statics of complex linkages. Comput. Aided Des. **38**(4), 300–314 (2006)
68. Rosenthal, D.: An order *n* formulation for robotic systems. J. Astronaut. Sci. **38**(4), 511–529 (1990)
69. Rosenthal, D.E., Sherman, M.A.: High performance multibody simulations via symbolic equation manipulation and Kane's method. J. Astronaut. Sci. **34**(3), 223–239 (1986)
70. Rossi, R., Isorce, M., Morin, S., Flocard, J., Arumugam, K., Crouzy, S., Vivaudou, M., Redon, S.: Adaptive torsion-angle quasi-statics: a general simulation method with applications to pro-

tein structure analysis and design. In: ISMB/ECCB (Supplement of Bioinformatics), vol. 23, pp. 408–417 (2007)

71. Scheraga, H.A., Khalili, M., Liwo, A.: Protein-folding dynamics: overview of molecular simulation techniques. Annu. Rev. Phys. Chem. **58**(1), 57–83 (2007)
72. Shahbazi, Z., Ilies, H., Kazerounian, K.: Hydrogen bonds and kinematic mobility of protein molecules. J. Mech. Robot. **2**(2), 021009 (2010)
73. Shillcocka, J.C., Lipowsky, R.: Equilibrium structure and lateral stress distribution of amphiphilic bilayers from dissipative particle dynamics simulations. J. Chem. Phys. **117**, 5048–5061 (2002)
74. Tozzini, V.: Coarse-grained models for proteins. Curr. Opin. Struct. Biol. **15**(2), 144–150 (2005)
75. Vaidehi, N., Jain, A., Goddard, W.A. III: Constant temperature constrained molecular dynamics: the Newton-Euler inverse mass operator method. J. Phys. Chem. **100**(25), 10508–10517 (1996)
76. Vereshchagin, A.F.: Computer simulation of the dynamics of complicated mechanisms of robot-manipulators. Eng. Cybern. **12**(6), 65–70 (1974)
77. Walker, M.W., Orin, D.E.: Efficient dynamic computer simulation of robotic mechanisms. J. Dyn. Syst. Meas. Control **104**(3), 205–211 (1982)
78. Wehage, R.A., Haug, E.: Generalized coordinate partitioning for dimension reduction in analysis of constrained dynamic systems. J. Mech. Des. **104**, 247–255 (1982)

Efficiency and Precise Interaction for Multibody Simulations in Augmented Reality

Lorenzo Mariti and Pier Paolo Valentini

Abstract In this chapter, an enhanced methodology for interactive, accurate, fast and robust multibody simulations using Augmented Reality is presented and discussed. This methodology is based on the integration of a mechanical tracker and a dedicated impulse based solver. The use of the mechanical tracker for the interaction between the user and the simulation allows to separate the processing of the data coming from the position tracking from those coming from the image collimation processing. By this way simulation results and visualization remain separated and the precision is enhanced. The use of a dedicated sequential impulse solver allows a quick and stable simulation also for a large number of bodies and overabundant constraints. The final result of this work is a software tool able to manage real time dynamic simulations and update the augmented scene accordingly. The robustness and the reliability of the system will be checked over two test cases: a ten pendula dynamic system and of a cross-lift mechanism simulation.

1 Introduction

During the last years, many investigations focused on the increasing trend of using Augmented Reality (AR) [1] to support a variety of engineering activities and to develop interactive tools in design. Augmented Reality has been used for supporting geometrical modeling [2], reverse engineering [3], assembly simulation [4–7], analysis [8].

The augmented reality deals with the combination of real world images and computer generated data. Most AR research is concerned with the use of live video imagery which is digitally processed and augmented by the addition of computer generated graphics. The purpose of the augmented environment is to extend the visual perception of the world, being supported by additional information and virtual objects. One of the most important feature of AR is the possibility to embed an high level of user's interaction with the augmented scene [9].

L. Mariti · P.P. Valentini (✉)
Department of Industrial Engineering, University of Rome "Tor Vergata", via del Politecnico, 1, 00133, Rome, Italy
e-mail: valentini@ing.uniroma2.it

J.-C. Samin, P. Fisette (eds.), *Multibody Dynamics*, Computational Methods in Applied Sciences 28,
DOI 10.1007/978-94-007-5404-1_8,

Some recent contributions showed an increasing interest in developing environments for simulating and reviewing physical simulations based on Augmented Reality [10–14]. They focused on the interactivity between the user and the augmented environment. By this way, the user is not a mere spectator of the contents of the augmented scene, but influences them. Interaction in dynamic simulations can concern both the definition of boundary conditions and initial parameters and the real time control of the process.

In a recent paper [15], P.P. Valentini and E. Pezzuti developed a methodology for implementing, solving and reviewing multibody simulation using augmented reality. According to their results, augmented reality facilitates the interaction between the user and the simulated system and allows a more appealing visualization of simulation results. For this reason, this approach has revealed to be suitable for didactical applications and teaching purposes as well.

On the other hand, the development of multibody simulations in augmented reality requires very fast solver in order to produce a smooth animation and an effective illusion. Valentini and Pezzuti proposed to use an optical marker to track the position of the user and interact with the objects in the scene. Moreover, they developed some simple examples to introduce the methodology.

Starting from the discussed background, this chapter aims to discuss two important enhancements of the work in [15] in order to improve the accuracy of user's interaction and to allow a robust simulation of large multibody systems. The accuracy in tracking the user is important to perform precise simulation that are required in many engineering applications. The use of optical marker is simple but in this case position tracking error highly depends on the resolution of the camera sensor and on the distance between the camera and the marker. Using standard USB cameras, this error can be some millimeters and can be unacceptable for precise applications, and unwanted flickering may occur due to the tracking algorithm limitations [16]. In order to improve the accuracy in tracking, in the present study we have included the use of a mechanical instrumented arm which is able to achieve a precision of about 0.2 mm in a working space of about 1.2 m of diameter. This enhancement is also important to separate the processing of the data coming from the position tracking from those coming from the image processing (for perspective collimation and visualization issues). By this way, higher precision in the analysis results can be achieved and only the graphical visualization is affected by optical imprecision.

The second enhancement which has been proposed and tested, is about the use of a dedicated solver for managing the integration of the equations of motion. The implemented solver makes use of the sequential impulse strategy [17] which allows a quick and stable simulation also for a large number of bodies, in presence of overabundant and unilateral constraints. According to some authors and applications (i.e. [18, 19]), this approach leads to a very fast and stable solution, but quite less accurate than global solution methods.

The chapter is organized as follows. First of all, a brief introduction about the state of the art of virtual engineering in augmented reality is presented, focusing on the aspects related to multibody simulations. Then, a description of hardware and software implementations is discussed, including the explanation of the iterative impulse solver and the details about the strategy for implementing interaction between the user and the scene. In the second part, two examples are presented and discussed.

2 Multibody Simulations in Augmented Reality Environments

The first and basic implementation of multibody simulations in augmented reality is about the possibility to project on a real scenario the results coming from a pre-computed simulation. It concerns the rendering on the augmented scene of all the objects involved in the simulation whose position is updated according to the results of the simulation.

This implementation is similar to that of the common post-processing software for visualizing graphics results. The only difference is in the introduction of the simulated system in a real context. The advantage is to perceive the interaction with the real world and check working spaces, possible interferences, etc.

Although it can be useful, this approach does not unveil all the potential of AR [15]. A more powerful way to enhance multibody simulation is to introduce an high level of interactivity. It means that the user does not only watch the augmented scene, but interacts with it. In this case, the solution of the equations of motion has to be computed synchronously to the animation in order to populate the scene with quickly updated information.

With this type of interaction, the user is active in the scene and can change the augmented contents by picking, pushing and moving objects and controlling the provided information and the environment behaves according to realistic physics laws. The interaction is carried out with advanced input/output devices involving different sensorial channels (sight, hear, touch, etc.) in an integrated way [20]. In particular, in order to interacts with digital information through the physical environment, the environment has to be provided of the so called Tangible User Interfaces (TUIs) [21–23].

In the most simple implementations, the patterned markers used for collimating real and virtual contents are used also as TUIs. In advanced implementation visual interaction is achieved by dedicated interfaces (mechanical, magnetic, optical, etc.). By this way, the image processing for the computation of the camera-world perspective transformation can be separated from the acquisition and processing of the user intent. Thanks to this split computation, it is possible to achieve a very precise interaction and simulation and a less precise (and more efficient) visual collimation. This means that the results of the simulations can be accurate and suitable for technical and engineering purposes. On the other hand, the small imprecisions in the optical collimation are limited to graphics display.

Starting from these considerations, the generic integration algorithm between multibody simulations and augmented reality presented in [15] can be modified separating the contributions of interaction and collimation. For this reason, an high-interactive generic multibody simulation in augmented reality can be implemented following five main steps:

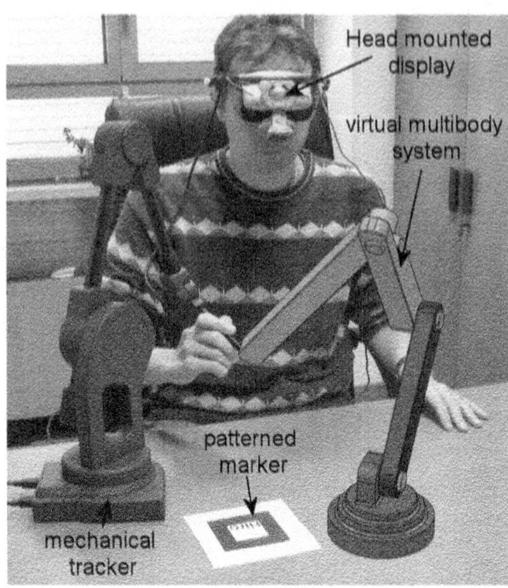

Fig. 1 Augmented reality hardware setup

1. Before the simulation starts, the geometries and topological properties (joints and connections) have to be defined (as for any multibody system);
2. The real scene has to contain information for collimating the real world to the virtual objects;
3. The real environment has to contain one or more TUIs for the acquisition of the user's intent of interacting with the scene;
4. During each frame acquisition, the user's intent has to be interpreted; the multibody equations have to be built and solved synchronously in order to compute the correct position of all the virtual bodies in the scene;
5. For each frame acquisition, virtual objects have to be rendered on the scene, after the numerical integration, in the correct position and attitude.

3 Implementation of the Augmented Reality Environment

3.1 Hardware Setup

For the specific purpose of this investigation, the implemented AR system (depicted in Fig. 1) includes an input video device Microsoft LifeCam VX6000 USB 2.0 camera, an Head Mounted Display (Emagin Z800) equipped with OLed displays, a Revware Microscribe GX2 mechanical tracker and a personal computer.

The Revware Microscribe is an instrumented arm (digitizer) which can be grabbed and driven by the user and possesses five degrees of freedom. It is able to acquire the real-time position of its tip stylus. The operating space is a sphere of about 1.2 m of diameter and the precision of tracking is up to 0.2 mm.

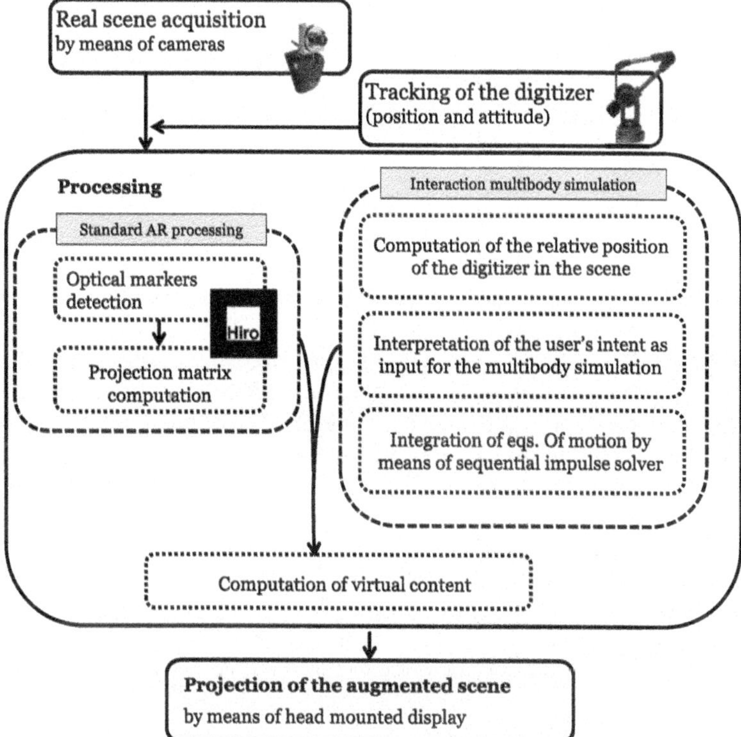

Fig. 2 Multibody simulation in augmented reality procedural scheme

3.2 Implementation and Interaction Strategy

In order to implement a AR environment suitable for multibody dynamics interactive simulation, we have chosen the following strategy (see Fig. 2). Before the simulation starts, the geometries and topological properties (joints and connections) have to be defined (as for any multibody simulation). Then, the real scene has to contain information for collimating the real world to the virtual objects. Usually this operation is performed by a patterned planar marker which is recognized by the processing units. For the scope it is necessary to perform the computation of the camera point of view and the corresponding perspective effect.

During each frame acquisition, the position of the user in the scene has to be acquired as well. This task can be performed using the mechanical tracker and computing the position and attitude of its end effector. This information can be interpreted and transferred as a spatial input for the simulation (user's intent to interact with the scene). Then, the multibody equations have to be solved in order to compute the correct position of all the virtual bodies in the scene, taking into account the user's intent.

After this computation, all the virtual objects have to be rendered on the scene in the correct position and attitude.

All the supporting software has been implemented using C++ programming language and Microsoft Visual Studio 2003 developing suite. Routines for image processing have been developed using the open source libraries named ARToolkit[1] which has been successfully used in many previous investigations. The libraries comprise a set of numerical procedures which are able to detect and recognize planar patterned markers in a video stream in real time. Using correlation techniques, the routines are also able to compute relative position and attitude between markers and camera with good precision for visual purposes. This computation is necessary for an accurate perspective collimation between virtual entities and real scene. The details about specific implementation and about the contents of the library can be found on the Internet site of the developers.[2]

The Microscribe GX2 has been integrated using the Microscribe SDK library that allows the real time access to position and attitude of each link of the instrumented arm. This library interprets the output coming from the rotation sensors of the five revolute joints of the mechanical arm and computes the position of the stylus tip by solving an open kinematic chain problem.

For managing complex geometries the OpenVrml library[3] has been included. All rendering tasks about virtual objects in the augmented scene have been performed using OpenGL library.

Details about the procedures for deducing and solving the equations of motion of the system under investigation have been provided in the next sections.

All these pieces of software have been integrated into a single simulation environment.

3.3 Collimation Procedure

The first step in the integration of the tracker in the augmented scene is the collimation between the information acquired by the instrumented device and that of the digital camera (see Fig. 3). The video stream acquired by the digital camera is elaborated by an image processing routine. It is able to recognize a patterned marker in the scene and to compute the corresponding transformation matrix between the camera and the real word. This matrix is used to project all the virtual contents in the augmented scene in the correct position and perspective.

The information acquired by the digitizer is concerned with the position and attitude of the end effector with respect to the reference frame fixed to the device itself.

In order to ensure the collimation between the data stream coming from the camera and that from the tracker, it is important to compute the relative transformation

[1]The ARToolkit libraries can be freely downloaded from the Internet site http://sourceforge.net/project/showfiles.php?group_id=116280.

[2]http://www.hitl.washington.edu/artoolkit/.

[3]The library can be freely downloaded from the Internet site http://openvrml.org/.

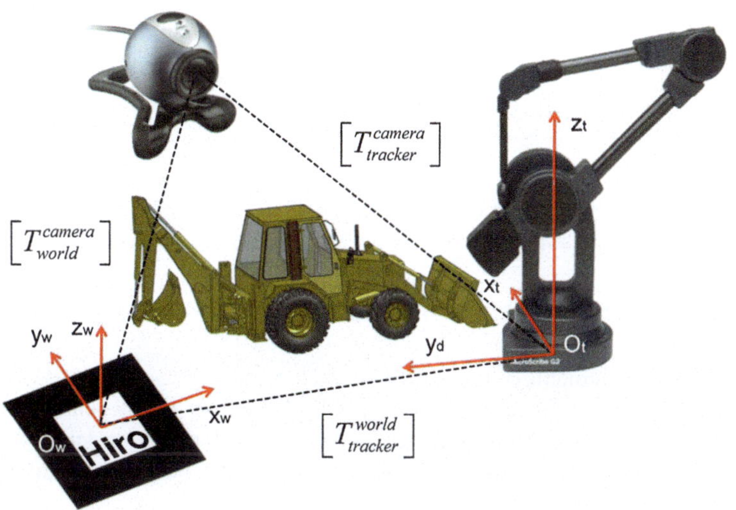

Fig. 3 Collimation between optical marker and mechanical tracking system

matrix between the tracker and the world (described by the marker). This calibration has to be performed only at the beginning of the application and it has to be repeated only if the relative position between the world marker and the digitizer changes.

The calibration procedure can be performed by picking with the tracker stylus a set not-aligned points (four no-coplanar points at least) at known positions with respect to the relative frame associated to the marker.

For expressing the coordinate transformation between points, it is useful to deal with homogeneous transformation matrices which include information on both rotation and translation parameters. A generic homogeneous transformation matrix can be expressed in the form:

$$[T] = \begin{bmatrix} [\text{Orientation}]_{3\times 3} & [\text{Position}]_{3\times 1} \\ 0 \quad 0 \quad 0 & 1 \end{bmatrix} \tag{1}$$

In the same way, a generic point can be expressed with the following coordinate vector:

$$\{P\} = \begin{Bmatrix} x & y & z & 1 \end{Bmatrix}^T \tag{2}$$

The coordinate transformation of a generic point P from the local coordinate system fixed to the digitizer to the world coordinate system attached to the marker can be written as:

$$\{P\}_{world} = \left[T_{world}^{digitizer}\right]\{P\}_{digitizer} \tag{3}$$

where:

$\{P\}_{world}$ is the vector containing the coordinate of the point P expressed in the world reference frame;

$\{P\}_{digitizer}$ is the vector containing the coordinate of the point P expressed in the local (tracker) reference frame.

Considering a collection of points P_1 P_2 ... P_n, we can build two matrices as:

$$[P]_{world} = \left[\{P_1\}_{world} \quad \{P_2\}_{world} \quad \dots \quad \{P_n\}_{world} \right] \tag{4}$$

$$[P]_{digitizer} = \left[\{P_1\}_{digitizer} \quad \{P_2\}_{digitizer} \quad \dots \quad \{P_n\}_{digitizer} \right] \tag{5}$$

In order to compute the matrix $[T_{word}^{digitizer}]$ we have to solve the following system of equations:

$$[P]_{world} = \left[T_{world}^{digitizer}\right][P]_{digitizer} \tag{6}$$

for the unknown elements of the matrix $[T_{word}^{digitizer}]$.

An homogeneous transformation matrix is defined by 6 independent parameters (three for the description of the rotation and three for the translation). For this reason, the system (6) has more equations than unknowns and the solution can be computed as:

$$\left[T_{world}^{digitizer}\right] = [P]_{world}^{-1^+}[P]_{digitizer} \tag{7}$$

where the $[P]_{world}^{-1^+}$ is the pseudo-inverse matrix of the $[P]_{world}$ matrix.

Due to numerical approximation or errors in acquisition, the orientation block of the computed matrix $[T_{word}^{digitizer}]$ can result not exactly orthogonal. Since it represents a rigid spatial rotation, it is important to correct this imprecision. For this purpose, we can operate a QR decomposition of this orientation block:

$$\left[\text{Orientation}_{word}^{digitizer}\right]_{3\times 3} = [R_1]_{3\times 3}[U_1]_{3\times 3} \tag{8}$$

where (due to the QR algorithm):

$[R_1]$ is an orthogonal matrix representing the corrected rotation;

$[U_1]$ is a matrix whose upper band contains the errors of approximation and the lower band has only zero elements. In case of a pure rotation (orientation block without errors) $[U_1] = [I]$.

Finally, in order to compute the transformation matrix between the digitizer and the camera $[T_{camera}^{digitizer}]$, useful to collimate the acquired points to the visualized ones, a matrix multiplication has to be performed:

$$\left[T_{camera}^{digitizer}\right] = \left[T_{word}^{digitizer}\right]\left[T_{camera}^{word}\right] \tag{9}$$

Figure 4 shows some snapshots acquired during a calibration procedure. The reference points are picked using a reference cube of known dimensions (80 mm × 80 mm × 80 mm).

3.4 User's Interaction

Once the position and the attitude of the tracker are correctly recorded, we have to define the methods to interact with the simulation.

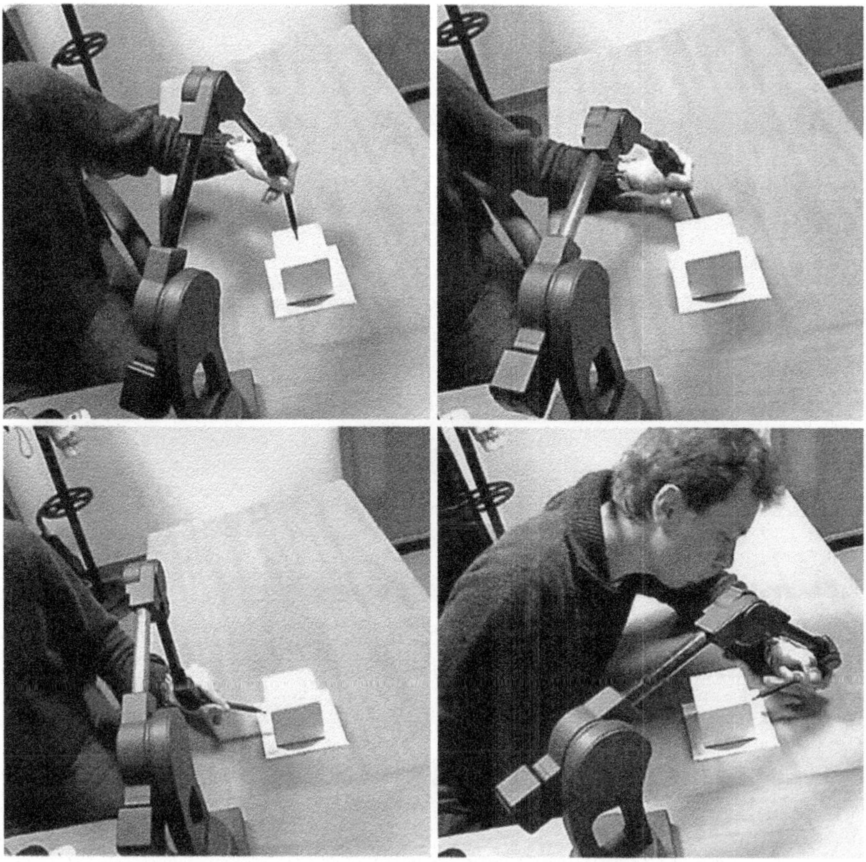

Fig. 4 Four snapshots taken during calibration procedure

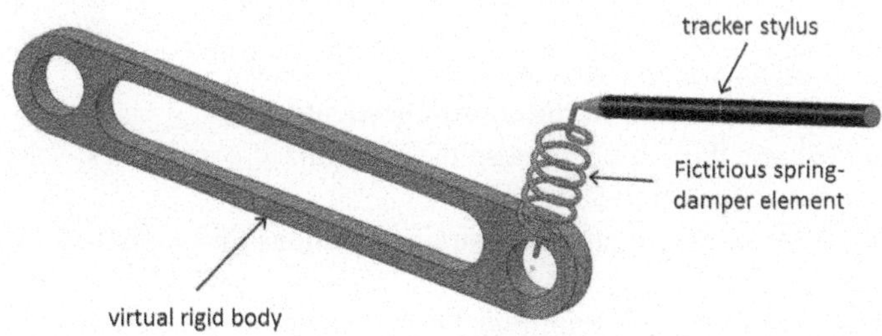

Fig. 5 The implementation of a fictitious spring for managing the user's interaction to the simulated bodies

A possible solution (see Fig. 5) is the use of a fictitious spring-damper element connected between the digitizer stylus tip and a point belonging to one of the rigid bodies in the simulation [24]. By this way, if the tip is moved in the scene, it affects the dynamics of that point that moves accordingly.

Mathematically, the use of a fictitious spring between the tracker stylus tip and a spline control point adds a term to the external force vector $\{F\}_e$ in equations of motion:

$$\{F\}_e = \{F\}_e - k_{f\_spring}\left\{d^{tracker\_tip}_{virtual\_body\_point}\right\} - c_{f\_spring}\left\{\dot{d}^{tracker\_tip}_{virtual\_body\_point}\right\} \tag{10}$$

where:

$\{d^{tracker\_tip}_{virtual\_body\_point}\}$ is the distance between the connected control point and the digitizer tip;
$\{\dot{d}^{tracker\_tip}_{virtual\_body\_point}\}$ is the derivative of $\{d^{tracker\_tip}_{virtual\_body\_point}\}$ with respect to time;
k_{f_spring} is the stiffness coefficient of the fictitious spring;
c_{f_spring} is the damping coefficient of the fictitious spring.

In order to prevent the dynamics of the system to be affected by the presence of an external (and not physical) elastic component the value of the stiffness parameter has to be chosen very stiff. Moreover, the use of a damping coefficient improves the stability of the system and prevents jittering in simulation.

4 Implementing the Sequential Impulse Solver

Given a collection of n_{body} rigid bodies, constrained by n_{joint} kinematic joints and subjected to a set of external forces, the equation of motion can be deduced following different approaches.

One of the most used multibody dynamics formulation is based on the Lagrange equations, obtaining a differential-algebraic system:

$$\begin{cases} [M]\{\ddot{q}\} - [\psi_q]^T\{\lambda\} = \{F_e\} \\ \{\psi\} = \{0\} \end{cases} \tag{11}$$

where:

$[M]$ is the inertia matrix of the collection of rigid bodies;
$\{\psi\}$ is the vector containing all the constraint equations;
$\{q\}$, $\{\dot{q}\}$ and $\{\ddot{q}\}$ are the vectors of absolute generalized coordinates, velocities and accelerations, respectively;
$[\psi_q]$ is the Jacobian matrix of the constraint equations (differentiated with respect to the generalized coordinates);
$\{\lambda\}$ is the vector of Lagrange multipliers associated to the constraint equations;
$\{F\}_e$ is the vector containing all the elastic and external forces (including the fictitious spring contribution).

In order to reduce the complexity of the solution, the constraint equations are often differentiated two times with respect to time and Eq. (11) is rearranged as:

$$\begin{bmatrix} [M] & [\psi_q]^T \\ [\psi_q] & [0] \end{bmatrix} \begin{Bmatrix} \{\ddot{q}\} \\ \{\lambda\} \end{Bmatrix} = \begin{Bmatrix} \{F_e\} \\ \{\gamma\} \end{Bmatrix} \tag{12}$$

where

$$\{\gamma\} = -\left([\psi_q]\{\dot{q}\}\right)_q \{\dot{q}\} - 2[\psi_{qt}]\{\dot{q}\} - \{\psi_{tt}\} \tag{13}$$

and the subscripts "q" and "t" denote the differentiation with respect to the generalized coordinates and time, respectively.

Both Eq. (11) and Eq. (12) allow to solve for the unknown generalized acceleration, velocities and positions taking into account all the constraint equations simultaneously. Of course, this approach can achieve accurate results with suitable DAE solver. On the other hand, a dynamic system with a lot of constraints increases the complexity of the problem and the computational effort to solve it. For this reason, the system in (12) can be rearranged for being suitable for the sequential impulse solution strategy.

There are two main steps in the impulse-based methodology. Firstly, the equations of motion are tentatively solved considering elastic and external forces but neglecting all the kinematic constraints. This produces a solution that is only approximated because the constraint equations are not satisfied. In a second step, a sequence of impulses are applied to each body in the system in order to correct its velocity according to the limitation imposed by the constraint. This second step is iterative. It means that a series of impulse is applied to the bodies until the constraint equations are fulfilled within a specific tolerance. It is important to underline that each impulse is applied independent from the others. By this way the constraint equations are not solved globally, but iteratively.

4.1 Solving the Equations of Motion

Following the approach introduced in the previous section, the sequential impulse formulation can be split into two main steps. The first one is about the solution of the equations of motion in (11) neglecting the constraint equations and constraint forces:

$$[M]\{\ddot{q}\}_{approx} = \{F_e\} \tag{14}$$

By this way, Eq. (14) can be solved for $\{\ddot{q}\}_{approx}$ that represent the vector of approximated generalized accelerations.

The values of the corresponding approximated generalized velocities and positions can be computed by linear approximation:

$$\{\dot{q}\}_{approx} = h \cdot \{\ddot{q}\}_{approx} \tag{15}$$

$$\{q\}_{approx} = h \cdot \{\dot{q}\}_{approx} \tag{16}$$

where h is the integration time step.

In order to correct the $\{\dot{q}\}_{approx}$ and fulfill the constraint equations, a series of impulses $\{P\}_{constraint}$ has to be applied to the bodies. Each impulse is computed imposing the fulfillment of the constraint equations written in terms of generalized coordinates. As well-known from the Newton's law, the application of the impulses causes a variation of momentum:

$$[M]\left(\{\dot{q}\}_{corrected} - \{\dot{q}\}_{approx}\right) = \{P\}_{constraint} \tag{17}$$

where $\{\dot{q}\}_{corrected}$ is the vector of generalized velocities after the application of impulses $\{P\}_{constraint}$.

The corrected velocities can be computed from Eq. (17) as:

$$\{\dot{q}\}_{corrected} = \{\dot{q}\}_{approx} + [M]^{-1}\{P\}_{constraint} \tag{18}$$

Considering that the impulses are related to the constraint equations, they can be computed as

$$\{P\}_{constraint} = [\psi_q]^T\{\delta\} \tag{19}$$

where $\{\delta\}$ is the vector of Lagrange multipliers associated to the impulses.

Since the effect of the impulses is to correct the generalized velocities and fulfill the kinematic constraints, the $\{\dot{q}\}_{corrected}$ has to satisfy the constraint equations written in terms of velocities:

$$\{\psi\} = \{0\} \quad \Rightarrow \quad \frac{\mathrm{d}\{\psi\}}{\mathrm{d}t} = [\psi_q]\{\dot{q}\} + \{\psi_t\} = \{0\} \tag{20}$$

$$[\psi_q]\{\dot{q}_{corrected}\} + \{\psi_t\} = \{0\} \tag{21}$$

Inserting Eq. (19) into Eq. (18) and substituting $\{\dot{q}\}_{corrected}$ into Eq. (21) we can obtain:

$$[\psi_q]\left(\{\dot{q}_{approx}\} + [M]^{-1}[\psi_q]^T\{\delta\}\right) + \{\psi_t\} = \{0\} \tag{22}$$

Equation (22) can be solved for $\{\delta\}$obtaining:

$$\{\delta\} = \left([\psi_q][M]^{-1}[\psi_q]^T\right)^{-1}\left([\psi_q]\{\dot{q}\}_{approx} + \{\psi_t\}\right) \tag{23}$$

Then, the impulses can be computed using Eq. (19) and the corrected values of generalized velocities using Eq. (18).

Since the impulses are computed sequentially, the global fulfillment of the constraint equations cannot be directly achieved. Some iterations are required. The computation of $\{\delta\}$, $\{P\}_{constraint}$ and $\{\dot{q}\}_{corrected}$ can be repeated till a tolerance on the fulfillment of Eq. (21) is reached or for a maximum number of time. Experience [17] shows that four or five iterations are sufficient to achieve an adequate tolerance. Since the constraints are imposed at velocity level, a stabilized formulation is required to control the constraints fulfillment at the position level. Details are provided in Sect. 4.3.

4.2 Computation of Reaction Forces

When simulating multibody dynamics, one of the most interesting results for engineers is the knowledge of the reaction forces, i.e. the forces exerted by the joints.

Using the sequential impulse formulation, the reaction forces cannot be computed directly but a preliminary computation is required. The problem is that impulses are applied sequentially, it means that each joint exerts more than one impulse during each time step and the various impulses have to be recollected.

In order to deduce a methodology for evaluating the reaction forces of the joints we have to introduce the concept of the accumulated impulse. It can be defined as the resultant impulse of each joint produced each time step and it can be computed as the sum of all the impulses exerted by the joint over the iteration.

In the solution of the equations of motion, the joint impulses are evaluated using Eq. (23) and Eq. (19). This computation is performed iteratively in order to reach a set of velocities congruent to kinematic joints. It means that at each iteration a new impulse vector $\{P\}_{constraint}$ is computed.

The accumulated Lagrange multipliers $\{\Delta\}$ of the impulses for each time step can be evaluated as:

$$\{\Delta\} = \sum_{iterations} \{\delta\} \tag{24}$$

The accumulated impulse $\{P_{tot}\}_{constraint}$ can be computed using Eq. (19) obtaining:

$$\{P_{tot}\}_{constraint} = [\psi_q]^T \{\Delta\} \tag{25}$$

The reaction forces $\{F\}_{constraint}$ exerted by the joints can be computed using the general relation between forces and impulses:

$$\{F\}_{constraint} = \frac{\{P_{tot}\}_{constraint}}{h} \tag{26}$$

4.3 Stability Issues

The use of the sequential impulse strategy is subjected to the use of constraint equations expressed in terms of generalized velocity. It means that the exact information about the kinematic joints may be lost during the integration process. In this case a position drift can be observed and stability problems may occur.

In order to enforce the constraints on position a stabilized formulation can be adopted. In this case, the constraint equations in Eq. (21) can be modified as:

$$[\psi_q]\{\dot{q}_{corrected}\} + \{\psi_t\} - \frac{\beta}{h}\{\psi\} = \{0\} \tag{27}$$

where β is a scalar chosen in the range 0–1.

Table 1 Geometrical and inertial parameters of the first example

Parameter	Value
Pendulum length	50 mm
Pendulum mass	0.1 kg
Pendulum principal moments of inertia	$[21, 21, 0.5]$ kg mm^2

5 Examples of Simulation

In order to test the proposed methodology and both hardware and software integration, in this section two examples of implementation are presented and discussed.

5.1 Ten Pendula Dynamic System

The first simulated scenario is about a collection of 10 rigid pendula that move under the effect of gravity. All the pendula have the same geometry and inertial properties which have been summarized in Table 1.

The first pendulum is pivoted to a fixed frame by means of a point-to-point 3 d.o.f. joint. The other pendula are sequentially connected by mean of revolute joints (hinges). The user can interact with the scene by connecting a fictitious spring between the tracker stylus and the free end of the last pendulum. In particular, the user can decide when the connection between the tracker and the last pendulum has to be activated (simulating the clipping) or deactivated (simulating the release). This scenario is also important to test the methodology with event based changes in the equations of motion.

In order to achieve stable and correct results, the solution strategy has to be able to manage rapid changes in force definition. The simulation has been performed with a fixed time step of 0.01 s.

Per each video frame, 4 integration steps are computed and the augmented scene is updated accordingly.

Figure 6 reports a series of four snapshots taken during the run of the simulation. The rigid body collection is real-time rendered along with the simulation. In the first part of the simulation (snapshot A of Fig. 6) the pendula are free to move and they are in an equilibrium position (aligned along the vertical direction).

Then, the user locates the tracker near the last pendulum tip and activates the fictitious spring connection (snapshot B). From this moment, the tip of the last pendulum moves subjected to this connection.

When the user moves the tracker, the tip of the pendulum follows it. It is important to notice that the motion of all the rigid body collection is continuously simulated according to the external action of the gravity and the driving force due to the user's presence.

When the user decides to release the fictitious spring connection (snapshot C), the collection of pendula moves subjected to gravity force only and it oscillates around the equilibrium position (snapshot D).

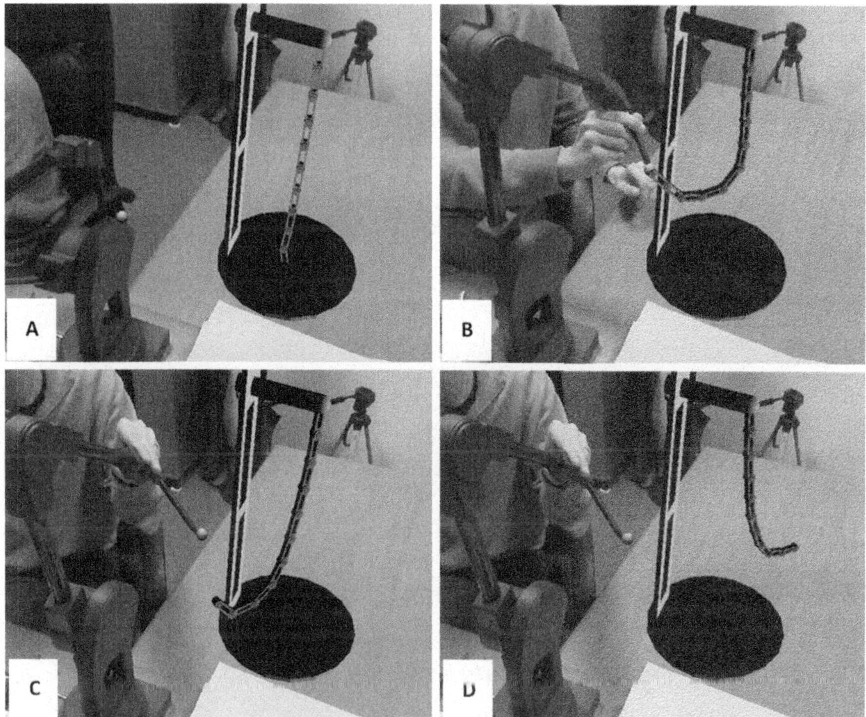

Fig. 6 Snapshots of the simulation of the first example

The accuracy and the stability of the simulation have been checked computing the error in the fulfillment of the constraint equations and the overall energy of the system (potential, kinetic and elastic). The norm of the constraint equations expressed using position variable is always lower than 10^{-6} m except for ten time steps when it reaches $1.6 \cdot 10^{-5}$ m. The variation of the overall energy of the systems (energy loss) is below 1 %. In the computation of the overall energy, the contribution of the external action of the user has been taken into account by evaluating the corresponding power as the dot product between the reaction force of the fictitious spring and the velocity of the digitizer stylus tip as:

$$Power_{user\_action} = \{F_{fictitious\_spring}\} \cdot \{v_{stylus\_tip}\} \tag{28}$$

where

$Power_{user\_action}$ is the power associated to the reaction force of the user's action;
$\{F_{fictitious\_spring}\}$ is the reaction force of the fictitious spring computed as in Eq. (10);
$\{v_{stylus\_tip}\}$ is the absolute velocity of the stylus tip.

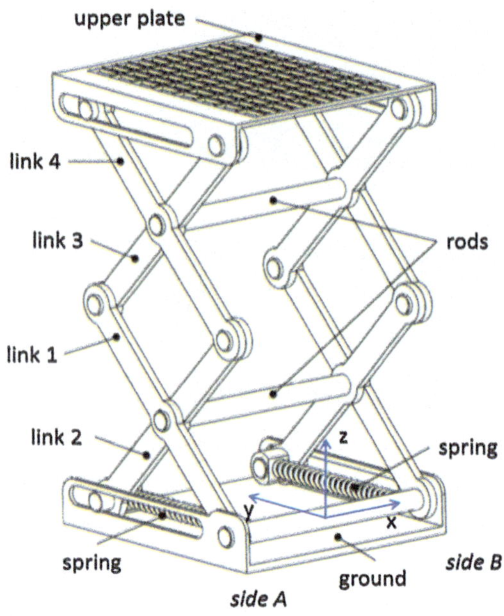

Fig. 7 Cross-lift device of the second example

5.2 Cross-Lift Dynamic Simulation

The second simulated scenario is about the dynamic motion of a cross-lift device which is comprised of 11 rigid bodies connected by 16 hinges and 4 slider joints (see Fig. 7). The links are connected by means of the following constraints:

- Two revolute joints between the frame and the first link on both side A and B of the mechanism;
- Two slider joints between the frame and the second link on both side A and B of the mechanism;
- Four revolute joints between adjacent links on each side of the mechanism (8 in total);
- A revolute joint between the third link and the upper plate on both side A and B of the mechanism;
- A slider joint between the fourth link and the upper plate on both side A and B of the mechanism;
- Four revolute joints connecting the two horizontal rods between the two side of the mechanism.

Two linear spring-damper elements act horizontally between the frame and the slider joint location of the second link on both side A and B of the mechanism. The gravity field acts downward along the vertical direction.

The example has been chosen in order to test the capabilities of the solver to deal with many overabundant constraints. Geometrical, inertial and elastic properties of the simulation have been summarized in Table 2.

Table 2 Geometrical, inertial and elastic parameters of the second example

Parameter	Value
Cross links length	300 mm
Distance between the two side of the mechanism (transverse rod length)	350 mm
Mass of the cross links	0.1 kg
Cross link principal moments of inertia	$[750, 750, 2]\ \mathrm{kg\,mm}^2$
Mass of the transverse rods	0.1 kg
Transverse rod principal moments of inertia	$[1021, 1021, 2]\ \mathrm{kg\,mm}^2$
Mass of the upper plate	0.1 kg
Upper plate principal moments of inertia	$[1021, 1021, 2]\ \mathrm{kg\,mm}^2$
Horizontal spring stiffness	40 N/mm

The user can interact with the scene by imposing a fictitious spring between the stylus tip and the middle point of the upper plate. In particular, the user can decide when the connection between the tracker and the upper plate has to be activated (simulating the clipping) or deactivated (simulating the release).

In the same way of the first example, the simulation has been performed with a fixed time step of 0.01 s. Per each video frame, 4 integration steps are computed and the augmented scene is updated accordingly.

Figure 8 reports a series of four snapshots taken during the run of the simulation. The cross-lift mechanism is real-time rendered along with the simulation.

In the first part of the simulation (snapshot A of Fig. 8) the mechanism is free to move and it is in an equilibrium position. Then, the user locates the tracker in the middle of the upper plate and activates the fictitious spring connection (snapshot B). From this moment, the upper plate vertical coordinates is controlled by the tracker and the mechanism moves subjected to this connection.

When the user moves the tracker, the upper plate follows it in the vertical direction (snapshot C), preserving the right connection with the other rigid bodies. It is important to notice that the motion of all the rigid body collection is continuously simulated according to the external action of the gravity, the springs and the driving force due to the user's presence.

When the user decides to release the fictitious spring connection, the lifting device moves freely subjected to gravity force and internal springs and it oscillates around the equilibrium position (snapshot D).

Also for this example, the accuracy and the stability of the simulation have been checked. Due to the presence of a lot of overabundant constraints, the norm of the constraint equations is higher than in the previous example. During the free motion (when the simulation run without the interaction of the user) it is lower than 10^{-6} m. When the user interacts by pushing and pulling the upper plate, the norm increases and reaches $6.1 \cdot 10^{-5}$ m. The variation of the overall energy of the systems is always below 2 %. As in the previous example, in the computation of the overall energy, the contribution of the external action of the user has been taken into account by

Fig. 8 Snapshots of the simulation of the second example

evaluating the corresponding power as the dot product between the reaction force of the fictitious spring and the velocity of the digitizer stylus tip using Eq. (28).

6 Conclusions

In this chapter, an enhanced methodology for interactive, accurate, fast and robust multibody simulations of mechanical systems using Augmented Reality has been presented and discussed. This methodology is based on the integration of a mechanical tracker and a dedicated impulse based solver.

In this context, the simulation of movement of mechanical systems in an Augmented Reality environment can be useful for projecting virtual animated contents into a real world. By this way, it is possible to build comprehensive and appealing representations of interactive simulations including pictorial view and accurate numerical results.

In particular, two important enhancements have been presented with respect to a previous implementations. First of all, it has been possible to improve the precision of the interaction between the user and the scene by means of a precise mechanical tracking instrumentation. This constitutes an important improvement if compared with the use of simple optical markers for tracking the user in the scene. In the latter case, the precision in tracking was affected by the resolution of the camera, while with a mechanical device, it is possible to separate the processing of the data coming from the position tracking, from those coming from the image collimation processing. By this way, the simulation input is independent from the visualization input and output.

The second important enhancement is the use of a dedicated solver based on the sequential impulse strategy in order to perform a fast and robust simulation.

According to this approach, the solution is based on the less computational demanding solution strategy. Following the implemented algorithm, the equations of motion are firstly tentatively solved considering elastic and external forces but neglecting all the kinematic constraints. This produces a solution that is only approximated because the constraint equations are not satisfied. In a second step, a sequence of impulses are applied to each body in the collection in order to correct its velocity according to the limitation imposed by the constraint. This second step is iterative and involves the application of a series of impulses to the bodies until the constraint equations are fulfilled within a specific tolerance.

The final result of this work is a tool able to manage real time dynamic simulation and to update the augmented scene accordingly. The robustness and the reliability of the system have been checked over two test cases: a ten pendula dynamic system and the dynamics of a cross-lift mechanism.

According to the proposed methodology, the user can directly control the simulation by a smooth visualization on the head mounted display.

The integration among Augmented Reality, dedicated solver and precise input tracker can be considered an advantage for the future development of a new class of multibody simulation software. Moreover, this integrated simulation environment can be useful for both didactical purposes and engineering assessments of mechanical systems.

References

1. Azuma, R., Baillot, Y., et al.: Recent advances in augmented reality. IEEE Comput. Graph. Appl. **21**(6), 34–47 (2001)
2. Gattamelata, D., Pezzuti, E., Valentini, P.P.: A CAD system in augmented reality application. In: Proc. of 20th European Modeling and Simulation Symposium, Track on Virtual Reality and Visualization, Briatico (CS), Italy (2008)
3. Valentini, P.P.: Augmented reality and reverse engineering tools to support acquisition, processing and interactive study of archaeological heritage. In: Virtual Reality. Nova Publ., New York (2011)
4. Raghavan, V., Molineros, J., Sharma, R.: Interactive evaluation of assembly sequences using augmented reality. IEEE Trans. Robot. Autom. **15**(3), 435–449 (1999)
5. Pang, Y., Nee, A.Y.C., Ong, S.K., Yuan, M.L., Youcef-Toumi, K.: Assembly feature design in an augmented reality environment. Assem. Autom. **26**(1), 34–43 (2006)
6. Valentini, P.P.: Interactive virtual assembling in augmented reality. Int. J. Interact. Des. Manuf. **3**, 109–119 (2009)

7. Valentini, P.P.: Interactive cable harnessing in augmented reality. Int. J. Interact. Des. Manuf. **5**(1), 45–53 (2011)
8. Gattamelata, D., Pezzuti, E., Valentini, P.P.: Virtual engineering in augmented reality. In: Computer Animation, pp. 57–84. Nova Publ., New York (2010)
9. Valentini, P.P.: Enhancing user role in augmented reality interactive simulations. In: Human Factors in Augmented Reality Environments. Springer, Berlin (2012)
10. Buchanan, P., Seichter, H., Billinghurst, M., Grasset, R.: Augmented reality and rigid body simulation for edutainment: the interesting mechanism—an AR puzzle to teach Newton physics. In: Proc. of the International Conference on Advances in Computer Entertainment Technology, Yokohama, Japan, pp. 17–20 (2008)
11. Chae, C., Ko, K.: Introduction of physics simulation in augmented reality. In: Proc. of the 2008 International Symposium on Ubiquitous Virtual Reality, ISUVR, pp. 37–40. IEEE Comput. Soc., Washington (2008)
12. Kaufmann, H., Meyer, B.: Simulating educational physical experiments in augmented reality. In: ACM SIGGRAPH Asia (2008)
13. Irawati, S., Hong, S., Kim, J., Ko, H.: 3D edutainment environment: learning physics through VR/AR experiences. In: Proc. of the International Conference on Advances in Computer Entertainment Technology, pp. 21–24 (2008)
14. Mac Namee, B., Beaney, D., Dong, Q.: Motion in augmented reality games: an engine for creating plausible physical interactions in augmented reality games. Int. J. Comput. Games Technol. **2010**, 979235 (2010)
15. Valentini, P.P., Pezzuti, E.: Interactive multibody simulation in augmented reality. J. Theor. Appl. Mech. **48**(3), 733–750 (2010)
16. Azuma, R.T.: Tracking requirements for augmented reality. Commun. ACM **36**(7), 50–51 (1993)
17. Mirtich, B.V.: Impulse-based dynamic simulation of rigid body systems. Ph.D. thesis, University of California, Berkeley (1996)
18. Schmitt, A., Bender, J.: Impulse-based dynamic simulation of multibody systems: numerical comparison with standard methods. In: Proc. Automation of Discrete Production Engineering, pp. 324–329 (2005)
19. Schmitt, A., Bender, J., Prautzsch, H.: On the convergence and correctness of impulse-based dynamic simulation. Internal report 17, Institut für Betriebs- und Dialogsysteme (2005)
20. Jaimes, A., Sebe, N.: Multimodal human-computer interaction: a survey. Comput. Vis. Image Underst. **108**(1–2), 116–134 (2007)
21. Ullmer, B., Ishii, H.: Emerging frameworks for tangible user interfaces. IBM Syst. J. **39**(3–4), 915–931 (2000)
22. Fiorentino, M., Monno, G., Uva, A.E.: Tangible digital master for product lifecycle management in augmented reality. Int. J. Interact. Des. Manuf. **3**(2), 121–129 (2009)
23. Slay, H., Thomas, B., Vernik, R.: Tangible user interaction using augmented reality. In: Proceedings of the 3rd Australasian Conference on User Interfaces, Melbourne, Victoria, Australia (2002)
24. Valentini, P.P., Pezzuti, E.: Dynamic splines for interactive simulation of elastic beams in augmented reality. In: Proc. of IMPROVE 2011 International Congress, Venice, Italy (2011)

Modelling of Contact Between Stiff Bodies in Automotive Transmission Systems

Geoffrey Virlez, Olivier Brüls, Nicolas Poulet, Emmanuel Tromme, and Pierre Duysinx

Abstract Many transmission components contain moving parts, which can come into in contact. For example, the TORSEN differentials are mainly composed of gear pairs and thrust washers. The friction involved by contacts between these two parts is essential in the working principle of such differentials. In this chapter, two different contact models are presented and exploited for the modelling of differentials. The former uses an augmented Lagrangian technique or a penalty method and is defined between two flexible bodies or between a rigid body and a flexible structure. The second contact formulation is a continuous impact modelling based on a restitution coefficient.

1 Introduction

Nowadays in automotive industries the requirements to reduce fuel consumption and environmental pollution are greatly increasing. Reducing the weight of the vehicle, lowering of mechanical losses and developing new hybrid electric propulsion systems are needed in order to reach this goal. Nevertheless these new vehicle designs should not alter the security and the comfort of the passengers. For instance,

G. Virlez (✉) · O. Brüls · E. Tromme · P. Duysinx
Department of Aerospace and Mechanical Engineering (LTAS), University of Liège,
Chemin des chevreuils, 1, B52/3, 4000 Liège, Belgium
e-mail: geoffrey.virlez@ulg.ac.be

O. Brüls
e-mail: o.bruls@ulg.ac.be

E. Tromme
e-mail: emmanuel.tromme@ulg.ac.be

P. Duysinx
e-mail: p.duysinx@ulg.ac.be

N. Poulet
JTEKT TORSEN Europe S.A., Rue du grand peuplier, 11, Parc industriel de Strépy,
7110 Strépy-Bracquegnies, Belgium
e-mail: npoulet@torsen.be

J.-C. Samin, P. Fisette (eds.), *Multibody Dynamics*,
Computational Methods in Applied Sciences 28,
DOI 10.1007/978-94-007-5404-1_9,

electronic control systems such as ABS or ESP involve additional automotive components and therefore tend to increase the global weight, but they highly improve the vehicle handling and allow to avoid accidents in a lot of situations. The mass reduction of structural parts can also lead to a higher flexibility, which can introduce vibrations and have an impact on the driving pleasure. Moreover, the comfort in the passenger cell can be affected by the reduction of acoustic isolation due to thinner structural panels.

In order to find a compromise between these antagonist criteria, the current trend addresses the development of reliable simulation tools to enhance automotive design processes.

Multibody simulation techniques are frequently used to model complex automotive systems. For instance, dynamic simulations of crankshaft or connecting rods have been carried out in [18] to analyse the mechanical losses; the study of deformations and stresses are available in [7, 17] for global multicylinder engines. The suspensions are also widely modelled using multibody tools [3]. The models are often composed of a mixed set of rigid and flexible bodies and enable to analyse the vehicle dynamic behaviour in case of maneuvers or braking (see for example [8] and [10]).

Models of transmission components is less mature because several complex physical phenomena are involved such as stick-slip, backlash between gear teeth, contact with friction, impact or hysteresis. The modelling of these nonlinear and discontinuous effects is not trivial and can lead to numerical problems during the simulations. The development of specific formulations is needed in order to manage these particular effects. The driveline devices such as clutch, gear box or differential highly interact together. They influence the driveline behaviour and also the whole vehicle performance. For example, the differential features can have a direct influence on the sizing of anti-roll bar and suspensions. Therefore, individual models of transmission components are often not sufficient and there is a need to have global drivetrain or even full vehicle models. In this way, the driveline modelling would allow the improvement of the performance not only of the transmission devices, but also of the other subsystems of the vehicle.

In automotive as in other fields of mechanics, many transmission components include contacts between different parts. These contacts inhibit the relative motion in one or several directions but they let free the motion in the other directions. The contact can be: bilateral or unilateral, rigid or flexible, frictional or frictionless. Several complex physical phenomena can be involved by contacts. For instance, if the relative velocity when the contact occurs is high for unilateral contact, the impact encountered can generate vibration waves in the body structure. Permanent plastic deformations can be induced [20]. The friction can also lead to stick-slip phenomena due to the difference between static and dynamic friction coefficients.

These accurate and efficient contact models are essential in order to get reliable drivetrain models. Gear boxes or differentials include numerous contacts which play a key role in the working principle of these mechanisms. It cannot be expected to set up a realistic dynamic model of this kind of transmission components without a good and reliable mathematical formulation of contacts.

In the literature, three main categories of contact modelling can be distinguished according to the behaviour considered to model the bodies subjected to contact: rigid-rigid contact [20], flexible-rigid contact [14] or flexible-flexible contact [23]. In the field of multibody systems dynamics, two different approaches are often used to formulate the contact condition: continuous contact modelling and instantaneous contact modelling. The continuous method does not need specific algorithmic tools to manage the impact phenomena. The contact forces are added in the equations of motion of the mechanism and a standard time integration scheme can be used to solve the complete system. The positions and velocities of all bodies vary continuously and it is not necessary to stop the time integration at the moment of contact establishment (see [15] for more details). With instantaneous contact models, the motion is divided into two periods, before and after the impact. While the displacements are continuous, a jump of the relative velocity is observed at the contact instant. Instantaneous contact formulations are often related to nonsmooth dynamic methods [1, 13]. The discontinuities in the velocity field require the use of special integration methods [5, 12, 16]. For instance, event-driven approaches require the interruption of the time integration at each impact whereas time-stepping methods discretize in time the complete multibody system dynamics including the unilateral constraints and the impact forces.

The objective of this chapter is to investigate two different unilateral contact formulations for modelling contacts in dynamic simulation of automotive drivetrains. For large models including numerous contacts, it can be very useful to use various contact conditions according to the detail level needed for each contact. For instance in case of a complete drivetrain model, accurate and fine contact formulations can be used inside the components modelled in detail whereas rougher contact conditions are sufficient for the transmission devices globally represented. This combination of contact formulations enables to obtain low CPU time consuming models for global applications. In order to be used with a classical integration scheme, two continuous contact formulations have been considered in this work. The first one is an accurate contact model defined between two flexible bodies or between a rigid body and a flexible body. This contact formulation uses an augmented Lagrangian approach or a penalty method. Each flexible body is represented by a finite element mesh, that notably enables to analyse the stresses on the contact surfaces. The second contact formulation is a simpler model defined between two rigid bodies. This method is based on the continuous impact theory and uses a restitution coefficient. The friction has been taken into account in both contact elements.

The application under study to validate the two contact elements is the TORSEN differential. This kind of limited slip differential is mainly composed of gear pairs and thrust washers. The axial force produced by the helical mesh leads to contact between the lateral circular faces of toothed wheels and the various thrust washers. The friction generated between these two bodies is at the source of the locking effects, specific to the operation of TORSEN differentials. A unilateral frictional contact model is then essential to model accurately and reliably these differentials.

In the sequel of this chapter, the nonlinear finite element approach for flexible multibody systems available in SAMCEF/MECANO [9] is briefly presented in Sect. 2. The two contact formulations are respectively presented in Sect. 3 for the rigid/flexible coupled interaction method and Sect. 4 for the continuous impact modelling. The working principle of TORSEN differential will be described in Sect. 5 and numerical results provided by the simulation of this transmission device are shown in Sect. 6 for both contact models.

2 Finite Element Method in Multibody System Dynamics

In this work, the chosen approach is based on the nonlinear finite element method for flexible multibody systems developed by Géradin and Cardona [9]. This method allows the modelling of complex mechanical systems composed of rigid and flexible bodies, kinematics joints and force elements. The degrees of freedom are the absolute nodal coordinates with respect to a unique inertial frame. Hence, there is no distinction between rigid and elastic coordinates which allows accounting in a natural way for many nonlinear flexible effects and large deformations. The Cartesian rotation vector combined with an updated Lagrangian approach is used for the parametrization of rotations. This choice enables an exact representation of large rotations.

The dynamics of a system including holonomic bilateral constraints is described by Eqs. (1) and (2). The modelling of unilateral contact conditions within this formulation is addressed in the next two sections.

$$\mathbf{M}(\mathbf{q})\ddot{\mathbf{q}} + \mathbf{g}^{gyr}(\mathbf{q}, \dot{\mathbf{q}}) + \mathbf{g}^{int}(\mathbf{q}, \dot{\mathbf{q}}) + \boldsymbol{\Phi}_{\mathbf{q}}^T (p\boldsymbol{\Phi} + k\boldsymbol{\lambda}) = \mathbf{g}^{ext}(t) \tag{1}$$

$$k\boldsymbol{\Phi}(\mathbf{q}, t) = \mathbf{0} \tag{2}$$

where $\mathbf{q}$, $\dot{\mathbf{q}}$ and $\ddot{\mathbf{q}}$ are the generalized displacements, velocities and acceleration coordinates, $\mathbf{M}(\mathbf{q})$ is the mass matrix, $\mathbf{g}^{gyr}$ is the vector of gyroscopic and complementary inertia forces, $\mathbf{g}^{int}(\mathbf{q}, \dot{\mathbf{q}})$ is the vector of the internal forces, e.g. elastic and dissipations forces and $\mathbf{g}^{ext}(t)$ is the vector of the external forces. According to the augmented Lagrangian method, the constraint forces are formulated by $\boldsymbol{\Phi}_{\mathbf{q}}^T (p\boldsymbol{\Phi} + k\boldsymbol{\lambda})$ where $\boldsymbol{\lambda}$ is the vector of Lagrange multipliers related to algebraic constraints $\boldsymbol{\Phi} = 0$; k and p are respectively a scaling and a penalty factor to improve the numerical conditioning.

Equations (1) and (2) form a system of nonlinear differential-algebraic equations. The solution is evaluated step by step using a second order accurate time integration scheme. For this study, the Chung-Hulbert scheme, which belongs to the family of the generalized α-method, has been used (see [2, 6]). At each time step, a system of nonlinear algebraic equations has to be solved using a Newton-Raphson method.

3 Coupled Iterations Method for Node to Surface Contact Element

For multibody simulation techniques based on a finite element approach, three geometrical configurations can be used to define the contact elements: node to node,

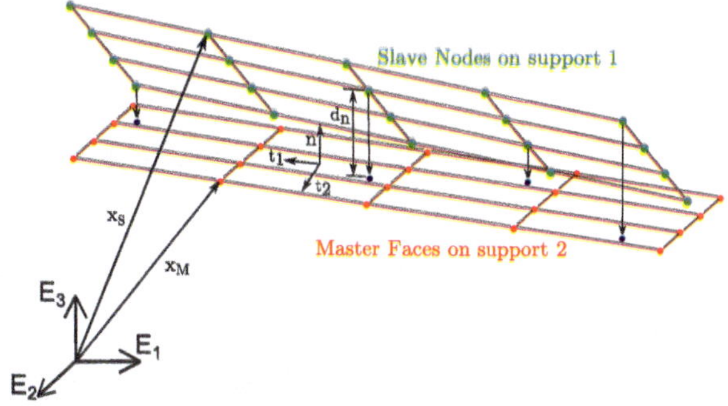

Fig. 1 Contact condition—projection of slave node on master surface

node to surface and surface to surface contact elements. The last configuration being often related to mortar algorithms [11, 19].

In this section, a node to surface contact element, based on the coupled iteration method, is presented. In contrast to uncoupled iterations, the contact problem is solved at the same time as the other nonlinearities and no distinction is done between the degrees of freedom linked by contact and the other ones. An augmented Lagrangian approach is used to define the kinematic constraints related to each contact condition. The resulting system of equations is solved simultaneously for the displacements and Lagrange multipliers.

This formulation is suitable for implicit nonlinear analysis and is able to model contacts between a rigid structure and a flexible part (flexible/rigid contact) or between two flexible parts (flexible/flexible contact). Contact elements are created between the nodes on the contact surface of the first support and a flexible facet of a finite element (in case of flexible/flexible contact) or a rigid master surface (in case of flexible/rigid contact) on the second support. Bilateral but also unilateral contacts can be represented. In this last case, the kinematic constraints are active in case of effective contact and inactive when the two bodies are separated. Once the set of active constraints has been determined, the equations of motion have the structure of Eqs. (1) and (2). The contact algorithm can be decomposed in two steps.

The first step is a geometrical step, which consists in searching the projection of each slave node on the master surface (Fig. 1), computing the normal distance (d_n) between the node and the surface, and measuring the displacement variations ($\Delta u_1, \Delta u_2$) in the tangent directions during the current time step.

The virtual variations δd_n, $\delta \Delta u_1$, $\delta \Delta u_2$ can be expressed as functions of the variations of nodal unknowns $\mathbf{q}$, which are the displacements of the slave nodes, and displacements and rotations of the node linked to the rigid surface (flexible-rigid contact) or the displacements of the nodes of the facet (flexible-flexible contact).

$$\delta d_n = \mathbf{n}^T \mathbf{B} \delta \mathbf{q} \tag{3}$$

$$\delta \Delta u_1 = \mathbf{t}_1^T \mathbf{B} \delta \mathbf{q} \tag{4}$$

$$\delta \Delta u_2 = \mathbf{t}_2^T \mathbf{B} \delta \mathbf{q} \tag{5}$$

where $\mathbf{n}$ is the normal, $\mathbf{t}_1$, $\mathbf{t}_2$ are the tangents to the surface, $\mathbf{B} = \partial(\mathbf{x}_S - \mathbf{x}_M)/\partial \mathbf{q}$, $\mathbf{x}_S$ and $\mathbf{x}_M$ being respectively the slave and master node positions expressed in the absolute reference frame.

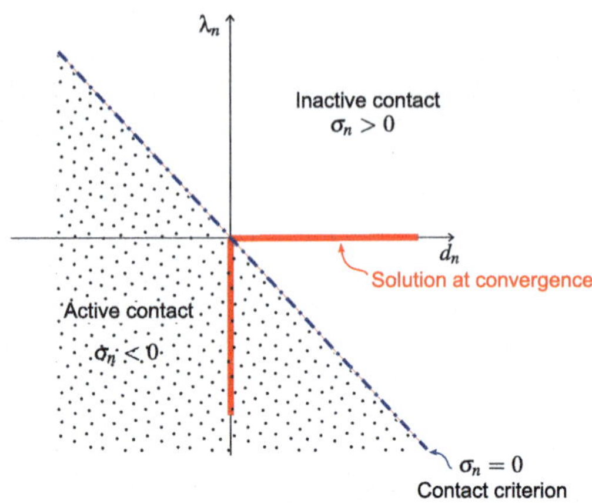

Fig. 2 Contact criterion and solution

The second step sets the contact conditions whose expression depends on the current status of the contact. For unilateral contacts, the case of an *active* or an *inactive* contact can be distinguished in the normal direction. The friction coefficient and the normal force enable to determine if the status is *stick* or *slip* in tangential directions.

In order to assess the contact status, two simple tests based on the quantities σ_n and σ_t are carried out. These quantities are defined as

$$\sigma_n = k\lambda_n + pd_n \tag{6}$$

$$\sigma_{t_1} = k\lambda_{t_1} + p\Delta u_1 \tag{7}$$

$$\sigma_{t_2} = k\lambda_{t_2} + p\Delta u_2 \tag{8}$$

$$\sigma_t = \sqrt{\sigma_{t_1}^2 + \sigma_{t_2}^2} \tag{9}$$

where λ_i are three Lagrange multipliers: one for the contact (λ_n) and two for the friction ($\lambda_{t_1}, \lambda_{t_2}$). k is a scaling factor and p is a regularization parameter.

For the behaviour in the normal direction, the bold line in Fig. 2 represents the solution of the contact condition at convergence of the differential-algebraic system of equation (Eqs. (1) and (2)): either the normal distance d_n being zero or the Lagrange multiplier λ_n being zero. During the iterations of the Newton-Raphson procedure, the system is not at the equilibrium and the contact criterion σ_n allows to determine if the contact condition related to an *active* (see Sect. 3.1) or an *inactive* (Sect. 3.2) contact has to be used. The dotted line $\sigma_n = 0$ divides the space in Fig. 2 in two zones. If $\sigma_n < 0$, the node is not considered in contact and if $\sigma_n > 0$, the node is considered in contact. The convergence property of this algorithm depends on the slope $(-p/k)$ of the line $\sigma_n = 0$. The scaling factor k being constant, the choice of the regularization parameter p influences the convergence in some particular situations but the solution at convergence will not depend on this regularization parameter (cf. bold line in Fig. 2).

3.1 Inactive Contact

If the contact criterion σ_n is positive, there is no contact nor friction forces applied on the nodes. The three Lagrange multipliers are set to zero. At the element level, the internal forces are computed by:

$$\delta \mathbf{q}^{*T} \mathbf{F}_{int} = -(\delta\lambda_n k\lambda_n + \delta\lambda_{t_1} k\lambda_{t_1} + \delta\lambda_{t_2} k\lambda_{t_2}) \tag{10}$$

with:

$$\mathbf{q}^* = \begin{Bmatrix} \mathbf{q} \\ \boldsymbol{\lambda} \end{Bmatrix} \tag{11}$$

3.2 Active Contact

When σ_n is negative, the contact is active and the kinematic constraint can be expressed by $\phi = d_n$. The virtual work principle enables to calculate the internal forces at the element level:

$$\delta \mathbf{q}^{*T} \mathbf{F}_{int} = \delta d_n (p d_n + k\lambda_n) + \delta\lambda_n k d_n \tag{12}$$

At the equilibrium, the kinematic constraint is satisfied ($\phi = 0$) and therefore the normal distance d_n is equal to zero. The Lagrange multiplier λ_n can be interpreted as the contact force divided by the scaling factor k. This last parameter is often chosen equal to the stiffness of the structural elements in order to have the same order of magnitude in the various terms of the iteration matrix of the Newton-Raphson process.

When the contact is active, two status are available for the friction behaviour: either the node is sticking to the surface or the node is sliding. The friction criterion σ_t defined in Eq. (9) is used to determine the friction status.

3.3 Sticking Friction

If σ_t is smaller than the normal force σ_n multiplied by the friction coefficient μ, the node is sticking on the master face. The friction coefficient is a constant value, no distinction between the static and kinetic friction coefficient is made. Two kinematic constraints equal to the variations of tangential displacements ($\phi_1 = \Delta u_1$ and $\phi_2 = \Delta u_2$) are introduced. As for the contact in the normal direction, the kinematic constraints are equal to zero after convergence ($\phi_1 = \Delta u_1 = 0$, $\phi_2 = \Delta u_2 = 0$). The internal forces related to friction forces for each contact element in a sticking situation are computed from the virtual work.

$$\begin{aligned} \delta \mathbf{q}^{*T} \mathbf{F}_{int} = {} & \delta \Delta u_1 (p \Delta u_1 + k\lambda_{t_1}) + \delta \Delta u_2 (p \Delta u_2 + k\lambda_{t_2}) \\ & + \delta\lambda_{t_1} k \Delta u_1 + \delta\lambda_{t_2} k \Delta u_2 \end{aligned} \tag{13}$$

3.4 Sliding Friction

When $\sigma_t > \mu|\sigma_n|$, the slave node is sliding on the master surface. The element forces are computed from:

$$\delta\mathbf{q}^{*T}\mathbf{F}_{int} = \{\delta\Delta u_1 \quad \delta\Delta u_2 \quad \delta\lambda_{t_1} \quad \delta\lambda_{t_2}\}\left\{\begin{array}{c} \mu|\sigma_n|v_1 \\ \mu|\sigma_n|v_2 \\ (\mu|\sigma_n|v_1 - k\lambda_{t_1})k/p \\ (\mu|\sigma_n|v_2 - k\lambda_{t_2})k/p \end{array}\right\} \tag{14}$$

In this case, the friction Lagrange multipliers are parallel to the variation of sliding displacements $(\Delta u_1, \Delta u_2)$. The iteration matrix for a sliding friction is not symmetrical and it is recommended to use a non-symmetric solver.

3.5 Penalty Method

The augmented Lagrangian approach presented above is sometimes not able to find a solution or can encounter great difficulties to converge in case of large discontinuities during the simulation. For instance, if the relative normal velocity of the colliding bodies is high at the contact establishment, an impact phenomenon occurs. This contact formulation cannot manage the strong discontinuity and the rebonds after the first impact have often erratic magnitudes and frequencies. Likewise, the switching between the stick and the slip is highly nonlinear for high speed systems.

In order to improve the convergence of the algorithm, the augmented Lagrangian method can be replaced by a pure penalty method. In contrast to the Lagrangian approach where the contact is infinitely rigid, the penalty allows a small penetration between the two bodies that slightly relaxes the discontinuity. The penalty function can be linear or nonlinear and can be seen physically as a finite stiffness that is active in compression but not in traction. To have a smoother response, it can also be useful to account for damping in the contact model. The first step consisting of the projection of the slave nodes on the master faces is unchanged compared with Lagrangian approach.

A regularization is often used to avoid the discontinuity when the sign of the relative sliding velocity shifts (see Fig. 3). The regularized friction coefficient μ_R can be defined in several ways with sometimes complex functions. In this study, a simple quadratic function is used:

$$\mu_R(\dot{\xi}) = \begin{cases} \mu\left(2 - \frac{|\dot{\xi}|}{\varepsilon_v}\right)\frac{\dot{\xi}}{\varepsilon_v} & |\dot{\xi}| < \varepsilon_v \\ \mu\frac{\dot{\xi}}{|\dot{\xi}|} & |\dot{\xi}| \geq \varepsilon_v \end{cases} \tag{15}$$

where $\dot{\xi}$ is the relative sliding velocity, μ is the friction coefficient and ε_v is the regularization tolerance.

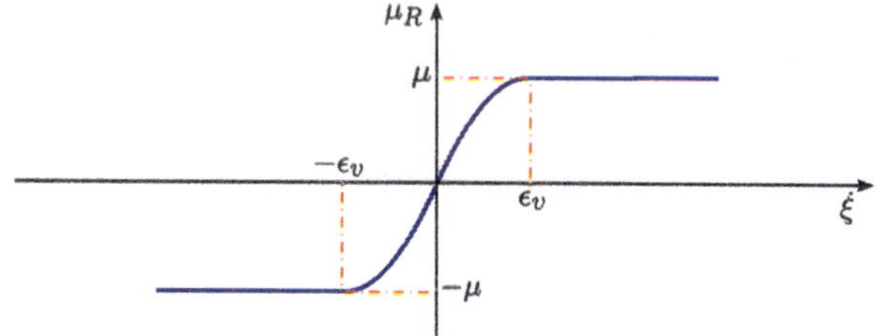

Fig. 3 Regularized friction coefficient

4 Continuous Impact Modelling

The continuous contact modelling is based on a continuous contact law that uses a restitution coefficient. During impacts between rigid bodies, some kinetic energy is lost. Indeed, impacts can initiate wave propagation in the bodies which absorb parts of the kinetic energy until they vanish owing to material damping. High stresses might also occur near the impact point and involve plastic deformation, which also contributes to kinetic energy loss, as well as visco-elastic material behaviour. Macro-mechanically, these various sources of kinetic energy loss are often summarized and expressed by a coefficient of restitution. The loss of kinetic energy described by the coefficient of restitution depends on the shapes and material properties of the colliding bodies as well as on their relative velocities. However, the restitution coefficient cannot be computed within the multibody system simulation. It has to be roughly estimated from experience, measured by experiments or determined by numerical simulations on a fast time scale [20].

There exist different definitions for the coefficient of restitution: kinematic (e_N), kinetic (e_P) or energetic (e_E):

$$e_N = -\frac{\dot{g}_{n_e}}{\dot{g}_{n_s}} \tag{16}$$

$$e_P = \frac{\Delta P_r}{\Delta P_c} = \frac{\int_{t_c}^{t_e} F\,\mathrm{d}t}{\int_{t_s}^{t_c} F\,\mathrm{d}t} \tag{17}$$

$$e_E^2 = -\frac{T_r}{T_c} = -\frac{\int_{t_c}^{t_e} F\dot{g}_n\,\mathrm{d}t}{\int_{t_s}^{t_c} F\dot{g}_n\,\mathrm{d}t} = -\frac{\int_{h_c}^{h_e} F\,\mathrm{d}t}{\int_{h_s}^{h_c} F\,\mathrm{d}t} \tag{18}$$

where $\dot{g}_{n_s}$ and $\dot{g}_{n_e}$ are respectively the relative velocity between the two bodies in normal direction before and after impacts; the time intervals $[t_s, t_c]$ and $[t_c, t_e]$ correspond to the compression and restitution phases; ΔP_c and ΔP_r are the impulse during the compression and restitution phases; T_c and T_r are the deformation energies during the compression and restitution phases; F is the contact force and $h = -g_n$ is the penetration allowed between the two bodies.

An impact with $e = 1$ means no energy loss (complete elastic contact), whereas $e = 0$ corresponds to a total loss of energy (plastic or inelastic contact), $0 \le e \le 1$.

These three forms of the restitution coefficient are equivalent unless the configuration is eccentric and the direction of slip varies during impact or if the bodies are

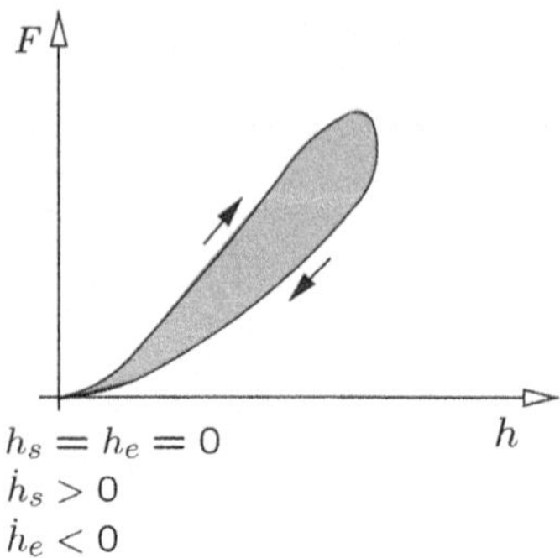

Fig. 4 Force law for continuous impact modelling

rough. Some differences can also appear in case of frictional contact or if several impacts occur simultaneously (see [21] for more details).

4.1 Force Law

A penalty approach is used for this continuous contact model whereby a small penetration h is allowed. The contact force is computed from this local penetration by a force law.

$$F(h,\dot{h}) = kh^n + ch^n\dot{h} \tag{19}$$

where k is the contact stiffness and c is a damping parameter.

In order to avoid a jump at the beginning of the impact and tension force at the end of the impact, the classical viscous damping term $(c\dot{h})$ has been multiplied by h^n.

As depicted in Fig. 4, this force law yields a hysteresis loop with $h_s = h_e = 0$ for the force-penetration curve. The enclosed area represents the kinetic energy loss during impact.

The parameters k and c have to be chosen in order to have realistic values for the impact duration, the local penetration and the kinetic energy loss. One way to set the damping parameter consists in formulating this coefficient as a function of the restitution coefficient. According to the contact configuration, various expressions are available in the literature (see for example [15]). For the contact considered in this work between gear wheels and washers, the expression (20) seems relevant and yields to a good approximation of the kinetic energy loss for large kinematic restitution coefficients ($e > 0.8$).

$$c = \frac{3(1-e^2)}{4}\frac{k}{\dot{h}_s} \tag{20}$$

where $\dot{h}_s$ is the relative normal velocity between bodies at the contact beginning. The force (Eq. (19)) applied on the two bodies while there are in contact as well as the contribution of this contact element to the global iteration matrix of the system have to be specified in the contact subroutine. In order to compute the tangent stiffness matrix and the damping matrix included in the iteration matrix of the contact element, the incremental form of the virtual work principle can be used.

$$\delta\,\mathrm{d}W = \delta\,\mathrm{d}h\;F(h,\dot{h}) + \delta h\,\mathrm{d}F\,(h,\dot{h}) \tag{21}$$

$$\delta\,\mathrm{d}W = \delta\,\mathrm{d}h\;F(h,\dot{h}) + \delta h\left(\frac{\partial F}{\partial h}\,\mathrm{d}h + \frac{\partial F}{\partial \dot{h}}\,\mathrm{d}\dot{h}\right) \tag{22}$$

This last expression has to be handled in order to obtain Eq. (23) and to identify the tangent stiffness matrix and damping matrix.

$$\delta\,\mathrm{d}W = \delta\mathbf{q}^T \underbrace{\frac{\partial \mathbf{F}_{int}}{\partial \mathbf{q}}}_{\mathbf{K}_T}\,\mathrm{d}\mathbf{q} + \delta\mathbf{q}^T \underbrace{\frac{\partial \mathbf{F}_{int}}{\partial \dot{\mathbf{q}}}}_{\mathbf{C}_T}\,\mathrm{d}\dot{\mathbf{q}} \tag{23}$$

Here $\mathbf{q}$ is the vector of nodal degree of freedom used by the contact element. In our current implementation, this vector contains the position parameters of the node located at the centre of the contact surface of two bodies candidate to contact ($\mathbf{q}^T = \{x_A\ y_A\ z_A\ x_B\ y_B\ z_B\}$).

In summary, the tangent stiffness and damping matrices can be expressed in the following form:

$$\delta\,\mathrm{d}W = \delta\mathbf{q}^T \underbrace{\left[\left(-\frac{F}{h} + \frac{\partial F}{\partial h} - \frac{\partial h}{\partial \dot{h}}\frac{\dot{h}}{h}\right)\mathbf{n}\mathbf{n}^T + \frac{F}{h}\mathbf{I} + \frac{\partial F}{\partial \dot{h}}\mathbf{n}\frac{\dot{\mathbf{x}}_{AB}^T}{h}\mathbf{I}\right]}_{\mathbf{K}_T}\mathrm{d}\mathbf{q}$$

$$+\,\delta\mathbf{q}^T \underbrace{\left[\frac{\partial F}{\partial \dot{h}}\mathbf{n}\mathbf{n}^T\right]}_{\mathbf{C}_T}\mathrm{d}\dot{\mathbf{q}} \tag{24}$$

with $\mathbf{x}_{AB}^T = \{x_B - x_A\ y_B - y_A\ z_B - z_A\}$, the vector between nodes A and B; $\mathbf{n}$ is the normal direction to the contact surface.

4.2 Friction Force

The friction force produced by the contact between the two rigid bodies can be easily added to the contact element presented in the previous section. Its magnitude is given by Eq. (25) where F_{norm} is equivalent to the contact force (Eq. (19)).

$$F_{fr} = \mu_R |F_{norm}| \tag{25}$$

This friction force is applied on a point M located at the middle of the segment AB and its direction is aligned with the tangential velocity vector $\mathbf{v}_t$.

5 Description of TORSEN Differentials

The two essential functions of a differential are to transmit the motor torque to the two output shafts and to allow a difference of rotation speed between these two outputs. In a vehicle, this mechanical device is particularly useful in turn when the outer wheels have to rotate quicker than the inner wheels to ensure a good handling.

The main drawback of a conventional differential (open differential) is that the total amount of available torque is always split between the two output shafts with the same constant ratio. In particular, this is a source of problems when the driving wheels have various conditions of adherence. If the motor torque exceeds the maximum transferable torque limited by road friction on one driving wheel, this wheel starts spinning. Although they do not reach their limit of friction, the other driving wheels are not able to transfer more torque because the input torque is often equally split between the two output shafts.

The TORSEN differentials significantly reduce this undesirable side effect. This kind of limited slip differential allows a variable distribution of motor torque depending on the available friction of each driving wheel. For a vehicle with asymmetric road friction between the left and right wheels, for example, right wheels are on a slippery surface (snow, mud, ...) whereas left wheels have good grip conditions, it is possible to transfer an extra torque to the left lane. That allows the vehicle to move forward whereas it would be hardly possible with an open differential. However, the overall driving torque cannot be applied on one output shaft while no load is exerted on the second shaft. When the difference between the 2 output torques becomes too large, the differential unlocks and lets different rotation speeds but keeps the same constant torque ratio.

When a TORSEN differential is used, the torque biasing is always a precondition before any difference of rotation speed between the two output shafts. Contrary to viscous coupling, TORSEN (a contraction of Torque-Sensing) is an instantaneous and pro-active process which acts before wheel slip.

The differential can be used either to divide the drive torque into equal parts acting on the traction wheels of the same axle, or to divide the output torque from the gearbox between the two axles of four-wheels drive vehicles. This second application is often called the transfer box differential or central differential.

As depicted in Fig. 5, the TORSEN differential contains a housing in two parts as well as several gear pairs and thrust washers. Due to the axial force produced by the helical mesh, several gear wheels can move axially and enter in contact with the various thrust washers fixed on the case or housing. The friction encountered by this relative sliding is at the origin of the locking effect of TORSEN differentials. The second important contribution to the limited slip behaviour is due to the friction between the planet gears and the housing holes in which they are inserted. When one axle tries to speed up, all encountered frictions tend to slow down the relative rotation and involve a variable torque distribution between the output shafts. The biasing on the torque only results from the differential gearing mechanical friction.

This limited slip differential has four working modes which depend on the direction of torque biasing and on the drive or coast situation. According to the

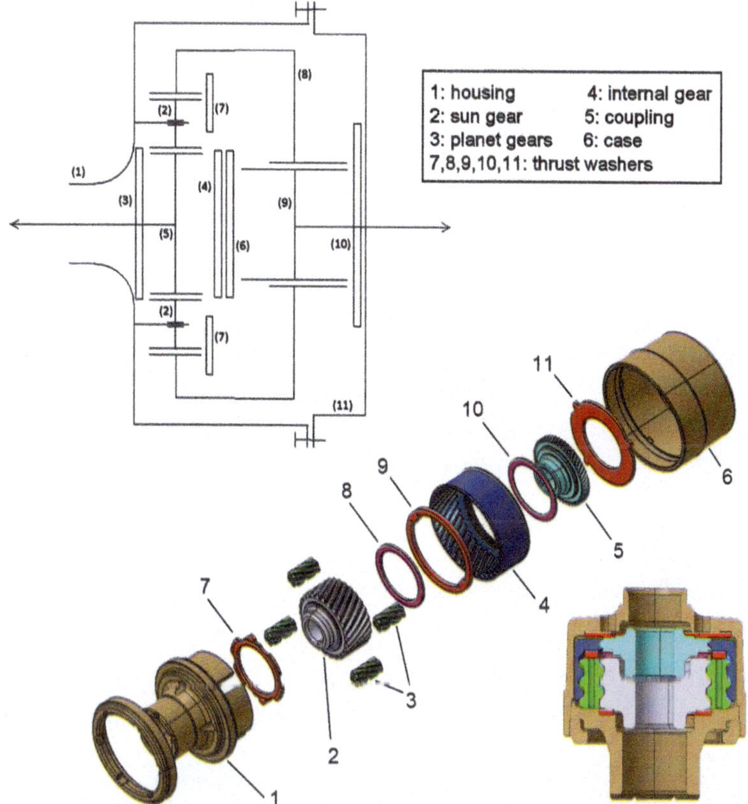

Fig. 5 Kinematic diagram, exploded view and cut-away view of type C TORSEN differential

considered mode, the gear wheels rub against one or the other thrust washers which can have different friction coefficients and contact surfaces.

6 Numerical Results

The two main kinematic constraints needed to model TORSEN differentials are gear pairs and contact conditions. The formulation used to model each gear pair is available for describing flexible gear pairs in 3 dimensional analysis of flexible mechanism. This gear element is developed in [4] and is a global kinematic joint defined between two physical nodes: one at the centre of each gear wheel which is represented as a rigid body. Nevertheless the flexibility of the gear mesh is accounted for by a nonlinear spring and damper element inserted along the instantaneous normal pressure line. Several specific phenomena in gear pairs which influence significantly the dynamic response of gears are also included in the model: backlash, mesh stiffness fluctuation, misalignment, friction between teeth. Contacts between the thrust

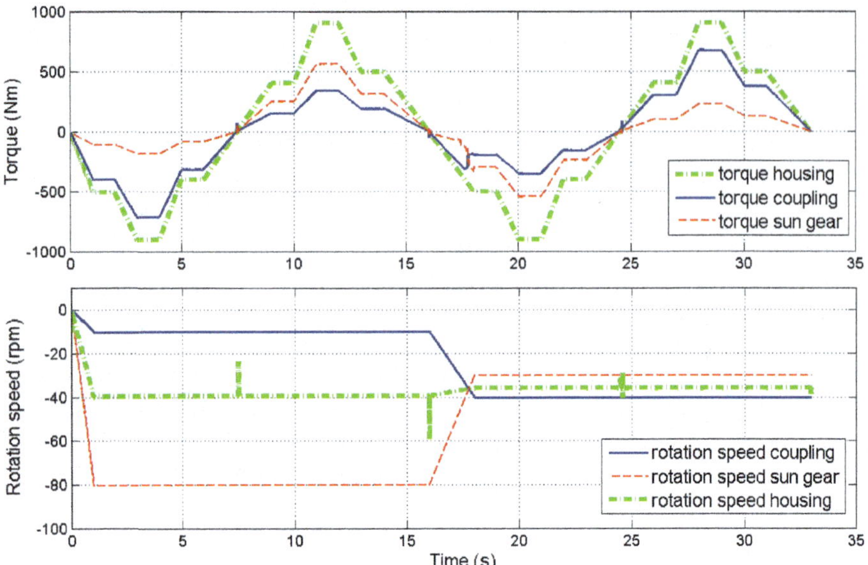

Fig. 6 Time evolution of torque and rotation speed of housing and output shafts of type C TORSEN on vehicle configuration

washers and the circular lateral faces of gear wheels have been successively modelled by the two contact formulation presented in this chapter.

A dynamic analysis of the type C TORSEN has been carry out when this component is considered in the vehicle configuration. The other driveline devices such as driveshafts are not represented in this work. Equivalent loads are applied on the differential inputs and outputs: a torque is applied on the housing whereas the rotation speed of sun gear and coupling are prescribed. The housing torque is equivalent to the driving force coming out from the gear box and transmitted to the differential housing through the propeller shaft. The prescribed rotation speeds can be seen as a measure of the adherence properties on the front and rear axles.

The time evolution of these loads has been chosen in order to study the four working modes of the differential during the same simulation and the transient behaviour at the switching time between two modes. Besides, in order to test the robustness of the model, the torque on the housing is applied with fast increasing or decreasing phases interrupted by steady state periods (see Fig. 6).

Some numerical results specific to the contact formulations studied in this chapter are presented in Sect. 6.1 for the coupled iterations method and in Sect. 6.2 for the continuous impact modelling.

A previous work addresses the development of full type B and type C TORSEN differential models. See Ref. [22] for more details on the construction of theses models and their global validation by comparison with experimental data.

6.1 Flexible-Rigid Contact

In the differential model, the five thrust washers are modelled with finite elements whereas the gear wheels are represented by rigid bodies. Therefore, the flexible-rigid version of the coupled iterations method for node to surface contact (Sect. 3) is used.

Owing to the high axial velocity of the gear wheels when the differential switches from one working mode to another one, the augmented Lagrangian approach has to be replaced by a full penalty method in order to allow the convergence of the integration algorithm as explained in Sect. 3.5. A damping force has also been used, besides the linear penalty function. The penalty function enable small interferences between the gear wheel and the thrust washer whereas the damping tends to slow down the impact velocity. The chosen damping function is a linear function with respect to the penetration velocity. The damping force has even been introduçed before the effective contact to anticipate the latter and reduce the shock phenomenon. This anticipating damping has been used to facilitate the convergence but has also a physical meaning. Indeed, in real operation the film of lubricating oil between the contacting surfaces tends to slow down the bodies before the contact and then plays the role of a damper. The damping coefficient used in this model should be identified with the oil properties to have a realistic damping behaviour. Nevertheless, in this work the damping coefficient has been chosen in order to allow the convergence but without a reference with the physical properties. The contact stiffness value used to enable the small penetration has also been set only to ensure the convergence. If this value is too large, the discontinuity is not sufficiently relaxed and if the value is too small the penetration of the two contacting bodies is too high which is in contradiction with the prescribed contact condition. A regularization of the friction coefficient has been needed to avoid a large discontinuity when the relative rotation between gear wheel and thrust washer changes direction.

Figure 7 illustrates the contact pressure of all contact elements introduced in the model when the differential is in the *drive to rear mode*. The time evolution of contact pressure is depicted in Fig. 8 where it can be observed that at each time and for each working mode, three contact elements are active and two are inactive. The contact between thrust washers #8 and #10 (cf. Fig. 5) is the only contact element always active. The *drive modes* ([0; 7] s and [16; 24] s) involve the contacts between the sun gear and the washer #7 and between the internal gear and the washer #11. On the other hand, the contacts between the coupling and the washer #11 and between the internal gear and washer #9 are active for the *coast modes* ([8; 15] s and [25; 32] s). The friction is taken into account in the five contacts. Figure 10 shows the spatial distribution of power dissipated by friction. The analysis of stresses in the thrust washers can be also provided by the simulation (Fig. 9). All these numerical data are useful to design the TORSEN differential. For instance, the internal and external radius of thrust washers or the friction coefficient of the contact can be adapted according to the locking effect wanted for each working mode. The washer thickness could also be modified thanks to the stress analysis. Further, the inclusion

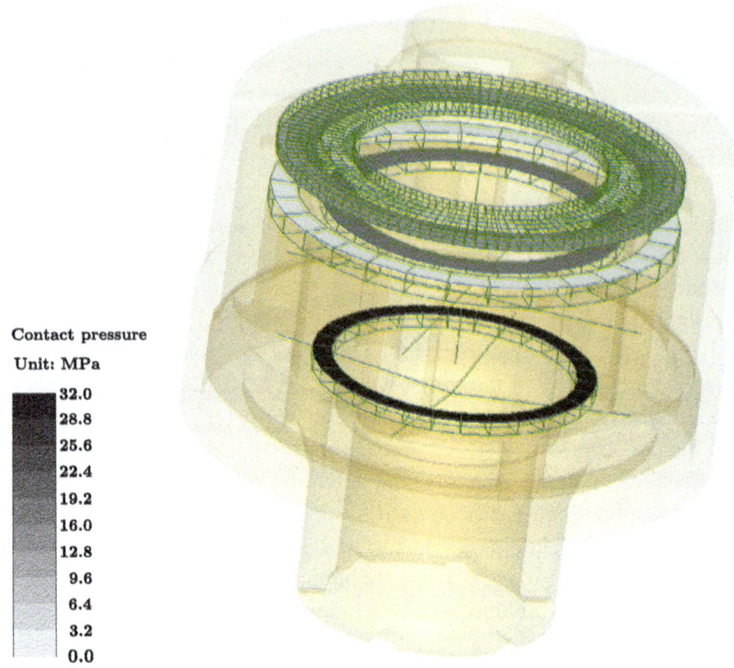

Fig. 7 Contact pressure on thrust washers ($t = 3.5$ s: *Drive to rear mode*)

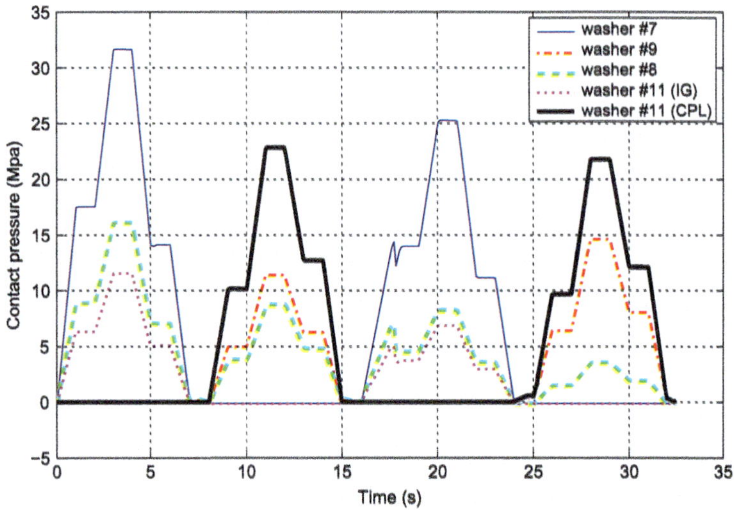

Fig. 8 Time evolution of the contact pressure in thrust washers

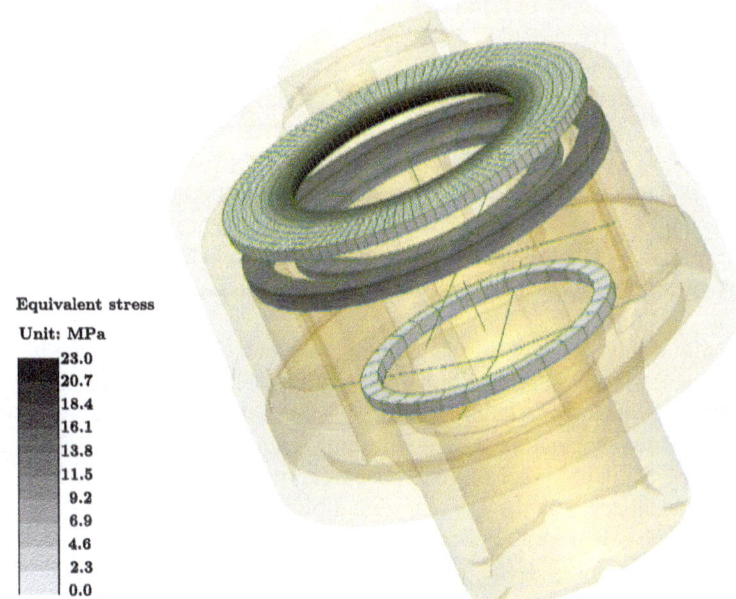

Fig. 9 Stress in the thrust washers ($t = 11.6$ s: *Coast to front mode*)

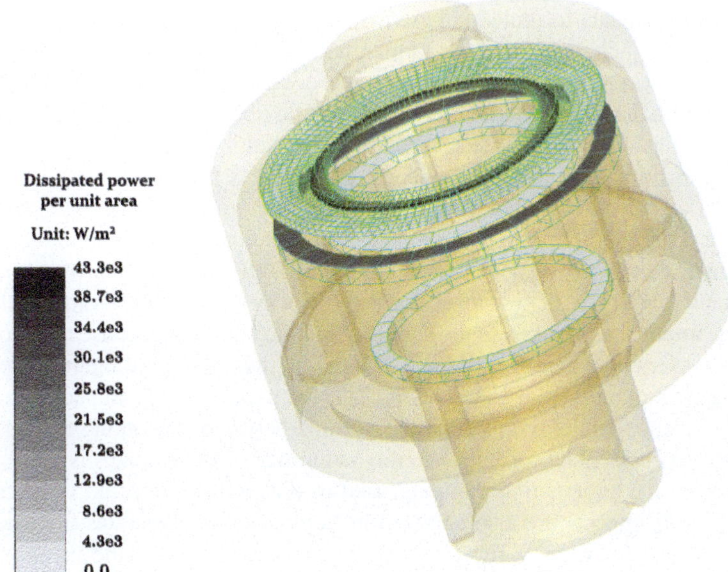

Fig. 10 Power dissipated by friction between gear wheels and thrust washers ($t = 28.2$ s: *Coast to rear mode*)

of this differential model in a full vehicle where the car body, suspensions, drive-shafts or tyres are represented, would allow to study the interaction between the differential and the vehicle dynamics.

A particular attention must be put on the meshing of the thrust washers. The kind of finite element used can influence the convergence properties and the results accuracy. It could be noticed that considering the thrust washers with a volume behaviour is better than with shell finite elements. Furthermore, it is better if the contact surface is composed of quadratic elements which are thereafter extruded to obtain hexahedron elements. In order to avoid tetrahedral elements in the mesh, the geometry of the thrust washers has been slightly simplified to obtain a perfect ring shape.

The main disadvantage of this rigid/flexible contact element is the high computational time needed owing to the important number of configuration parameters required by this contact formulation. Indeed, this contact condition requires to model at least one of two contacting bodies with finite elements. For instance, in case of volume elements, the number of generalized coordinates is more than three times the number of nodes plus the Lagrange multipliers linked to the contact element attached to each slave node. For some applications, it is not always necessary to account for the flexibility of the bodies in contact. In this case, this contact element increases the size and unnecessarily complicates the model.

6.2 Rigid-Rigid Contact

The frictional contact formulation based on the continuous impact theory, described in Sect. 4 of this chapter, has been implemented in the user element framework of SAMCEF/MECANO. In order to test this new contact element and compare its performance with the coupled iterations method for flexible-rigid contact, the TORSEN differential has also been considered as the application system. The same kind of loading and limit conditions have been used whereas the five contact conditions have been replaced by the new contact elements.

Figure 11 shows the time evolution of the torque on the sun gear and coupling when the *drive to rear mode* of the differential is active (first quarter of the simulation depicted in Fig. 6). These torques can be seen as the reaction torque on the differential output shafts linked to the sun gear and the coupling because their rotation speed is prescribed whereas a torque is applied on the differential input (housing).

The spikes on curves of Fig. 11 represent the shocks due to impacts when the gear wheels move quickly at the switching time between to working modes. With the rigid/flexible contact formulation, the magnitude of these spikes is almost vanished due to the anticipating damping used to enable the algorithm convergence. The effects of impacts are also depicted in Fig. 12 which illustrates the axial displacement of the sun gear. The rebonds of gear wheels against the various thrust washers occur at each change of differential working mode. The magnitude and the frequency of these rebonds depends on the restitution coefficient and the contact stiffness used for each contact element.

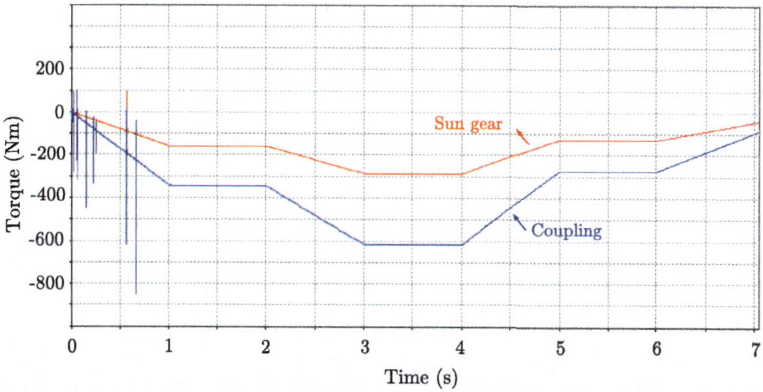

Fig. 11 Reaction torque on the sun gear and the coupling

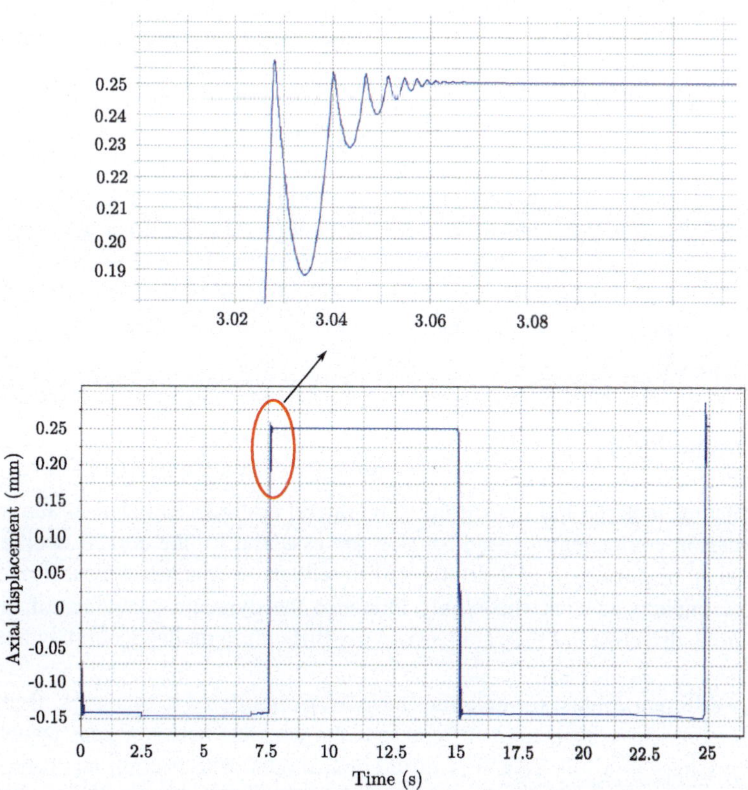

Fig. 12 Axial displacement of the sun gear ($e = 0.8$)

This rigid/rigid contact condition based on the continuous impact modelling is suitable for global models of complex automotive components due to its robustness and implementation simplicity. Besides the modelling of TORSEN differential, this contact formulation could be easily extended to model the contact between plates in clutch or between synchronization devices in gear boxes.

Compared with the previous contact element, the main advantage is the reduction of the computation time by a factor of five. Moreover, the transient behaviour before and after the impact is better represented because it is not mandatory to add an anticipating damping before the impact. The amount of numerical damping needed to enable the convergence is less than for the previous contact formulation. A Hilbert-Hughes-Taylor integration scheme can be used whereas the Chung-Hulbert method with the maximum of numerical damping allowed was necessary for the rigid/flexible contact. The contact stiffness and the restitution coefficient are the only two parameters needed for this contact element. Their value have to be determined according to several criteria: material properties of the two contacting bodies, geometry of the contact surfaces, ... Most of the time, in order to find accurate and reliable values for theses parameters, some physical experiments have to be carried out. For intricate configurations, the experimental measures often require a complicated set-up and expensive instruments. Another way to fix the restitution coefficient and contact stiffness addresses the achievement of detailed numerical simulations. These later are carried out on a faster time scale than the multibody model simulation within the contact condition is included. Nevertheless, in both case these operations are time consuming and request a lot of resources. This is one drawback of this contact formulation based on the restitution coefficient. Currently, neither experiments nor detailed numerical simulations have been carried out for the TORSEN differential model. For the five contact elements, the contact stiffness has been chosen according to the Young's modulus of the steal used for the gear wheels and thrust washers. The restitution coefficient has been prescribed to a value commonly used for contact between two metallic bodies ($e = 0.8$).

7 Conclusions

This chapter is about the modelling of unilateral contacts included in automotive transmission components. These contacts are essential for the correct operation of mechanical devices such as gear boxes, differentials or clutches. Physical phenomena like impacts or stick-slip highly influence the dynamic behaviour of the full automotive driveline but they are particularly difficult to model accurately and efficiently.

Two different contact formulations have been considered in this work. Both contact elements have been used to represent the contacts between gear wheels and thrust washers in a full TORSEN differential model. The friction has been taken into account in all contact elements because friction torques are fundamental for the locking effect of this kind of limited slip differential.

The first one uses the coupled iterations method and is able to model contact between two flexible bodies modelled with finite elements or between a rigid body and a flexible body. During the first step of this contact algorithm, each slave node on the contact surface of the first body is projected on the master face of the second body submitted to the contact condition. From this projection, an associated distance sensor in the normal and tangential directions is created. The relative displacement in theses directions allows to determine the contact status: inactive or active contact, sticking or sliding contact. The second step addresses the definition of the contact condition according to the contact status. An augmented Lagrangian approach is used to express the three kinematic constraints. However if impacts occur, some convergence problems can appear. For the simulation of the TORSEN differential, the augmented Lagrangian approach has to be replaced by a full penalty method in order to have a robust model. A dynamic simulation including the four working modes has been performed and has been globally validated by comparison with experimental data [22].

The second contact formulation implemented is defined between two rigid bodies and is based on the continuous impact theory. With this method, a restitution coefficient is used to account for the kinetic energy loss during the impact process. The contact force is determined by an impact law which depends on the contact stiffness, the restitution coefficient and the relative local penetration and penetration velocity between the two rigid bodies. This second formulation is more robust to represent the transient behaviour close to the impact and enables to greatly reduce the number of degrees of freedom as well as the computing time. However, in contrast to the first contact formulation (flexible/flexible or rigid/flexible), the analysis of deformations and stresses in the contacting bodies is not available with this global contact model.

For global applications like full automotive drivetrain systems, the two contact formulations presented in this chapter could be used together inside the same model. For instance, the coupled iterations method between flexible bodies could be used for the contacts included within the transmissions devices accurately modelled. While for the driveline components coarsely modelled, the rigid-rigid contact formulation could be used to represent these contact conditions in a more global way. Further, the two contact models could be extended in order to capture specific phenomena such as stick-slip often encountered in mechanical transmission devices.

Acknowledgements The author, Geoffrey Virlez would like to acknowledge the Belgian National Fund for Scientific research (FRIA) for its financial support.

References

1. Acary, V., Brogliato, B.: Numerical methods for nonsmooth dynamical systems. Lect. Notes Appl. Comput. Mech. **35**, 540 (2008)
2. Arnold, M., Brüls, O.: Convergence of the generalized-α scheme for constrained mechanical systems. Multibody Syst. Dyn. **18**(2), 185–202 (2007)

3. Blundell, M., Harty, D.: The Multibody Systems Approach to Vehicle Dynamics. Elsevier Butterworth-Heinemann, Amsterdam (2004)
4. Cardona, A.: Flexible three dimensional gear modelling. Eur. J. Comput. Mech. **4**(5–6), 663–691 (1995)
5. Christensen, P., Klarbring, A., Pang, J., Strömberg, N.: Formulation and comparison of algorithms for frictional contact problems. Int. J. Numer. Methods Eng. **42**, 145–173 (1998)
6. Chung, J., Hulbert, G.: A time integration algorithm for structural dynamics with improved numerical dissipation: the generalized-α method. J. Appl. Mech. **60**, 371–375 (1993)
7. Drab, C., Engl, H., Haslinger, J., Offner, G.: Dynamic simulation of crankshaft multibody systems. Multibody Syst. Dyn. **22**, 133–144 (2009)
8. Fichera, G., Lacagnina, M., Petrone, F.: Modelling of torsion beam rear suspension by using multibody method. Multibody Syst. Dyn. **12**, 303–313 (2004)
9. Géradin, M., Cardona, A.: Flexible Multibody Dynamics. Wiley, New York (2001)
10. Gonçalves, J.P., Ambrosio, J.A.C.: Optimization of vehicle suspension systems for improved comfort of road vehicles using flexible multibody dynamics. Nonlinear Dyn. **34**, 113–131 (2003)
11. Hartmann, S., Ramm, E.: A mortar based contact formulation for non-linear dynamics using dual Lagrange multipliers. Finite Elem. Anal. Des. **44**, 245–258 (2008)
12. Hjiaj, M., Feng, Z.Q., De la Saxcé, G., Mróz, Z.: Three-dimensional finite element computations for frictional contact problems with non-associated sliding rule. Int. J. Numer. Methods Eng. **60**, 2045–2076 (2004)
13. Jean, M.: The non-smooth contact dynamics method. Comput. Methods Appl. Mech. Eng. **177**, 235–257 (1999)
14. Kogut, L., Etsion, I.: Elastic-plastic contact analysis of a sphere and a rigid flat. J. Appl. Mech. **69**, 657–662 (2002)
15. Lankarani, H., Nikravesh, P.: Continuous contact force models for impact analysis in multibody analysis. Nonlinear Dyn. **5**, 193–207 (1994)
16. Lens, E., Cardona, A.: A nonlinear beam element formulation in the framework of an energy preserving time integration scheme for constrained multibody systems dynamics. Comput. Struct. **86**, 47–63 (2008)
17. Ma, Z.D., Perkins, N.C.: An efficient multibody dynamics model for internal combustion engine systems. Multibody Syst. Dyn. **10**, 363–391 (2003)
18. Potenza, R., Dunne, J., Vulli, S., Richardson, D.: A model for simulating the instantaneous crank kinematics and total mechanical losses in a multicylinder in-line engine. Int. J. Eng. Res. **8**(4), 379–397 (2007)
19. Puso, M.: A 3d mortar method for solid mechanics. Int. J. Numer. Methods Eng. **69**, 657–662 (2004)
20. Seifried, R., Schiehlen, W., Eberhard, P.: The role of the coefficient of restitution on impact problems in multi-body dynamics. Proc. IMechE, Part K: J. Multi-body Dyn. **224**, 279–306 (2010)
21. Stronge, W.J.: Impact Mechanics. Cambridge University Press, Cambridge (2000)
22. Virlez, G., Brüls, O., Poulet, N., Duysinx, P.: Simulation of differentials in four-wheel drive vehicles using multibody dynamics. In: Proceedings of the ASME 2011 International Design Engineering Technical Conferences and Computers and Information in Engineering Conference, IDETC/CIE 2011, Washington, DC, USA, August 29–31, 2011
23. Yang, B., Laursen, T.A.: A mortar-finite element approach to lubricated contact problems. Comput. Methods Appl. Mech. Eng. **198**, 3656–3669 (2009)

The manufacturer's authorised representative in the EU is Springer Nature Customer Service Centre GmbH, Europaplatz 3, 69115 Heidelberg, Germany. If you have any concerns regarding our products, please contact ProductSafety@springernature.com

Printed and bound by CPI Group (UK) Ltd, Croydon, CR0 4YY
15/07/2026
02167627-0001